ELECTRIC
DRIVES

SECOND EDITION

ELECTRIC DRIVES

SECOND EDITION

Ion Boldea
Fellow IEEE
University Politehnica
Timisoara, Romania

S. A. Nasar
Life Fellow IEEE
University of Kentucky
Lexington, KY

Taylor & Francis
Taylor & Francis Group

Boca Raton   London   New York   Singapore

A CRC title, part of the Taylor & Francis imprint, a member of the
Taylor & Francis Group, the academic division of T&F Informa plc.

Published in 2006 by
CRC Press
Taylor & Francis Group
6000 Broken Sound Parkway NW, Suite 300
Boca Raton, FL 33487-2742

International Standard Book Number-10: 0-8493-4220-1 (Hardcover)
International Standard Book Number-13: 978-0-8493-4220-2 (Hardcover)
Library of Congress Card Number 2005049133

Library of Congress Cataloging-in-Publication Data

Nasar, S. A.
    Electric drives / S.A. Nasar, Ion Boldea.-- 2nd ed.
        p. cm.
        Boldea's name appears first on the earlier edition.
    Includes bibliographical references and index.
    ISBN 0-8493-4220-1 (alk. paper)
    1. Electric driving. I. Boldea, I. Electric drives. II. Title.

TK4058.B64 2005
621.46--dc22   2005049133

Taylor & Francis Group
is the Academic Division of T&F Informa plc.

**Visit the Taylor & Francis Web site at
http://www.taylorandfrancis.com**

**and the CRC Press Web site at
http://www.crcpress.com**

PREFACE TO SECOND EDITION

The electric drive technologies have grown and matured notably in the last 10 years and prompted this second edition of our book.

Our main problem was how to introduce briefly the required new knowledge without confusing the existing users (faculty and engineers in R&D and manufacturing), while providing for usable new knowledge.

This is what we came up with:

- We added with almost all existing chapters recent literature and a few paragraphs, but as much as possible at the end of the chapter.
- We have introduced new paragraphs on a.c. brush series motor and its drive, on capacitor-split inductor motor and its drive, on single-phase PMSM and on single-phase switched reluctance motors and on tooth wound PMSM in their existing chapters, with numerical examples to facilitate a feeling of magnitudes for the new knowledge.
- For the main a.c. drives, with induction and synchronous motors, we have introduced case studies with ample simulation and test results on advanced motion sensorless control with direct torque and flux control (DTFC) with space vector modulation (SVM) to demonstrate very large speed range (about 1000/1 for IM and 200/1 for the synchronous motor) without signal injection for state estimation.
- We added a complete new (last) chapter on "Control of electric generators". Control at constant and variable speed through power electronics is approached, for excited synchronous, PM synchronous, wound rotor induction and switched reluctance generator, with sample results and applications from standard and distributed power systems (with renewable energy conversion) through standby and standalone to vehicular applications (automotive, railroad, marine and aircraft).

We felt that this is a fast growing breed of technology with large worldwide markets and its place is here with electric drives as electric machines are reversible. However, fast, robust active and reactive power control in constant and variable speed generators exhibits notable peculiarities that need separate treatment in contrast to the so-called regenerative braking of electric drives:

- To the 8 Matlab/Simulink close loop power electronics computer programs: 3 for the IM, 3 for the PMSM, one for the d.c. brush motor and one for SRM we added two more:
  - The doubly fed (wound rotor) induction generator control through a rotor connected bidirectional a.c., d.c., a.c. PWM converter that is used for pump storage hydro and wind energy conversion today.
  - The single phase PMSM standard and low switch count power electronics drive for low power low cost applications.
- We decided not to add to existing proposed problems (as is standard in text books) because today's reader is very busy and because, instead he

has now 10 Matlab/Simulink computer programs for 10 different close-loop basic and up to date electric drives to exercise with, make changes in order to deepen his expertise in the field on his own.

- We do believe that the book may be used both as an undergraduate course (one part of it) and a graduate course (the other part of it) with the instructor free to choose his chapters and the corresponding paragraphs of interest.
- Additionally, the book has strong attributes of an up to date monograph that should inspire future work for the MS and PhD students, R&D engineers and new faculty.
- The CD interactive version contains the whole text, selected slides and the solution's manual for the instructor and the 10 Matlab/Simulink simulated close-loop power electronics drives.
- The 10 Matlab/Simulink programs are marked in the paper book by the Matlab icon!
- Special thanks are due to our PhD Student Cristian Ilie Pitic who diligently computer edited this second edition.

Ion Boldea,
Timisoara, Romania,

PREFACE TO FIRST EDITION

Industrial motion, torque, speed, position, control is paramount in raising productivity and quality and in reducing energy and equipment maintenance costs in all industries.

Electric drives share most of industrial motion control applications. Variable speed is inherent in modern electric drives. Such drives contain various high performance motors, power electronic converters and digital control systems. They are provided with input power filters and comply with electomagnetic interference (EMI) standards. The international markets for electric drives are rising at a 10% annual rate and already enjoy a multibillion U.S. dollars (USD) market worldwide.

This book presents a comprehensive view of modern (variable speed) electric drives. It may be considered as both a practical textbook and as an up-to-date monograph. So the readership spectrum is expected to be broad, from senior and graduate students to R&D and plant engineers who are interested in getting in depth and panoramic knowledge on various electric drive solutions in terms of topology, performance, design elements, digital simulation programs and test results, practical issues in industrial drives.

In order to follow the book, prior thorough knowledge of electric machinery or power electronics is not required. But two semesters of electric circuits and one of linear systems are required.

Various brushless motor (of induction, permanent magnet (PM) or reluctance, switched reluctance types) drives as well as d.c. brush motor drives are dedicated special chapters. Separate chapters are devoted to practical issues in pulse width modulation converter drives and to various large power drives.

There are plenty of solved numerical examples and proposed problems throughout the book. Also digital simulation results from 8 Matlab/Simulink programs are included.

Book core

The book is structured into 14 chapters. The first chapter deals with energy conversion in electric drives, application range, saved energy pay-back, typical loads, motion/time profiles or multiquadrant operation, etc.

Electric motor types (configurations) for electric drives are described in Chapter 2. Conventional a.c. motors, d.c. brush motors, and power electronic converter dependent motors are all included.

Chapter 3 synthesizes the main power electronic converters for electric drives. Power electronic switches and various diode rectifiers are treated in some detail while phase-controlled rectifiers, d.c. choppers, voltage-source PWM inverters, current-source inverters and cycloconverters are only briefly reviewed.

Topologies, performance and state space equations of d.c. brush motors are treated in Chapter 4.

Chapters 5 and 6 treat, almost exclusively through explained numerical examples, various rectifier and d.c. chopper fed brush motor drives, respectively. A strong feeling of phenomena and magnitudes is obtained this way. A digital simulation program in Matlab/Simulink treats 4 quadrant d.c. chopper brush motor drives.

Chapter 7 deals with closed-loop motion (torque, speed, position) control in electric drives. Cascade motion controllers, state space controllers as well as non-linear controllers such as sliding mode, fuzzy logic and neuro-fuzzy systems are reviewed and exemplified on d.c. chopper brush motor drives.

Chapter 8 treats the construction, performance, characteristics and space-phasor models for induction motors in electric drives.

Chapter 9 deals extensively with most aspects of modern induction motor drives. Various vector control methods, direct torque and flux control (DTFC) with or without motion sensors are treated in detail, including 3 Matlab-Simulink digital simulation programs. Sensorless control via scalar control methods (V / f with slip compensation) is also included.

Chapter 10 presents PM and reluctance (or hybrid) synchronous motors and their characteristics and d-q (space phasor) models for drives.

Chapter 11 is dedicated to vector (d-q) control and direct torque and flux control (DTFC) of PM and reluctance synchronous motors. Besides sinusoidal, trapezoidal control is included. Three Matlab/Simulink programs are provided to explore the performance of such drives during steady state and transients.

Chapter 12 deals with switched reluctance motors and drives. General and high grade (servo) drive solutions are presented both with and without motion sensors. A Matlab/Simulink program deals with digital simulations for both motoring and regenerative braking operations of such drives.

Chapter 13 includes practical issues with PWM converter drives (all drives presented so far). Input (grid) current harmonics, long power cable effects, bearing currents and remedies are presented in considerable detail.

Chapter 14 treats representative large power drives which use GTOs and thyristors. Three level voltage source inverter and cycloconverter excited synchronous motor drives are followed by current source converter synchronous motor drives and finally by sub- and hyper-synchronous four-quadrant induction motor cascade drives.

Ways to use the book

The book may be considered as a two-semester textbook where the first 7 chapters represent the first semester. It may also be used, through selective paragraphs (chapters), as a one-semester textbook.

Finally the advanced reader may study paragraphs or chapters of his own choice, skipping the basics and jumping directly to new, inspiring knowledge of the subject matter.

The CD interactive version

The CD version, also available through CRC Press, contains:
- the whole text (for browsing);
- problem solutions;
- selected slides (to extract) for the teacher;
- 8 digital simulation Matlab/Simulink programs which are highly interactive and user-friendly. They refer to closed loop d.c. chopper drives (1), vector control and DTFC (sensorless) induction motor drives (3), PM or reluctance synchronous motor drives with rectangular current and vector control (3), and switched reluctance motor drives (1). These digital simulation programs written in a known programming medium, provided with comprehensive helps, may constitute a special vehicle to an in-depth independent study of most representative modern electric drives.

Acknowledgement

The authors express their gratitude to Mr. Jánosi Loránd, their Ph.D. student who worked out the 8 digital simulation programs and computer-edited the entire book.

The rather new philosophy of this book of electrical drives, which does not require prior knowledge of electric machines and power electronics, and includes comprehensive digital simulations and an interactive CD version of the book, is bound to produce, besides beneficial consequences, some secondary effects, also.

Consequently, the authors will welcome readers' feedback.

Timisoara, Romania
Lexington, KY, USA

CONTENTS

Chapter 1

ENERGY CONVERSION IN ELECTRIC DRIVES

1.1. ELECTRIC DRIVES - A DEFINITION

An electric drive is an industrial system which performs the conversion of electrical energy to mechanical energy (in motoring) or vice versa (in generator braking) for running various processes such as: production plants, transportation of people or goods, home appliances, pumps, air compressors, computer disc drives, robots, music or image players, etc.

About 50% of electrical energy is used in electric drives today.

Electric drives may run at constant speed (Figure 1.1) or at variable speed (Figure 1.2).

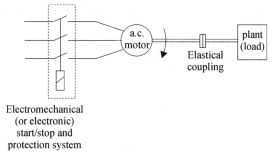

Figure 1.1. Constant speed electric drive

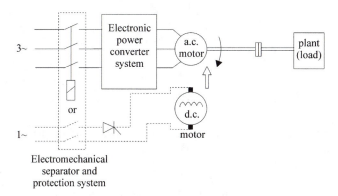

Figure 1.2. Variable speed electric drive

The constant speed electric drive contains the electric (alternating current) motor, the mechanical coupling, the mechanical load (plant) and the electromechanical (or electronic) start/stop and protection system. Today about (75-80)% of all electric drives still run at constant speed because there is not much need for speed control except for starting, stopping and protection.

However there is a smaller (20-25)% group of applications, with very fast annual expansion rate where the torque and speed must be varied to match the mechanical load.

A typical variable speed electric drive (Figure 1.2) contains, in addition, a power electronic converter (PEC) to produce energy conservation (in pumps, fans, etc.) through fast, robust and precise mechanical motion control as required by the application (machine tools, robots, computer-disk drives, transportation means, etc.)

Even though constant speed electric drives could constitute a subject for a practically oriented book, in this book we will refer only to variable speed electric drives which make use of power electronic converters (PECs). For brevity, however, we use the simple name of electric drives.

1.2. APPLICATION RANGE OF ELECTRIC DRIVES *later*

A summary of the main industrial applications and power range of electric drives is shown in Figure 1.3.

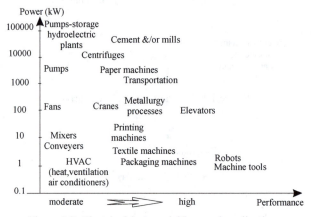

Figure 1.3. Electric drives-variable speed applications

Singular applications such as pumped-storage hydroelectric plants are now built for unit powers of 100MW or more. High performance means, in Figure 1.3, wide speed range, fast and precise response in speed, or position control.

Traditionally, for variable speed, d.c. brush motors have been used for decades [1] but a.c. motors [2,3] have been catching up lately (since 1990) as shown in Figure 1.4. This radical shift is mainly due to the rapid progress in power electronic converters for a.c. motors [4].

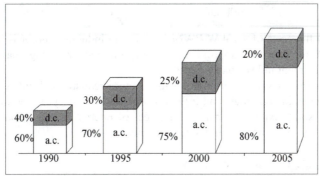

Figure 1.4. A.c. versus d.c. electric drives market dynamics

For variable speed, a.c. motors require both variable voltage amplitude and frequency while d.c. brush motors require only variable d.c. voltage. As the mechanical commutator of the motor does, the frequency changes itself.

A.c. motors are in many cases brushless and have a higher torque (power) density (Nm/kg or kW/kg) than d.c. brush motors and lower initial and maintenance costs. Today power electronic converters (PECs) for both d.c brush or brushless motors, for reversible speed applications, have comparable prices. The PEC costs are higher than the motor costs but this ratio (C) decreases as the power level increases

$$C = \frac{\text{PEC price}}{\text{a.c. motor price}} = 5 \div 2 \qquad (1.1)$$

in general from the kW to MW power range. So what could justify introducing the PEC (Figure 1.2) which is more expensive than the motor?

In most applications the energy savings pay off the additional investment in the PEC in less than 5 years for applications with energy savings of only 25% for only a 1 to 3 speed variation ratio but 24 hours a day operation, above 10kW rated power level. The larger the power the lower the revenue period for a given energy savings percentage.

1.3. ENERGY SAVINGS PAY OFF RAPIDLY

Let us illustrate the energy savings potential pay off for the additional expenses in variable speed drives implied by the presence of the PEC system. Consider a real case when a motor pump system of 15kW works 300 days a year, 24 hours a day and pumps 1200m³ of water per day. By on/off and throttling control only, the system uses 0.36kWh/m³ of pumped water to keep the pressure rather constant for variable flow rate.

Adding a PEC, in the same conditions, the energy consumption is 0.28kWh/m³ of pumped water, with a refined pressure control.

Let us consider that the cost of electrical energy is 8 cents/kWh.

The energy savings per year S is:

$$S = 1200 \cdot 300(0.36 - 0.28) \cdot \$0.08 / \text{year} = \$2304 / \text{year} \qquad (1.2)$$

Now the costs of a 15kW PWM-PEC for an induction motor is less than $8000. Thus, to a first approximation, only the energy savings pay off the extra investment in less than 4 years.

After this coarse calculation let us be more realistic and notice that the energy costs, interest rates and inflation rise slowly every year and taxes also lower to some extent the beneficial effect of energy saving through PEC, contributing to an increase in the revenue period.

Example 1.1. The revenue time.

Let us consider that S dollars (1.2) have been saved in the first year on energy losses through the introduction of a PEC system, and denote by i the interest rate and by i_p the power cost yearly increase.

Thus the effective interest rate per year i_E is

$$i_E = \frac{1+i}{1+i_p} - 1 \qquad (1.3)$$

The net present value (NPV) of losses for an n-year period is

$$NPV = S \cdot \frac{(1+i_E)^n - 1}{i_E \cdot (1+i_E)} \qquad (1.4)$$

We may now consider the influence of taxes and inflation on these savings

$$NPV^E = NPV_e + NPV_d \qquad (1.5)$$

where NPV_e is the net present value of energy savings and NPV_d is the net present value of depreciation on premium investment (straight line depreciation is assumed). With T the tax range

$$NPV_e = NPV \cdot (1 - T) \qquad (1.6)$$

$$NPV_d = \frac{NPV_e}{n} \cdot \frac{\left[(1+i)^n - 1\right]}{i \cdot (1+i)^n} \cdot T \qquad (1.7)$$

With S = $2304 as the first year savings, for a period of n = 5 years, with i = 10%, i_p = 5% and T = 40% we obtain gradually

from (1.3):
$$i_E = \frac{1+0.1}{1+0.05} - 1 = 0.0476$$

from (1.4):
$$NPV = 2304 \cdot \frac{1.0476^5 - 1}{0.0476 \cdot 1.0476^5} = \$10047.7$$

from (1.6):
$$NPV_e = 10047.7 \cdot (1 - 0.4) = \$6028.6$$

from (1.7): $NPV_d = 6028.6 \cdot \dfrac{(1+0.1)^5 - 1}{0.1 \cdot (1+0.1)^5} \cdot 0.4 = \$1828.8$

Finally the premium investment that can be expanded to achieve $2034 energy savings in the first year for a period of 5 years is, from (1.5),

$$NPV^E = 6028.6 + 1828.8 \approx \$7857.$$

This is roughly $8000 (the costs of the PEC), so more exact calculations have led to an increase of the revenue duration from 4 to 5 years.

Considering that lifetime of PEC drives is more than 10-15 years, the investment pays off well.

1.4. GLOBAL ENERGY SAVINGS THROUGH PEC DRIVES

So far the energy savings produced by the PEC in variable speed drives have been calculated for the drive only-PEC and motor.

If we consider the costs to produce and transport the saved electric energy to the consumer (drive) location, the rentability of energy savings increases dramatically. Figure 1.5 shows the energy flow from the power plant to the motor-pump system for constant speed operation and 10kW of useful power, when using throttling to control (reduce) the flow rate.

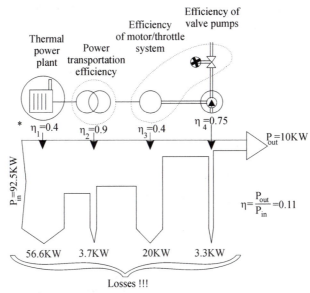

*with recent combined cycle gas turbines $\eta_1 = 0.6$
Figure 1.5. Primary energy consumption for throttle / motor / pump system

The total efficiency from the primary energy source is only 11%. This is poor energy utilization. When a PEC is introduced to produce the same useful power a much better energy utilization ratio is obtained (Figure 1.6).

The energy utilization factor (η) is doubled by the presence of PEC drives for variable speed. So, to the energy savings value in the drive, we should add the savings along the entire chain of energy conversion and distribution, from the power plants to the consumers.

Variable speed with PEC drives, applied today to 15-20% of all electric motor power in developed countries is expected to reach 30-40% by the year 2010, with an annual extension rate of 7-8%. If we add to this aspect the beneficial effects of motion control by PEC drives on the quality of products and manufacturing productivity, we get the picture of one of the top technologies of the near and distant future with large and dynamic markets worldwide (already 4 billion USD in 2004).

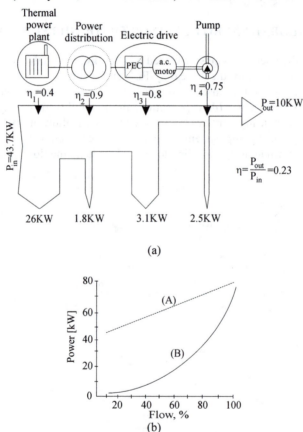

Figure 1.6. a) Primary energy consumption for PEC/motor/pump systems; b) Input power for a 73.6 KW pump motor (A) with output throttle value, (B) with variable voltage/frequency 100 KVA power electronics converter.

1.5. MOTOR/MECHANICAL LOAD MATCH

The role of electric drives is, in fact, to match the electric motor to the mechanical load and to the electrical power grid.

The mechanical load is described by shaft torque/speed or torque/time plus speed/time or position/time relationships.

1.5.1. Typical load torque/speed curves

Typical load torque/speed curves are shown in Figure 1.7. They give a strong indication of the variety of torque/speed characteristics. Along such curves the mechanical power required from the motor varies with speed.

The base speed (unity speed in Figure 1.7) corresponds to continuous duty rated torque (and power) and rated (maximum) voltage from the PEC.

$$\text{Power} = T_{load} \cdot \Omega_r \qquad (1.8)$$

To match the required speed/torque (and power) envelope, the motor and PEC should be carefully chosen or designed. However, the motor to mechanical load match should be provided not only for steady state but also during transients such as drive acceleration, deceleration or short overload periods. The transients require higher torques, in general below base speed, and both the motor and the PEC have to be able to withstand it.

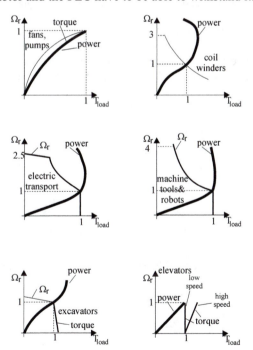

Figure 1.7. Typical load speed/torque, speed/power curves

1.6. MOTION/TIME PROFILE MATCH

For some applications such as robots and electric transportation for given load inertia and load torque/time, a certain position/time (and) or speed/time profile is required. An example is shown in Figure 1.8. There are acceleration, cruising and deceleration intervals. They are strong grounds for the design (sizing) of the electric drives. The torque varies with time and so does the motor current (and flux linkage level). The electric, magnetic and thermal loadings of the motor and the electric and thermal loading of the PEC are definite constraints in a drive specification.

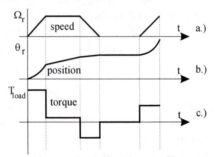

Figure 1.8. Motion/time profile:
a.) speed; b.) position; c.) required load torque.

Here is an illustrative example.

Example 1.2. The direct drive torque/time curve.

A direct drive has to provide a speed/time curve such as in Figure 1.9 against a constant load torque of $T_L = 10$Nm, for a motor load inertia $J = 0.02$ kgm^2.

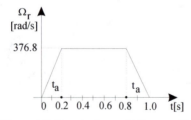

Figure 1.9. Required speed / time profile

Neglecting the mechanical losses lets us calculate the motor torque (T_e)/time requirements.

The motion equation for a direct drive is

$$T_e(t) = J \cdot \dot{\Omega}_r(t) + T_L(t) \qquad (1.9)$$

For the linear speed/time (acceleration-deceleration) zones the speed derivative is

$$\dot{\Omega}_r = \pm \frac{\Omega_{r\,max}}{t_a} = \pm \frac{376.8}{0.2} = \pm 1884 rad/s^2 \qquad (1.10)$$

For the constant speed (cruising) zone $\dot{\Omega}_r = 0.0$.

Consequently, the torque requirements from the motor for the three zones are:

$$T_e = \begin{cases} 1884 \cdot 0.02 + 10 = 37.68 + 10 = 47.68 Nm; & \text{for } 0 \le t \le 0.2s \\ 0 + 10 = 10 Nm; & \text{for } 0.2 \le t \le 0.8s \\ -1884 \cdot 0.02 + 10 = -37.68 + 10 = -27.68 Nm; & \text{for } 0.8 \le t \le 1s \end{cases} \qquad (1.11)$$

The motor torque/time requirements of (1.11) are shown in Figure 1.10.

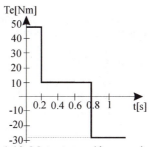

Figure 1.10. Motor torque/time requirements

It is known that a theoretical-sudden motor torque is not practical with a limited available source power. However, a quick torque variation is required.

Advanced PEC drives today manage a zero to rated torque ramping in only 1-5 ms. Lower values correspond to low powers (1kW), while higher values refer to larger powers (hundreds of kW or more). Many electric drives use a mechanical transmission for speed multiplication or for speed reduction by a ratio. The motor and load inertia are "coupled" through the mechanical transmission. Let us illustrate this aspect through a numerical example.

Example 1.3. Gear-box drive torque/time curve.

Let us consider an electric drive for an elevator with the data shown in Figure 1.11.

The motor rated speed n_n = 1550rpm. The efficiency of the gearing system is η = 0.8.

Let us calculate the total inertia (reduced to motor shaft), torque and power without and with counterweight.

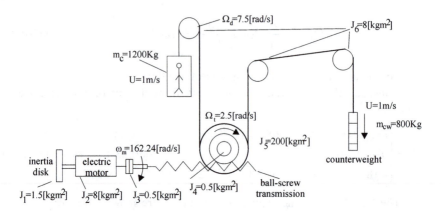

Figure 1.11. Elevator electric drive with multiple
mechanical transmissions and counterweight

First the motor angular speed ω_m is

$$\omega_m = 2 \cdot \pi \cdot n_n = 2 \cdot \pi \cdot \frac{1550}{60} = 162.22 \text{ rad/s} \qquad (1.12)$$

The gear ratios may be defined as speed ratios Ω_t / ω_m for $J_4 + J_5$ and Ω_d $/\omega_m$ for J_6 (Figure 1.11).

Consequently the inertia of all rotating parts J_r, reduced to the motor shaft (Figure 1.11), is

$$J_r = J_1 + J_2 + J_3 + (J_4 + J_5) \cdot \frac{\Omega_t^2}{\omega_m^2} + J_6 \cdot \frac{\Omega_d^2}{\omega_m^2} =$$

$$= 15 + 8 + 2 + (0.5 + 200) \cdot \left(\frac{2.5}{162.22}\right)^2 + 8 \cdot \left(\frac{7.5}{162.22}\right)^2 = 25.062 \text{ kgm}^2 \qquad (1.13)$$

For the cabin and the counterweight, the inertia reduced to motor shaft (J_e) is

$$J_e = (m_c + m_{cw}) \cdot \frac{u^2}{\omega_m^2} = (1200 + 800) \cdot \frac{1^2}{166.22^2} = 0.07238 \text{ kgm}^2 \qquad (1.14)$$

Thus the total inertia J_t is

$$J_t = J_r + J_e = 25.062 + 0.07238 = 25.135 \text{ kgm}^2 \qquad (1.15)$$

In absence of counterweight the law of energy conservation leads to

$$T_{em} \cdot \omega_m \cdot \eta = m_c \cdot g \cdot u \qquad (1.16)$$

Consequently the motor torque, T_{em}, yields

$$T_{em} = \frac{1200 \cdot 9.81 \cdot 1}{162.22 \cdot 0.8} = 90.71 \, \text{Nm} \qquad (1.17)$$

The motor electromagnetic power P_{em} is

$$P_{em} = T_{em} \cdot \omega_m = 90.71 \cdot 162.22 = 14715 \, \text{W} \qquad (1.18)$$

On the other hand in the presence of a counterweight (1.16) the power balance becomes

$$T_{em}' \cdot \omega_m \cdot \eta = (m_c - m_{cw}) \cdot g \cdot u \qquad (1.19)$$

$$T_{em}' = \frac{(1200 - 800) \cdot 9.81 \cdot 1}{162.22 \cdot 0.8} = 30.71 \, \text{Nm} \qquad (1.20)$$

So the motor electromagnetic P'_{em} is

$$P_{em}' = T_{em}' \cdot \omega_m = 30.71 \cdot 162.22 = 4905 \, \text{W} \qquad (1.21)$$

It should be noted that the counterweight alone produces a 3 to 1 reduction in motor electromagnetic power yielding important energy savings. On top of this, for the acceleration-deceleration periods, the PEC drive adds energy savings and produces soft starts and stops for a good ride quality. To do so in tall buildings, high speed quality elevators with up to 1/1000 motor speed variation (control) range are required. This is why in Figure 1.3 elevators are enlisted as high performance drives.

1.7. LOAD DYNAMICS AND STABILITY

Load dynamics in an electric drive with rigid mechanical coupling between motor and load is described by the equation

$$J_t \cdot \frac{d\Omega_r}{dt} = T_e - T_{friction} - T_{load} \qquad (1.22)$$

where J_t is the total inertia of motor and load reduced to motor shaft, T_e is the motor electromagnetic torque, T_{load} is the actual load torque, and $T_{friction}$ is the total friction torque of the motor/transmission subsystem. The friction torque, $T_{friction}$, has quite a few components

$$T_{friction} = T_S + T_C + T_V + T_W \qquad (1.23)$$

where T_S is the static friction torque (at zero speed); T_C is Coulomb friction torque (constant with speed); T_V is viscous friction torque (proportional to speed) and T_W is windage friction (including the ventilator braking torque, proportional to speed squared)

$$T_V = B' \cdot \Omega_r \qquad (1.24)$$

$$T_W = C \cdot \Omega_r^2 \qquad (1.25)$$

The friction torque components are shown in Figure 1.12. Only to a first approximation we may write

$$T_{friction} = B \cdot \Omega_r \qquad (1.26)$$

In this latter case (1.22) becomes

$$J \cdot \frac{d\Omega_r}{dt} = T_e - T_{load} - B \cdot \Omega_r \qquad (1.27)$$

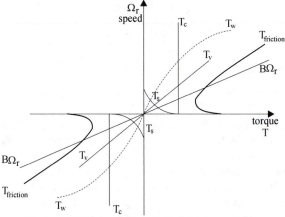

Figure 1.12. Components of friction torque, $T_{friction}$

If T_e and T_{load} are constant, the solution of (1.27) is

$$\Omega_r(t) = \Omega_{r\,final} + B \cdot e^{-t/\tau_m} \qquad (1.28)$$

where $\Omega_{r\,final}$ is the final, steady state, speed and $\tau_m = J/B$ is the so-called mechanical time constant, and A is a constant to be determined from initial conditions. Equation (1.28) reflects a stable aperiodic response. As

$$\Omega_r = \frac{d\theta_r}{dt} \qquad (1.29)$$

(1.27) becomes

$$J \cdot \frac{d^2\theta_r}{dt^2} + B \cdot \frac{d\theta_r}{dt} - T_e = -T_{load} \qquad (1.30)$$

Clearly, for stable operation the transients in θ_r must die out. As B is generally small, stability is obtained if the motor torque T_e is of the form

$$T_e \approx -C_e \cdot \frac{d\theta_r}{dt}; \quad C_e > 0 \qquad (1.31)$$

or
$$T_e \approx -C_e \cdot \frac{d\theta_r}{dt} - C_i \cdot \theta_r; \quad C_e, C_i > 0 \tag{1.32}$$

with (1.31)-(1.32), (1.30) becomes

$$J \cdot \frac{d^2\theta_r}{dt^2} + (B + C_e) \cdot \frac{d\theta_r}{dt} + C_i \cdot \theta_r = -T_{load} \tag{1.33}$$

Now it is evident that the θ_r transients will be stable.

Note: For the synchronous machine θ_r is in fact the angle between the e.m.f. and the terminal voltage, which is constant for steady state.

Let us see if the torque/speed curves of basic d.c. brush and a.c. motors satisfy (1.33). Typical torque/speed or torque/angle curves for these machines are shown in Figure 1.13.

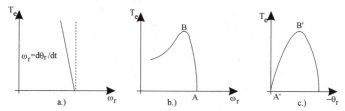

Figure 1.13. Mechanical characteristics: a) d.c. brush motor with separate excitation, b) induction motor, c) synchronous motor

Condition (1.13) is fulfilled by the d.c. brush motor as

$$-C_e = \frac{dT_e}{d\left(\dfrac{d\theta_r}{dt}\right)} < 0 \tag{1.34}$$

The torque decreases with speed steadily. For the induction motor (Figure 1.13b) only along the zone AB

$$\frac{dT_e}{d\omega_r} < 0 \tag{1.35}$$

Finally for the synchronous motor (Figure 1.13c) only along the zone A´B´

$$-C_i = \frac{dT_e}{d\theta_r} < 0 \tag{1.36}$$

The above discussion serves only to signal the problem of electric drive dynamics and stability. The presence of PECs frees the mechanical characteristics from the forms in Figure 1.13 obtained for constant voltage (and frequency). Consequently, stable response in position, speed, or torque may be artificially obtained through adequate control of voltage and frequency at the motor electrical terminals.

Two simple examples follow:

Example 1.4. D.c. brush motor drive stability.

A permanent magnet d.c. brush motor with the torque speed curve: $\Omega_r = 200.0 - 0.1 \cdot T_e$ drives a d.c. generator which supplies a resistive load such that the generator torque / speed equation is $\Omega_r = 2T_L$. We calculate the speed and torque for the steady state point and find out if that point is stable.

Solution:

Let us first draw the motor and load (generator) torque speed curves in Figure 1.14.

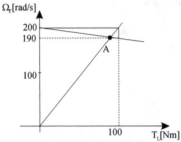

Figure 1.14. D.c. brush motor load match

The steady state point, A, corresponds to constant speed and B = 0 in (1.27). Simply, the motor torque counteracts the generator braking torque

$$T_L = T_e \tag{1.37}$$

Using the two torque speed curves we find

$$\Omega_{rA} = 200 - 0.1 \cdot \frac{\Omega_{rA}}{2} \tag{1.38}$$

and thus

$$\Omega_{rA} = \frac{200}{1 + 0.1/2} = 190.476 \text{ rad/s} \tag{1.39}$$

and
$$T_{eA} = T_{LA} = \frac{\Omega_{rA}}{2} = \frac{190.476}{2} = 95.238 \text{ Nm} \tag{1.40}$$

The static stability is met if

$$\left(\frac{\partial T_e}{\partial \Omega_r}\right)_A < \left(\frac{\partial T_L}{\partial \Omega_r}\right)_A \tag{1.41}$$

In our case from the two torque/speed curves

$$-10 < \frac{1}{2} \tag{1.42}$$

and thus, as expected, point A represents a situation of static equilibrium.

Example 1.5. Induction motor drive stability.
An induction motor has the torque speed curve given by

$$T_e = \frac{2T_{eK}}{\dfrac{S}{S_K} + \dfrac{S_K}{S}} ; \; S = \frac{\omega_1 - \omega_r}{\omega_1} ; \; S_K = 0.1; \; T_{eK} = 20 \text{ Nm} \tag{1.43}$$

where S - slip; ω_1 - primary frequency; ω_r - rotor electrical speed ($\omega_r = \Omega_r.p$; p - number of winding pole pairs). This motor drives a d.c. generator with a resistive load whose torque / speed curve is $T_L = C.\omega_r$. Neglecting the mechanical losses (B = 0) we decide to check out static stability conditions for $T_e = 10 \text{Nm}$.
Solution:
Let us draw the two mechanical characteristics (Figure 1.15).

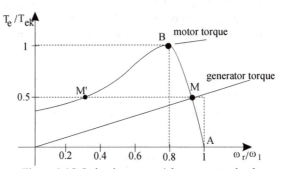

Figure 1.15. Induction motor / d.c. generator load

The slip value for $T_e = 10 \text{Nm}$ may be found from

$$10 = \frac{2 \cdot 20}{\dfrac{S}{0.2} + \dfrac{0.2}{S}} \tag{1.44}$$

The solutions of (1.44) are

$$S_1 = 0.0436; \text{ point M in figure } 1.15$$

$$S_2 = 0.7464; \text{ point M' in figure } 1.15 \tag{1.45}$$

The static stability depends on the sign of $\partial T_e / \partial \omega_r$

$$\frac{\partial T_e}{\partial(\omega_r/\omega_1)} = \frac{-\partial T_e}{\partial S} = \frac{2T_e}{\left(\dfrac{S}{S_k}+\dfrac{S_K}{S}\right)^2} \cdot \left(\frac{1}{S_k}-\frac{S_K}{S^2}\right) \qquad (1.46)$$

So for
$$S < S_K;\ \ \frac{\partial T_e}{\partial(\omega_r/\omega_1)} < 0;\ \text{stable zone} \qquad (1.47)$$

and for
$$S > S_K;\ \ \frac{\partial T_e}{\partial(\omega_r/\omega_1)} > 0;\ \text{unstable zone} \qquad (1.48)$$

As at point M $S_1 = 0.0536 < S_K = 0.2$, M corresponds (as expected) to static stability conditions while at point M' $S_2 = 0.7464 > S_K = 0.2$; M' is within the instability zone.

1.8. MULTIQUADRANT OPERATION

An electric drive may be required to provide for forward and reverse motion with rapid (regenerative) braking in both directions. Motor operation implies torque in the direction of motion. In regenerative braking the torque is opposite to the direction of motion, and the electric power flow in the motor is reversed also (negative). These possibilities are summarized in Table 1.1 and in Figure 1.16.

Table 1.1.

Mode of operation	Forward motoring	Forward regenerative braking	Reverse motoring	Reverse regenerative braking
Speed, ω_r	+	+	-	-
Torque, T_e	+	-	-	+
Electric power flow	+	-	+	-

Positive (+) electric power flow means electric power drawn from the PEC by the motor while negative (-) refers to electric power delivered by the motor (in the generator mode) to the PEC.

The PEC has to be designed to be able to handle this bi-directional power flow. In low and medium power PECs (up to hundreds of kW) with slow braking demands the generated power during the braking periods is interchanged with the strong filter capacitor of PEC or d.c. (dynamic) braking is being used.

For d.c. dynamic braking the kinetic energy of the motor-load system is converted to heat in the motor rotor. For fast and frequent generator brakings the PEC has to handle the generated power either by a controlled braking resistor or through bidirectional power flow. All these aspects will be discussed in some detail later in subsequent chapters.

For a fast speed response, modern variable speed drives may develop a maximum transient torque up to base speed ω_b and maximum transient power up to maximum speed, provided that both the motor and the PEC can handle these powers.

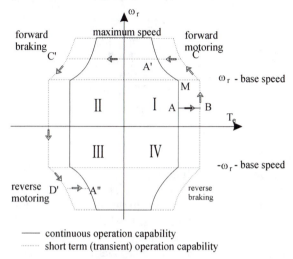

Figure 1.16. Electric drives with four quadrant operation

A rapid increase in speed from point A to point A′ (Figure 1.16) is managed along the ABCA′ path which remains in the first quadrant. A slow speed increase from A to A′ is performed along the path AMA′. For a speed reversal (from A to A″ the trajectory goes along the path AA′C′D′ through the second quadrant (regenerative braking). Such speed transitions indicate the complexity of energy conversion and transfer between the motor, PEC and power source in a multiquadrant electric drive.

Note: So far we discussed only rotary motor electric drives. However for each rotary motor there is a linear counterpart. Torques and angular speeds will be replaced by thrust and linear speed, respectively, in linear motor drives [5]. Linear motor, however, are not treated in this book.

1.9. PERFORMANCE INDEXES

The electric drives performance indexes defined in what follows are divided into three main categories: energy conversion indexes, drive response indexes and costs and weight indexes.

Electric drives perform the conversion of electrical energy to mechanical energy or vice versa through the use of magnetic energy as a storage medium. A typical electric drive more detailed structure contains the following: an electric motor, a static power converter, feedback sensors (or observers) and a digital (or analog or hybrid) motion controller (Fig. 1.17)

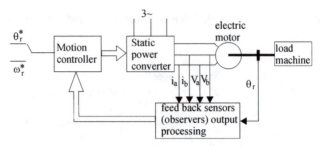

Figure 1.17. Electric drive basic topology

(a) Power efficiency (for steady state)

Energy conversion in an electric drive takes place both in the static power converter and in the electric motor. The energy flows from the static power converter to the electric machine in the motor regime (Fig. 1.18a), and vice versa for regenerative braking (Fig. 1.18b).

Energy conversion is accompanied by losses: conduction and commutation losses in the static power converter (p_{conv}), conduction, core, and mechanical losses in the motor (p_{mot}).

For steady state, the rating of energy conversion from electrical to mechanical or vice versa may be derived through the power efficiency η_p defined for the motor (η_{pm}) or for the drive (η_{pd}):

$$\eta_{pm} = \frac{P_{out}}{P_{out} + \sum p_{mot}} \tag{1.49}$$

$$\eta_{pd} = \frac{P_{out}}{P_{out} + \sum p_{mot} + \sum p_{conv}} \tag{1.50}$$

As speed is variable in electric drives, the power efficiency is relevant for base speed (ω_b), maximum speed (ω_{max}), and rated torque for these two situations.

Base speed ω_b is the speed for which full voltage is required to produce the peak design continuous operation torque T_{ek}. Above base speed the voltage remains constant and, in a.c. motors, only the frequency increases. In many servo drives $\omega_b = \omega_{max}$, that is, the peak torque T_{ek} is available up to maximum speed when the static power converter rating is defined for maximum rather than for base speed.

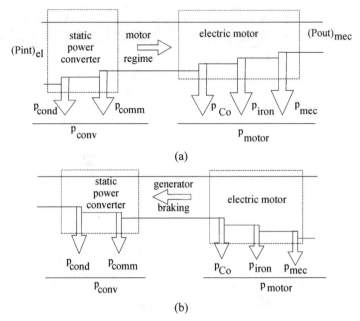

Figure 1.18. Energy conversion in electric drives;

a) motor regime, b) generator braking.

(b) Energy efficiency (for mechanical transients)

In some applications the variable speed drives undergo very frequent mechanical transients (speed and torque transients): hybrid and electric cars.

To assess the energy conversion efficiency for such cases the energy efficiency η_E for the motor (η_{Em}) and for the drive (η_{Ed}) are defined:

$$\eta_{Em} = \frac{W_{out}}{W_{out} + W_{motor}} \qquad (1.51)$$

$$\eta_{Ed} = \frac{W_{out}}{W_{out} + W_{motor} + W_{conv}} \qquad (1.52)$$

Where W_{out} is the output (useful) energy, W_{motor} are the motor total energy losses, and W_{conv} are the converter total energy losses.

The energy efficiency may be an important optimization criterion for the design of the drives (motor, converter and controller) with frequent mechanical transients (robotics, urban transit drives, etc.).

(c) Losses / torque [W/Nm] ratio.

A more flexible energy conversion index is the ratio between the motor losses and the motor torque. The W/Nm ratio may refer to motor winding power losses only. In this case the W/Nm ratio is most adequate for prolonged low speed operation, as the core losses at low speed are small.

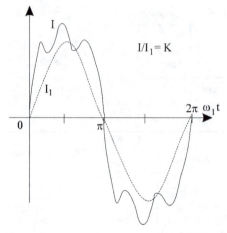

Figure 1.19. Alternating motor phase current in the field weakening zone

(six-pulse inverter regime)

For high speed however, all losses in the motor have to be considered in the W/Nm, in order to assess the heating of the motor fairly.

(d) RSM kW/kVA ratio

The RSM kW/kVA ration for a.c. motors is in general computed for the fundamental components and represents, in fact, the displacement power factor DPF:

$$DPF = (kW / kVA)_{motor} = \frac{P_{in}}{3V_1 I_1} \tag{1.53}$$

For given torque, the higher the displacement power factor, the smaller the current drawn by the motor for given terminal voltage and frequency. The power factor DPF thus has considerable influence on stator winding losses, and finally on power efficiency.

(e) Peak kW/kVA ratio

As most variable speed servodrives make use of MOSFET or IGBT transistors, the static power converters are sized according to the peak current and voltage values.

For a.c. motors, at high speeds, in the weakening zone, the motor phase currents depart considerably from sinusoidal waveforms.

The active input power P_{in} is:

$$P_{in} = \frac{3}{\pi} V_0 I \frac{1}{K} DPF \tag{1.54}$$

Where V_0 is the d.c. source voltage at the input of the power converter and I is the peak flat topped phase current; K is the ratio between the actual peak current and the peak value of the current fundamental (Fig. 1.19), in general $k \leq (1.1 - 1.15)$.

$$\text{peak } kW / kVA = \frac{P_{in}}{S_1} = \frac{3}{\pi} V_0 I \frac{1}{K} \frac{DPF}{6 V_0 I} = \frac{DPF}{2\pi K} \tag{1.55}$$

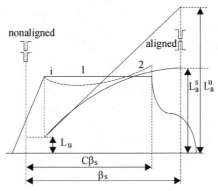

Figure 1.20. Current waveforms in a switched reluctance motor:

1-ideal (chopped) current; 2-actual current at full voltage (no chopping)

On the other hand, the switched reluctance motor (Chapter 12) has unipolar pulse currents with one phase working at a time (Fig. 1.20).

The same performance index may be defined for the switched reluctance motor [6] as:

$$\text{peak } kW / kVA = \frac{\beta_s N_r Q}{8\pi} \tag{1.56}$$

where β_s is the stator pole/stator pole pitch ratio ($\beta_s \cong 0.4$); N_r is number of rotor poles.

$$Q \approx C \left(2 - \frac{C}{s} \right) \tag{1.57}$$

where C is the ratio between the phase turn-on angle and the stator pole angle.

In general, at start (zero speed) C = 1 and decreases with speed ($C \cong 0.65$ at base speed).

$$s = \frac{\lambda_u - 1}{\lambda_u \sigma - 1}; \quad \lambda_u = \frac{L_a^u}{L_u} \approx (6 \div 10); \quad \sigma = \frac{L_a^s}{L_a^u} = (0.3 \div 0.4) \qquad (1.58)$$

L_u, L_a^u, L_a^s are the phase inductance values shown in Fig. 1.20.

The peak apparent power, S_1 is:

$$S = 2 \cdot m \cdot V_0 \cdot I \qquad (1.59)$$

m is the number of phases (m = 3, 4 in general), V_0 is the d.c. source voltage; I – peak phase current value.

In the kW power range, the peak kW/kVA ratio is in the interval 0.55 - 0.65 for flat top ideal current, even with a considerable magnetic core saturation level (σ = (0.3 – 0.4)), suggesting a reasonable peak kVA in the static power converter rating [1].

(f) Peak torque / inertia ratio

As already mentioned, most electric drives provide the peak torque T_{ek} up to base electric speed ω_b only. In some servodrives the base speed is equal to maximum speed.

The ratio between the peak torque T_{ek} and the rotor inertia J represents the maximum ideal acceleration a_{max} (radian / sec^2) up to the base electric speed ω_b:

$$a_{max} = \frac{T_{ek}}{J} [rad/s^2] \qquad (1.60)$$

Conversely, the time t_a required to reach the base speed ω_b under peak torque and no-load may be defined:

$$t_a = \frac{\omega_b / p}{a_{max}} = \omega_b \frac{J}{p T_{ek}} [ms]; p - pole \ pairs \qquad (1.61)$$

Both, a_{max} and t_a tends to become catalogue data for high performance variable speed drives.

(g) Field weakening speed range (ω_{max}/ω_b)

Many applications require constant power over a wide range of speeds. For such applications the motor's full voltage is reached at base speed while above base speed the flux is gradually reduced to allow motoring torque.

The ratio between maximum and base speeds (ω_{max}/ω_b) is a performance index because it shows the capability of the drive to reach high speeds avoiding overheating.

Ratio ω_{max} / ω_b of 1.5 - 2 is characteristic for permanent magnet synchronous and induction motors. Higher values require special measures and traded converter and motor overrating.

(h) Variable speed ratio: $\omega_{max}/\omega_{min}$

The speed close loop control range, that is, $\omega_{max}/\omega_{min}$, is an indication of the drive capability to accommodate different applications.

Electric drives with $\omega_{max}/\omega_{min} > 200$ are considered to perform wide speed range control and require a position (or at least speed) feedback sensor for precision position (or speed) control.

On the other hand, when $20 < \omega_{max}/\omega_{min} < 200$ the speed range control is medium and the position and speed feedback signals for speed control are obtained through observers.

Speed errors of 2 to 3 % are allowable for such drives. As the motion (position and speed) sensor costs are important they should be replaced by motion observers but this is feasible only in applications with medium variable speed range.

For $\omega_{max}/\omega_{min} < 20$ open loop control with some speed overcompensation under load is applied with so called V/f a.c. drives. V/f a.c. drives are the most common as for fans and pumps applications they do the job properly at rather low costs (less than 200 USD / kVA).

(i) Torque rise time (t_{Tek})

In fast response drives it is important to provide for fast torque variation. Torque transients depend on speed, motor parameters, converter type, input voltage level, and method of control.

As, inevitably, the torque rise time decreases with speed due to increased back e.m.f., it seems that the rise time of torque from zero to T_{ek}, t_{Tek}, with the flux linkage already present in the machine, at zero speed, may constitute a reliable performance index.

In fast response drives t_{Tek} is in the range of 2-6 ms. For a.c. (brushless) motor drives such fast torque response is obtained through the field orientation (transvector) control of induction, PM- synchronous, and synchronous reluctance motors.

(k) Torque ripple ratio ($\Delta T_e/T_{erated}$)

The pulsations in torque depend on motor type, parameters, and also on torque control strategy. They may be assessed through the ratio of torque ripple peak to rated torque ($\Delta T_e/T_{erated}$). The rated torque is defined here as the steady-state continuous torque at base speed. The rated torque pulsations influence the torque, speed and position control precision, vibrations and noise. Consequently, they should be as low as possible in a trade off between performance and costs.

(l) Thermal limitation (ϑ_{motor})

Direct drives for position or wide speed range are in direct mechanical contact with the load (machine tools, for example). In order to avoid mechanical deformations in the shaft of the load, in such cases, the temperature of the motor ϑ_{motor} should be [7]:

$$\vartheta_{motor} < 20^0 C + \vartheta_{ambient} \qquad (1.62)$$

This is a serious limitation in motor design.

For most drives the motor temperature is limited by the electric insulation class (B, E or F).

(m) Noise level, L_{noise} (dB)

The noise radiated by a variable speed drive is due to both the motor and the static power converter. The noise level accepted depends on the application. As an example, for machine tools, [2]:

$$L_{noise} \approx 70 + 20\log(P_n / P_{n0}); \quad P_{n0} = 1kW; \quad P_n = (1 \div 10)kW \qquad (1.63)$$

where P_n is the rated power of the motor.

(n) Motion control precision and robustness

Motion control means torque, speed and position close loop control.

Torque control is the most demanding, as the torque feedback sensor should account for the core losses. In general, motion control precision may be measured as torque, speed or position error:

$$\Delta T_e; \Delta \omega_r (rpm); \Delta \theta_r (^0) \qquad (1.64)$$

The torque error is defined in relative values and the speed error in rpm, while the position error is given in degrees.

For speed and position control, the torque loop is present, sometimes as a current limiter.

Robustness of control is defined as the sensitivity of the drive response (in torque, speed, position) with respect to motor parameters, inertia, and load torque variations. It is the degree of immunity of the drive response to drive parameters detuning.

Between robustness and quickness of speed control there is a conflict in the sense that, for robustness, available quickness of response has to be partially sacrificed.

Advanced motion controllers such as self-tuning, model reference, or variable structure, fuzzy reasoning controllers are proposed to increase response robustness.

Robustness indexes could be defined as: torque error (ΔT_e), for a given parameter detuning ΔP_{ar} ($\Delta T_e / \Delta P_{ar}$), or speed rise time ratio from zero to base speed for J and 2J, respectively, or for zero and rated load torque.

(o) Dynamic stiffness

Dynamic stiffness is the ratio between perturbation torque ($\Delta T_{perturbation}$) and controlled variable error (Δx) versus frequency of torque perturbation. It is a kind of dynamic robustness to torque perturbations:

$$DS = \frac{\Delta T_{perturbation}}{\Delta x}$$

(p) Specific costs and weights

The overall costs C_{total} of an electric drive may be considered as a cost criterion:

$$C_{total} = C_{equip} + C_{loss} + C_{maint} \tag{1.65}$$

where:

C_{equip} is the cost of motor, converter, sensors, and of controller;

C_{loss} is the capitalized energy loss cost of the drive over its lifetime;

C_{maint} is the maintenance cost.

The relative importance of C_{loss} increases with drive power (or torque) and should be accounted for in any computation attempt of total cost break down.

In the equipment cost C_{equip}, the cost of motor, static power converter, sensors (observers), and motion controllers have a relative importance depending notably on the motor power, with the motor cost's importance increasing with power level. Here the net present worth cost (including inflation, premium on investments dynamics) for the foreseeable life of the drive is to be considered.

The specific weights are considered also as performance criteria:

$$\text{motor specific weight} = \frac{\text{peak torque}}{\text{weight}}, [\text{Nm/Kg}] \tag{1.66}$$

$$\text{converter specific weight} = \frac{\text{peak apparent power}}{\text{weight}}, [\text{kVA/Kg}] \tag{1.67}$$

The first weight criterion (1.66) is a strong comparison criterion between different motors for drives. The second refers, from the same point of view, to the static power converter since the peak apparent power is a design criterion for the MOSFET or IGBT PWM inverters, widely applied to servodrives. For thyristor inverters the RMS apparent power should be considered, instead of peak apparent power.

1.10. SUMMARY

/ •   Modern electric drives perform electrical to mechanical energy conversion at variable speeds. They contain power electronic converters (PEC) which modify the voltage (and frequency) with high efficiency. [9 - 11]

•   The introduction of PECs in electric drives (15 – 20 % of all drives in 2004) is justified by energy savings or process control (productivity and quality) performance.

• *later* D.c. brush motor-PEC drives are now more and more replaced by a.c. motor-PEC drives as the motors are more rugged, less expensive for about the same PEC costs and comparative performance, especially for reversible motion control applications.

•   Energy savings with PEC drives provide a revenue period of 5 years or less from 10 kW up while global energy savings (from the power plant to the application site) are spectacular.

/ •   Motor torque has to match the load torque during steady state while stable transients are produced by closed loop motion control in PEC drives. Load dynamics and stability are priority issues in PEC drives.

/ •   PEC drives provide, in general, motoring in both directions of motion. Also, they are capable of fast regenerative braking in both directions of motion, provided the PEC can retrieve the generated power back to the power source. When this is not possible, for slow dynamics (braking), the energy is dumped into the motor rotor or in an external, controlled, braking resistor attached to the PEC.

•   The PECs have a limited voltage ceiling V_b for which the motor produces continuous rated power P_b at the so-called base speed ω_b.

•   Most PEC drives can work above ω_b for constant power P_b (and constant voltage V_b) up to $\omega_{max} = (2\text{-}4)\,\omega_b$.

•   A set of coherent / practical energy conversion, drive response quality and cost / weight performance for modern (power electronics) electric drives has been introduced.

1.11. PROBLEMS

1.1.   A PEC drive saves S = \$500 of energy in the first year of operation. For a 4 year (n = 4) money return period calculate the price of PEC investment if the interest rate i = 8%, the energy costs yearly raise i_p = 4% and the tax range T = 35%.

1.2.   Calculate the global power requirements for 100kW of useful (mechanical) power of an electric drive at 60% of rated speed if the power plant efficiency is η_1 = 40%, energy transportation is η_2 = 90% for two cases:

    a.   with a conventional drive at η_{34} = 60% total efficiency;
    b.   with a PEC drive at η'_{34}= 85% total efficiency.

1.3. An electric drive uses a mechanical transmission between the motor and load with a reduction ratio a = 1/10. The motor inertia J_m = $0.02kgm^2$ while the load-transmission inertia J_L = $2kgm^2$. Mechanical losses are neglected and the load torque T_L = 200Nm. For the speed/time curve in Figure 1.9, calculate the motor torque requirements after reducing the load inertia to motor shaft.

1.4. The base speed and power of a PEC a.c. motor drive are ω_b = 367 rad/s and P_b = 100kW. The motor has 4 poles (2p = 4). Calculate the motor torque T_{eb} at base speed and power and the torque for constant power P_b for ω_{max} = $3\omega_b$. What is the torque value and sign for generator braking at ω_b and delivering P_b with zero motor losses?

1.12. SELECTED REFERENCES

1. *P. Sen*, Thyristor D.C. drives, John Wiley & Sons, 1981.
2. *W. Leonhard*, Control of electric drives, Springer-Verlag first edition (1985), second edition (1996), third edition (2001).
3. *G.K. Dubey*, Power semiconductor controlled drives, Prentice Hall, 1989.
4. *N. Mohan, T.M. Undelund, W.P. Robbins*, Power electronics, John Wiley & Sons, 1989, 1995.
5. *I. Boldea, S.A. Nasar,* Linear electric actuators and generators, Cambridge Univ. Press 1997.
6. *T.J.E. Miller*, "Converter volt – ampere requirements of the switched reluctance motor drive", IEEE Trans., vol.IA-21, no. 5, pp. 1136-1144.
7. *W. Weck, G. Ye*, "Catalog of servodrives for NC machines", ETZ, vol. 111, no. 6, pp. 282-285, 1990 (in german).
8. *BT Boulter*, "Selecting the right drive", IEEE – IA – Applications Magazine, vol. 10, no. 4, 2004, pp. 16-23.
9. **B.K. Bose,** Modern power electronics and drives, Prentice Hall, 2001.
10. **R. Krishnan,** Electric motor drives, Prentice Hall, 2001.
11. **H.A. Toliyat, S. Campbell,** DSP – based electromechanical motion control, CRC Press, Florida, 2004.

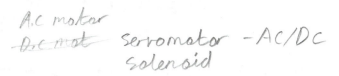

Chapter 2

ELECTRIC MOTORS FOR DRIVES

In this chapter we will introduce typical electric drive configurations and various electric motors for electric drives.

2.1. ELECTRIC DRIVES - A TYPICAL CONFIGURATION

A modern electric drive [1-8] capable of controlled variable speed is made of some important parts (Figure 2.1) such as:

- the electric motor;
- the power electronic converter (PEC);
- the electric and motion sensors;
- the drive controller;
- the command interface.

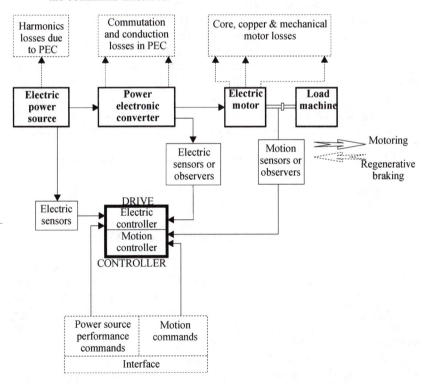

Figure 2.1. Electric drive basic topology

The drive controller may be considered to be made of motion-speed and/or position-controller, more or less the same for all types of electric drives, and the electric controller which refers to current and voltage (or flux linkage and torque) control within the PEC.

Electric sensors (observers) refer to voltage, current, flux as measured (or calculated) state variables while motion sensors (observers) mean position and/or speed and torque as measured (or calculated) state variables.

The electric controller has, in general, electric sensor (observer) inputs from both the power source and the PEC output. The motion controller handles, in general, only motion sensor (observer) outputs.

On the other hand, the electric controller commands are related to power source side energy conversion performance commands (unity power factor, harmonics elimination) while the motion controller commands relate to motion (speed, position, torque) control commands handled through an interface from a local digital controller or from a remote process control host computer. As expected, both electric and motion controllers are blended together in the drive controller hard and soft.

By now both the drive controller and the interface are carried out by high performance digital signal processors (DSPs) which can handle several MFLOP/s, not an excessive feature with modern electric drives.

In what follows in this chapter, we will briefly characterize the candidate electric motors used for drives. The various losses occurring in an electric drive (Figure 2.1) will be dealt with in relation to specific motors and PECs in subsequent chapters. The sensors will be described in some detail when applied to various electric drives. The motion controllers, more or less common to all electric drives, will have a separate chapter devoted to them in relation to d.c. brush motor drives.

Finally, electric controllers tend to be more specific and will be treated during discussions of applications for various electric drives.

And now a short classification of electric motors for electric drives follows.

2.2. ELECTRIC MOTORS FOR DRIVES

In essence, all existing types of electric motors may be accompanied by PECs with digital motion controllers to produce high performance electric drives. The ideal speed-torque curve for motion control is a straight descending line. This fact and the ease to produce a variable d.c. voltage source for a wide range of speed control, had made the d.c. brush motor the favorite variable speed drive up to 1960. Since then the development of reliable price-competitive PECs with variable a.c. voltage and frequency capability and the linearization of the speed-torque curve through the so-called VECTOR CONTROL have contributed to the take over of the variable speed drive applications by a.c. motors. Let us start with the d.c. brush motor.

2.3. D.C. BRUSH MOTORS

A typical topology of a d.c. brush motor with stator PM (or electromagnetic) excitation and a rotor comprising the armature winding and the mechanical commutator with brushes is shown in Figure 2.2a. The mechanical commutator is, in fact, an electromechanical d.c.-a.c. bidirectional power flow power converter as the currents in the rotor armature coils are a.c. while the brush-current is d.c.

Figure 2.2c shows an axial airgap PM d.c. brush motor with a printed, winding, ironless-disk rotor and mechanical commutator with brushes. PM excitation, especially with the nonmagnetic disk rotor, yields extremely low electric time constants L/R (around or less than 1 ms in the sub kW power range). Thus quick response in current (torque) is expected, though the current (torque) harmonics are large unless the switching frequency in the PEC is not high enough.

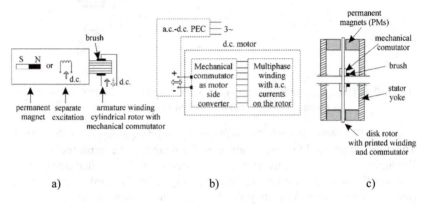

a)   b)   c)

Figure 2.2. D.c. brush motor: a.) with cylindrical rotor;

b.) equivalence of d.c. motor to a.c. motor plus motor side power converter;

c.) with PMs and disk rotor.

Unfortunately the mechanical commutator, though not bad in terms of losses and power density, has serious commutation current and speed limits and thus limits the power per unit to 1-2 MW at 1000 rpm and may not be at all acceptable in chemically aggressive or explosion-prone environments.

The PEC, as a d.c. variable voltage source (Figure 2.2b), when capable of four quadrant operation (positive and negative d.c. voltage and d.c. current output), is as complicated and as expensive as the PEC for a.c. motors. Moreover, fast speed reversal is quite a problem.

Still, the d.c. brush motors, especially with PM excitation, are and will be used for a good period of time, in numerous applications, where 1-2 quadrant operation suffices at low powers and moderate speeds.

2.4. CONVENTIONAL A.C. MOTORS

By conventional a.c. motors we mean traveling field motors — induction, synchronous with electromagnetic excitation and reluctance synchronous — that may be used either at on-line start (constant speed) or variable speed (with PEC) applications. Consequently, the rotors of all conventional a.c. motors have squirrel cages for asynchronous starting. Among them the induction motor (Figure 2.3a) is mostly used for variable speed drives.

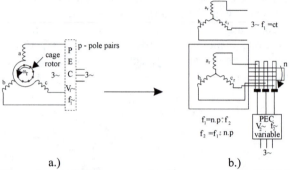

a.)   b.)

Figure 2.3. Three phase induction motor: a.) with a cage rotor and PEC in the stator; b.) with slip-rings wound rotor and PEC on the rotor.

A PEC system controls the voltage V_1 and frequency f_1 in the stator. This is a full power PEC and is adequate for wide speed control (below 1/2). For high power applications with limited speed control, a limited power PEC supplies the wound rotor through the slip rings (Figure 2.3b) with adequate (low) voltage V_2 and frequency f_2 while the stator is line-fed at constant V_1 and f_1.

Thus the sizing and costs of PEC drives are drastically reduced though, for full torque starting, a special arrangement is required as the PEC cannot handle directly the rather large starting rotor currents and voltages. Still the slip ring-brush mechanical system poses notable problems of maintenance and may not be used in hostile environments.

The cage-rotor induction motors cover wide power ranges from 0.5kW to 10MW while even higher powers may be handled through wound-rotor configurations.

Electromagnetically excited conventional synchronous motors have three-phase a.c. windings on the stator and a d.c. excited cage rotor with salient or nonsalient poles (Figure 2.4).

A conventional synchronous motor when used in variable speed drives requires two PECs - one full power a.c.-a.c. PEC in the stator and a low power (1-5%) a.c.-d.c. PEC that supplies, in general through slip rings and brushes, the excitation windings on the rotor. A coordinated control of the two PECs provides for speed (active power) and reactive power control and for efficient wide speed range control in high power applications (MW and tens of MW range).

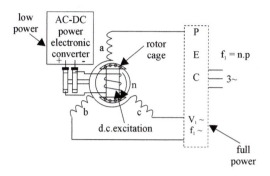

Figure 2.4. Conventional salient pole rotor synchronous motor fed

through a full power PEC in the stator

Why do, above 2 MW, synchronous motors predominate over induction motors? Because the reactive power control through excitation is possibly easier with synchronous motors, and thus the requirements in this regard from the full power PECs are lower. Consequently, the PECs are simpler and less expensive and the overall costs of the drive are lower, though the synchronous motor, for comparable power and speed, is more expensive than the induction motor. The efficiency of the synchronous motor above 2 MW is also, in general, higher than that of the cage-rotor induction motor.

Conventional reluctance synchronous motors (RSMs) have a cylindrical stator with three a.c. windings and a windingless rotor with a moderate orthogonal axis magnetic saliency up to 4 (6) to 1. High magnetic saliency is obtained with multiple flux barriers (Figure 2.5)

A cage on the rotor is sometimes maintained for stability purposes when the motor is fed through a full power a.c.-a.c. PEC with a rigid relationship between frequency f_1 and motor speed n (n = f_1/p).

The conventional RSMs are to some extent (up to 100kW) used in low dynamics variable speed drives with open loop speed control as the speed does not decrease with load. Consequently, control is simpler than with induction motors.

The main drawback of the conventional RSM is the low power factor and not high enough torque density (Nm/kg), which leads to higher kVA ratings of the PEC (by 15-20%).

Besides the conventional a.c. motors (which may also be used for line-start constant speed applications due to the action of the rotor cage), in the last decades, new motor configurations totally PEC-dependent, have been introduced. The main types of PEC dependent motors are described below.

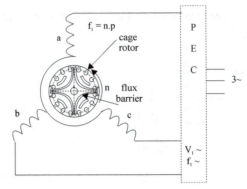

Figure 2.5. Conventional reluctance synchronous motor

fed through a full power PEC in the stator

2.5. POWER ELECTRONIC CONVERTER DEPENDENT MOTORS

Evidently PEC dependent motors can not operate without PECs. They are, in general, <u>multiphase motors</u> (to limit torque pulsations) and provide for self-starting from any initial position. In essence this new breed of motors may be fed through unipolar or bipolar multiphase currents in tact with the rotor position. Also they have singly salient or doubly salient magnetic structures with or without permanent magnets (PMs) on the rotor.

PEC-dependent motors with PM-rotors [13] have evolved from synchronous motors by replacing the electromagnetic excitation with high energy PMs (Figure 2.6).

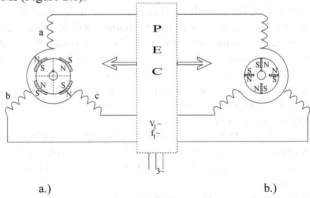

a.)   b.)

Figure 2.6. PM synchronous motor (PM-SM) - singly salient

a.) pole magnet rotors, b.) interpole magnet rotors

The absence of the rotor cage renders the PM-SM fully dependent on PECs as the stator frequency and rotor speed should be in synchronism at all moments.

The stator three-phase windings are concentrated (one coil/pole/phase) when rectangular bipolar current control (based on inexpensive rotor position sensors) is performed. For more refined performance (wider speed range (1/1000 range)) sinusoidal current control is performed. A more expensive precise position sensor (or observer) is needed.

For both rectangular and sinusoidal current control, nonoverlap windings have also been introduced recently (from 1 W to tens of kW power levels) to reduce copper losses and frame length and to reduce PM torque at zero current (cogging torque).

The pole magnet and interpole magnet rotors have both advantages and disadvantages as shown in the chapter on PM-SM motor drives.

The other category of PEC-dependent motors is the so-called stepper motors. In fact they are doubly salient multiphase motors with windingless rotor and unipolar position-dependent or position-independent current control through a.c.-d.c. (unipolar current) PECs (Figure 2.7).

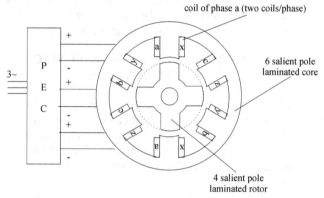

Figure 2.7. Three-phase switched reluctance 6/4 motor (SRM) - PEC dependent.

It is evident that the switched reluctance motor, SRM [14] — a commercial name for the power step motor with rotor position controlled slewing (continuous motion) — works on the principle that by sequentially supplying each phase with a current pulse when the inductance of that phase increases will produce positive torque (motoring). Regenerative braking is obtained if the phase is supplied when the inductance slope is negative. The phases are switched on and off based on position feedback (provided by a position sensor or observer) and thus torque and motion control "without loosing steps" is obtained.

The SRM has undergone more than two decades of development and seems ready now for marketing for powers from a few watts to megawatt levels, both for low and high dynamics applications. The SRM is, together with PM-SM, a competitive motor for high performance variable speed drives.

The reluctance or hybrid (PM-reluctance) stepper motors resemble the SRM and PM-SMs with rectangular current, but they rely on open-loop feed-forward (position independent) control with moderate dynamics and acceptable position precisioning by a notable increase in the number of stator and rotor poles and the number of phases (4-5 in general).

Stepper motors refer to low power positioning applications only. For more details on stepper motors, please see [15].

PM-SM and SRM drives will be devoted separate chapters in this book.

Other cageless rotor converter-dependent motors are spin-offs from electromagnetically-excited synchronous motors (SM), induction motors (IM), reluctance synchronous motors or doubly salient motors RSM (by adding PMs on the rotor or on the stator).

Despite of their severe torque pulsations, single phase PM synchronous (brushless) motors, as well as single phase SRM drives, configurations are being gradually introduced in low power (for home appliances or small electric actuators on cars) applications to cut costs by reducing the number of controlled power electronic switches in the static converter.

They will be discussed later, (briefly), in pertinent chapters.

2.6. ENERGY CONVERSION IN ELECTRIC MOTOR/GENERATORS

In all systems where mass is not destroyed or produced, the principle of energy conservation holds. So energy is not created nor destroyed, it is only converted from one form into another. This principle, together with the Faraday, Ampere, Gauss, Ohm and Newtonian mechanical laws and theory of electric circuits govern the electro-mechanical energy conversion in electric motor-generators.

The electromechanical energy conversion in electric motor – generators, implies a few forms of energy:

$$
\left\langle \begin{array}{l} \text{Energy form} \\ \text{electrical source}(\pm) \end{array} \right\rangle = \left\langle \begin{array}{l} \text{mechanical} \\ \text{energy}(\pm) \end{array} \right\rangle + \left\langle \begin{array}{l} \text{stored} \\ \text{magnetic} \\ \text{energy} \\ \text{increase} \end{array} \right\rangle + \left\langle \begin{array}{l} \text{energy} \\ \text{converted} \\ \text{to heat} \\ \text{(losses)} : (+) \end{array} \right\rangle
$$

$$(2.1)$$

The electric energy from the electric source and mechanical energy are considered positive for motoring and negative for generating.

The energy converted into heat has three main causes:

- Electric machine coils (copper) losses: p_{Co};
- Mechanical losses (gear friction and windage losses): p_{mec};
- Magnetic (hysteresis and eddy current) losses in the magnetic cores of the machine: p_{iron}.

Including the losses at their location we obtain:

$$\left\langle \begin{array}{c} \text{electric} \\ \text{energy} \\ -p_{Co} \end{array} \right\rangle = \left\langle \begin{array}{c} \text{magnetic energy} \\ \text{storage increase} \\ +p_{iron} \end{array} \right\rangle + \left\langle \begin{array}{c} \text{mechanical} \\ \text{energy} \\ +p_{mec} \end{array} \right\rangle \qquad (2.2)$$

So, in fact, the electrical to mechanical energy conversion in an electric motor implies an intermediate energy form for storage: the magnetic energy, mostly present in the magnetic field of the airgap between the rotor (mover) and the stator (fixed part) of the electric motor (Figure 2.8.).

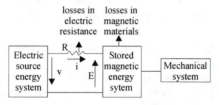

Figure 2.8. Energy conversion in electric motor / generators

The net electrical energy increment dW_e may be described in terms of voltage v, current i, electromotive voltage e.m.f. e:

$$dW_e = (v - Ri)i \cdot dt \qquad (2.3)$$

To realize electrical to mechanical energy conversion the coupling electromagnetic field in the machine has to produce a reaction to the electric circuit (to electric source voltage v), e:

$$-e = v - Ri; dW_e = -(e)i \cdot dt \qquad (2.4)$$

If the electric energy is transmitted to the coupling magnetic field in the machine through more than one electric circuit (phase) eqn. (2.4) contains more terms.

According to Faraday's law the e.m.f. e is:

$$e = -\frac{d_s\lambda}{dt} \qquad (2.5)$$

d_s/dt is the total time derivative of the flux linkage λ, which may contain two terms: a transformer (pulsational) term and a motion-caused term.

According to (2.4):

$$dW_e = i \cdot d_s\lambda, \quad dW_{mec} = T_e d\theta_r \qquad (2.6)$$

dW_{mec} is the mechanical energy increment.

Where T_e is the instantaneous electromagnetic torque developed by the electric machine and $d\theta_r$ the rotational motion angle increment.

From (2.4) – (2.5) and the energy conservation condition:

$$dW_e = id_s\lambda = dW_m + T_e d\theta_r \qquad (2.7)$$

dWm is the stored magnetic energy increment.

If the magnetic flux linkage is constant ($d_s\lambda = 0$) the energy increment from the electrical energy source is zero and thus the electromagnetic torque T_e is:

$$T_e = -\left(\frac{\partial W_m}{\partial\theta_r}\right) \qquad (2.8)$$

This situation occurs only when the rotor-stored mechanical energy of the motor is converted into losses in the shortcircuited (v = 0) electric generator mode.

The electromagnetic torque is negative and the machine is braked gradually close to standstill.

In most applications the case when there is electric energy transfer from (to) the electric source is more relevant as it defines the electric motor and electric generator operation modes.

In such cases (2.7) writes:

$$dW_m = id_s\lambda - T_e d\theta_r; i = -\frac{\partial W_m}{\partial\lambda} \qquad (2.9)$$

with λ and θ_r as independent variables.

We may choose the current i instead of flux linkage λ as independent variable. But in that case a new function – magnetic coenergy W'_m – may be defined as:

$$W'_m = i\lambda - W_m \qquad (2.10)$$

Differentiating this equation we obtain:

$$dW'_m = id\lambda + \lambda di - dW_m \qquad (2.11)$$

Or, with (2.9):

$$dW'_m = id\lambda + \lambda di - id\lambda + T_e d\theta_r \qquad (2.12)$$

And finally:

$$T_e = \left(\frac{\partial W'_m}{\partial\theta_r}\right)_{i=const}; \lambda = \frac{\partial W'_m}{\partial i} \qquad (2.13)$$

$$W_m = \int_0^{\lambda_0} id\lambda; W'_m = \int_0^{i_0} \lambda di \qquad (2.14)$$

Eqn (2.9) presupposes that the magnetic flux linkage λ varies in time and thus the electric energy exchange with the electric machine dW_e is now nonzero.

In accord with eqns. (2.8) and (2.13) the electromagnetic torque T_e is nonzero only when the magnetic field coenergy varies with respect to rotor position θ_r.

Let us consider a primitive switched reluctance motor (SRM) with two stator and two rotor poles (Figure 2.9).

We have, in fact, an inductance L (made of two coils in series or in parallel) with an airgap and a magnetic core. It is evident that when the rotor moves around, the magnetic permeance varies with θ_r.

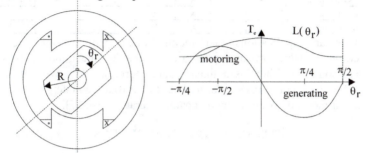

Figure 2.9. The 2 / 2 stator / rotor pole primitive SRM:
a) cross section, b) inductance L and torque T_e versus angle θ_r.

Approximately:

$$L(\theta_r) \approx L_0 + L_2 \cos 2\theta_r \qquad (2.15)$$

If the magnetic saturation is neglected, L_0 and L_2 are constant, the magnetic coenergy W'_m of this coil is:

$$W'_m = \int_0^{i_0} L(\theta_r)\, idi = \frac{i_0^2}{2} L(\theta_r) \qquad (2.16)$$

The electromagnetic torque T_e is:

$$T_e = \left(\frac{\partial W'_m}{\partial \theta_r} \right)_{i_0 = const} = \frac{i_0^2}{2}\frac{\partial L(\theta_r)}{\partial \theta_r} = -i_0^2 L_2 \sin 2\theta_r = -T_{em} \sin 2\theta_r \,(2.17)$$

The polarity of the current does not influence the torque polarity. Positive torque means motoring and negative torque implies generating. As the torque has positive and negative values with θ_r it follows that the 2/2 SRM may not start from any initial rotor position and thus special measures (a parking permanent magnet, for example) are needed to provide self starting from same (parking) position. Multiphase SRMs may start from any position.

2.7. SUMMARY

We summarize the motor candidates for variable speed electric drives in Table 2.1.

Table 2.1. Electric motors for drives

Motor type	Singly salient	Non salient	Doubly salient	Unipolar current	Bipolar current	Conventional (caged rotor)	Converter dependent (cageless rotor)
D.c. brush	X			X			X[1]
Induction		X			X	X	
Excited synchronous	X	X			X	X	X
ReluctanceSynchronous	X				X	X	X
PM synchronous	X	X	X		X		X
Switched reluctance		X	X				X
Stepper motor			X	X	X		X

There are applications that require linear motion and, provided competitive costs/performance rates are reasonable, linear motors become practical solutions. There is a linear counterpart for every type of rotary motor. Also their control is similar to that of their rotary counterparts.

Linear motor drives have their separate literature [16, 17].

2.8. SELECTED REFERENCES

1.   **W. Leonhard,** Control of electrical drives, First, Second & Third Edition, Springer Verlag, 1985, 1996, 2001.
2.   **P.C. Sen,** Thyristor D.C. drives, John Wiley & Sons, 1981.
3.   **G.K. Dubey,** Power semiconductor controlled drives, Prentice Hall, 1989.
4.   **P. Vas,** Vector control of a.c. machines, Oxford Univ. Press, 1990.
5.   **I. Boldea, S.A. Nasar,** Vector control of a.c. drives, CRC Press, 1992.
6.   **M.P. Kazmierkovski, H. Tunia,** Automatic control of converter-fed induction motor drives, Elsevier, 1994.
7.   **P.M. Trzynadlowski,** The field orientation principle in control of induction motors, Kluwer Acad. Press, 1994.
8.   **D.W. Novotny, T.A. Lipo,** Vector control and dynamics motor of a.c. drives, Clarendon Press, 1996.
9.   **S.A. Nasar, I. Boldea,** Electric machines steady state, CRC Press, 1990.

10. **S.A. Nasar, I. Boldea,** Electric machine dynamics and control, CRC Press, 1992.
11. **P. Krause,** Analysis of electric machinery, McGraw Hill, 1986.
12. **I. Boldea,** Reluctance synchronous machines & drives, O.U.P., 1996.
13. **S.A. Nasar, I. Boldea, L.E. Unnewehr,** Permanent magnet, reluctance and selfsynchronous motors, CRC Press, 1993.
14. **T.J.E. Miller,** Switched reluctance motors and their control, Magna Physics & Clarendon Press, 1993.
15. **T. Kenyo,** Stepping motors and their microprocessor controls, Clarendon Press, 1984.
16. **I. Boldea, S.A. Nasar,** Linear electric actuators and generators, Cambridge Univ. Press, 1997.
17. **I. Boldea, S.A. Nasar,** Linear motion electromagnetic devices, Taylor and Francis 2001.

---

[1] d.c. brush motors may be supplied through a series resistor and in this sense they are conventional (converter independent for variable speed) but this method is energy consuming and thus it is hardly used anymore.

Chapter 3

POWER ELECTRONIC CONVERTERS (PECs) FOR DRIVES

Power electronic converters (PECs) for drives are now a mature technology with notable and dynamic worldwide markets. Extraordinary research and development efforts are also devoted to the subject today and are likely to continue in the future.

There is a distinct rich literature on the subject [1]. Here we only summarize the main configurations and the voltage-current waveforms while more details will be given in subsequent chapters devoted to various motor drives.

PECs may be classified in many ways. Among them we may include the type of electronic switch used, the steps in power conversion, and current (voltage) output waveforms.

Let us first review the main power electronic switches (PESs).

3.1. POWER ELECTRONIC SWITCHES (PESs)

Power electronic switches (PESs) used in PECs undergo frequent commutations from a blocked to saturated (open) condition to provide high energy conversion rates (efficiency above 85-95%) over a wide range of voltages and, when it is the case, frequencies.

The blocked state is characterized by high voltage over the power terminal of the PES when the conducting current is practically zero. The saturated (open) state is defined by the presence of the conducting current while the voltage at the power terminals is low, in the 1-2V range.

The PESs may be classified as:

a. - uncontrolled;

b. - semicontrollable;

c. - fully controllable.

The <u>diode</u> is an uncontrolled PES (Figure 3.1a) whose state of conduction is determined solely by the direction of the current from anode (A) to cathode (K).

The ideal characteristic is shown in Figure 3.1b and shows that the diode conducts only for positive current, when the ideal voltage drop on it V_{AK} is zero. In reality V_{AK} is not zero but small (around 1-2V).

The diode is present in all PECs either as a rectifier or for protection. The diode does not have a control circuit or a driver for it.

The <u>thyristor</u> is a semicontrollable PES. Its state of conduction may be controlled only in the sequence blocked — saturated when in the command circuit (driver) gate (G) — cathode (K) a positive current is present and if the PES is traveled by a positive current from anode (A) to cathode (K) (Figure

41

3.2). The saturated state is kept even after the command signal is inhibited until the current in the power circuit (A-K) becomes zero.

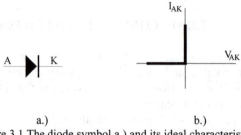

a.)   b.)
Figure 3.1.The diode symbol a.) and its ideal characteristic b.)

The thyristor is used especially in PECs having an interface with a.c. power grids at high power levels and low commutation frequencies (up to 300Hz in general).

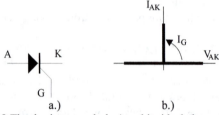

a.)   b.)
Figure 3.2.The thyristor symbol a.) and its ideal characteristic b.)

The GTO (gate turn off thyristor) (Figure 3.3) is a fully controllable PES. Its saturation is obtained as for the thyristor, but its blocking is accessible when a negative current i_G is applied to the command (driver) circuit.

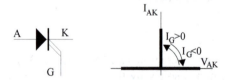

a.)   b.)
Figure 3.3. The GTO's symbol a.) and its ideal characteristic b.)

This PES is used for high power PECs and for interfacing with a.c. grids. Its switching frequency is higher than that of the thyristor but still below 1 kHz in general.

The bipolar junction transistor (BJT) (Figure 3.4) is also a fully controllable PES but at high commutation frequency and low and medium (up to tens of kW) powers. Its command — saturation — is obtained through its base (B) current with respect to the emitter (E) and is maintained only in the presence of the command current. When the command current becomes zero the BJT gets blocked. The commutation time is lower than for the thyristor and the GTO. The BJT may operate only in the first quadrant (Figure 3.4) as its power circuit collector (C)-emitter (E) may not withstand negative polarization voltages.

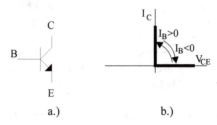

a.)   b.)

Figure 3.4. The bipolar junction transistor symbol

a.) and its ideal characteristic b.)

The <u>MOS transistor</u> is also a fully controllable PES. The MOS transistor — in contrast to the other PESs — works with voltage (rather than current) signals in the command (driver) circuit gate (G) - source (S). Consequently the command power is practically zero which notably simplifies the command circuitry.

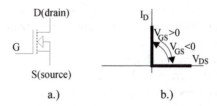

a.)   b.)

Figure 3.5. MOS Transistor symbol a.) and its ideal characteristic b.)

The MOS transistor has the lowest commutation time, allowing for very high switching frequency (tens of kHz); however, at low and moderate power and voltage levels.

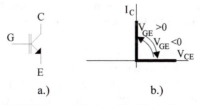

a.)   b.)

Figure 3.6. IGBT's symbol a.) and its ideal characteristic b.)

The IGBT (insulated gate bipolar transistor) (Figure 3.6) possesses the MOS command merits (voltage command signal), the low commutation time (up to 20kHz switching frequency) and power circuit assets (low saturation voltage, larger voltage (current) ratings) of the bipolar transistor (BJT).

In the voltage/current ratings, the switching frequencies are getting higher and higher and their conduction and commutation losses are becoming lower at a high pace. As an example, PECs with IGBTs have reached in 1996 500kVA per unit and, with four units in parallel and simultaneously controlled, 2MVA.

At the same time thyristors are less expensive and in the MW (and higher) power range they are still competitive in many applications and in particular topologies. GTOs tend to win rather quickly over the power range from thyristors.

For high power, medium voltage MOS-controlled thyristors (MCTs) have been recently (1992) introduced. MCTs are turned on and off by a short voltage pulse on the MOS gate with thousands of microcells in parallel on a chip to yield high current levels.

Since 1997 the insulated gate-commutated thyristors (IGCT) are available also for high voltage, high power (4500 V, 300 A per unit). The turn off negative gate current is also 300 A but is very short.

High switching frequency characterizes both MCTs and IGCTs.

Still in the labs, SiC-based higher voltage (15 kV, 300A) medium current static power switches with a few kHz switching frequency, seem likely to revolutionize the medium and large power electronics by 2010.

The PECs may be classified many ways. In what follows we will refer to their input and output voltage/current waveforms and distinguish:

- a.c.-d.c. converters (or rectifiers);
- d.c.-d.c. converters (or choppers);
- a.c.-d.c.-a.c. converters (indirect a.c.-a.c. converters) - 2 stages;
- a.c.-a.c. converters (direct a.c.-a.c. converters).

We should note that a.c.-d.c.-a.c. converters contain an a.c.-d.c. source side converter (rectifier) and a d.c.-a.c. converter called an inverter. These converters are mostly used with a.c. motor drives of all power levels.

For bidirectional power flow the rectifier should be capable of working as an inverter, while the inverter is in principle capable of working as a rectifier.

Direct (one stage) a.c.-a.c. converters (cycloconverters) are applied in large power a.c. drives while the a.c.-d.c.-a.c. converters are used within the whole power range.

A.c.-d.c. converters are either part of a.c.-d.c.-a.c. converters or they are used independently for driving d.c. brush motors.

D.c.-d.c. converters are either connected directly to batteries (d.c. sources in general) or they are fed from uncontrolled (diode) rectifiers for driving d.c. brush motors, in multiphase configurations.

As the diode rectifier is used in many PECs as a line side converter, it will be presented here in some detail. All other PECs will merely be introduced here. Further details are given in subsequent chapters on various electric motor drives.

3.2. THE LINE FREQUENCY DIODE RECTIFIER FOR CONSTANT D.C. OUTPUT VOLTAGE V_d

In most electric drives the power is provided by the local grid at 60 (or 50) Hz in single-phase or three-phase configurations. Single phase a.c. is available at low power in general in various buildings, while industrial power grid is three phase. We treat them in sequence.

The output voltage of the diode rectifier should be as ripple free as possible and thus it requires a rather large capacitor filter (Figure 3.7).

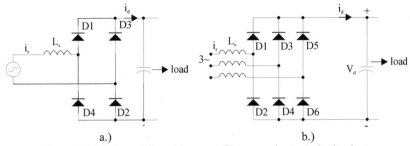

Figure 3.7. Diode rectifier with output filter capacitor: a.) single-phase,
b.) three-phase

Let us consider first a basic rectifier circuit (Figure 3.8) with instantaneous commutation and a line source inductance L_s providing constant V_d on no load.

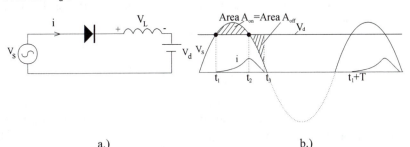

a.)   b.)
Figure 3.8. a.) Basic rectifier equivalent circuit and
b.) the voltage and current waveforms

The diode starts conducting when $V_s \geq V_d$ at t_1. At t_2, $V_s = V_d$ but, due to inductance L_s, the current goes into the diode until it dies out at t_3 such that $A_{on} = A_{off}$. In fact the integral of inductance voltage V_L from t_1 to t_1+T should be zero, that is, the average flux in the coil per cycle is zero

$$\int_{t_1}^{t_1+T} V_L \, dt = 0 = A_{on} - A_{off} \qquad (3.1)$$

As V_d is close to the maximum value $V_s\sqrt{2}$, the current i becomes zero prior to the negative (next) cycle of V_s.

Figure 3.8 illustrates only the positive voltage, that is, diodes D_1-D_2 conducting. For the negative V_s, D_3-D_4 are open and a similar current waveform is added (Figure 3.9).

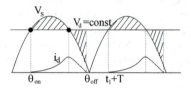

Figure 3.9. Single-phase rectifier - the waveforms

As long as the current i_d is non zero

$$V_L = L_s \cdot \frac{di_d}{dt} = \sqrt{2} \cdot V_s \cdot \sin \omega t - V_d \qquad (3.2)$$

$$\omega \cdot L_s \cdot i_d = \int_{t_1}^{t_1+T} \left(\sqrt{2} \cdot V_s \cdot \sin \omega t - V_d \right) d(\omega t); \quad \theta_{on} < \omega t < \theta_{off} \qquad (3.3)$$

For θ_{on} $$V_d = \sqrt{2} \cdot V_s \cdot \sin \theta_{on} \qquad (3.4)$$

For $\omega t = \theta_{off}$, $id(\omega t) = 0$ and from (3.3) we may calculate θ_{off} as a function of θ_{on}.

Finally the average coil flux linkage $L_s I_d$ is

$$L_s \cdot I_d = \frac{1}{\pi} \cdot \int_{\theta_{on}}^{\theta_{off}} L_s \cdot i_d(\omega t) \cdot d(\omega t) \qquad (3.5)$$

For given values of $L_s I_d$, iteratively θ_{on}, θ_{off} and finally V_d are obtained from (3.3)-(3.5) (Figure 3.10).

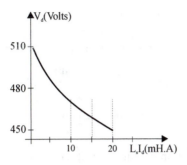

Figure 3.10. V_d versus L_sI_d

3.3. LINE CURRENT HARMONICS WITH DIODE RECTIFIERS

The line current has the same shape as i_d in Figure 3.9 but with alternate polarities (Figure 3.11). It is now evident that the line (source) current in a rectifier is rich in harmonics. This distortion from sinusoidal can be described by some distortion indexes.

Also, the current fundamental is lagging the source voltage by the displacement power factor (DPF) angle φ_1

$$DPF = \cos\varphi_1 \qquad (3.6)$$

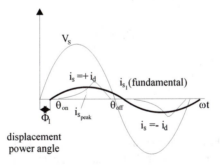

Figure 3.11. Source current shape

The source current r.m.s. value is I_s. Thus the apparent power magnitude S is

$$S = V_s \cdot I_s \qquad (3.7)$$

where V_s is the r.m.s. voltage value.

The power factor
$$PF = \frac{P}{S} \qquad (3.8)$$

where
$$P = V_s \cdot I_{s1} \cdot DPF \qquad (3.9)$$

So
$$PF = \frac{I_{s1}}{I_s} DPF \qquad (3.10)$$

A strong distortion in the line current will reduce the ratio I_{s1}/I_s, and thus a small power factor PF is obtained even if DPF is unity.

Now
$$I_s = \sqrt{I_{s1}^2 + \sum_{\upsilon=2}^{\infty} I_{s\upsilon}^2} \qquad (3.11)$$

The total harmonic current distortion THD (%) is

$$THD\% = 100 \frac{I_{dis}}{I_{s1}} \qquad (3.12)$$

where
$$I_{dis} = \sqrt{\sum_{\upsilon=2}^{\infty} I_{s\upsilon}^2} \qquad (3.13)$$

The peak current I_{speak} is also important to be defined as a relative value constant called the crest factor (CF)

$$CF = \frac{I_{speak}}{I_s} \qquad (3.14)$$

or the form factor (FF)

$$FF = \frac{I_s}{I_d} \qquad (3.15)$$

It has been shown that the displacement power factor DPF is above 0.9 but the power factor PF is poor if the source inductance L_s is small.

The ratio V_d / V_s only slightly decreases with an $L_s I_d / V_s$ ratio but both the crest factor (CF) and the form factor (FF) sharply decrease at low (below 0.03) $L_s I_d / V_s$ values. After that, however, they only slightly decrease with $L_s I_d / V_s$.

The distorted waveform of the source current suggests that a.c. filters are required to eliminate the harmonics pollution of the local power grid. The tendency is that every PEC drive which uses a diode rectifier on the source side be provided with a source harmonics filter. More on this aspect in Chapter 13.

<u>Example 3.1.</u>
A single-phase diode rectifier with constant e.m.f. is fed from an a.c. source with the voltage $V_s(t) = V_s\sqrt{2}\cdot\sin\omega t$ ($V_s = 120V$, $\omega = 367rad/s$). The discontinuous source current (Figure 3.11) initiates at $\theta_{on} = 60^0$ and becomes zero at $\theta_{off} = 150^0$. The source inductance is $L_s = 5mH$. Calculate the d.c. side voltage V_d and the waveform of the source current $i_d(\omega t)$.
According to Figure 3.9 from (3.3) we obtain

$$\int_{\theta_{on}}^{\theta_{off}}\left(\sqrt{2}\cdot V_s\cdot\sin\omega t - V_d\right)\cdot d(\omega t) = 0 \tag{3.16}$$

$$0 = \sqrt{2}\cdot V_s\cdot\left(\cos\theta_{on} - \cos\theta_{off}\right) - V_d\cdot\left(\theta_{off} - \theta_{on}\right) \tag{3.17}$$

From (3.17)

$$V_d = \frac{\sqrt{2}\cdot 120\cdot\left(\cos 60 - \cos 150\right)}{\frac{5}{6}\pi - \frac{1}{3}\pi} = 147V \tag{3.18}$$

Now from (3.3) again

$$i_d = 0; \quad 0 < \theta < \theta_{on}$$

$$i_d(\omega t) = \frac{1}{\omega L_s}\cdot\int_{\theta_{on}}^{\omega t}\left(\sqrt{2}\cdot V_s\cdot\sin\omega t - V_d\right)\cdot d(\omega t); \quad \text{for } \theta_{on} \le \theta \le \theta_{off} \tag{3.19}$$

$$i_d = 0; \quad \theta_{off} < \theta < 180^0$$

Consequently

$$i_d(\omega t) = \frac{\sqrt{2}\cdot V_s\cdot\left(\cos\theta_{on} - \cos\omega t\right) - V_d\cdot\left(\omega t - \theta_{on}\right)}{\omega L_s}$$

$$= \frac{\sqrt{2}\cdot 120\cdot\left(0.5 - \cos\omega t\right) - 147\cdot\left(367t - \pi/3\right)}{367\cdot 5\cdot 10^{-3}} \tag{3.20}$$

for $60^0 < \theta < 150^0$

Though not convenient to use, (3.19)-(3.20) allow for the computation of I_s (rms), peak current I_{speak}, fundamental I_1, TDH% (3.12), crest factor (3.14), average d.c. output current Id.

3.4. CURRENT COMMUTATION WITH $I_d = ct$ AND $L_s \neq 0$

For constant d.c. current $I_d = ct$ (Figure 3.12):

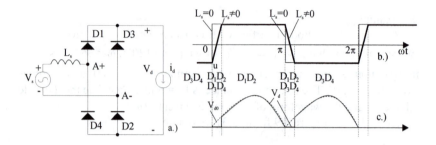

Figure 3.12. Current commutation in single-sided rectifier with I_d = ct.

a.) equivalent circuit; b.) source current; c.) rectified voltage

Ideally ($L_s = 0$), the source current will change stepwise from $-I_d$ to I_d at $\omega t = 0$ and $\omega t = \pi$ (Figure 3.12b). Due to the nonzero L_s, during commutation, all four diodes conduct and thus $V_d = 0$. For $\omega t < 0$, D_3D_4 conduct while after commutation ($\omega t > u$), only D_1D_2 are on.

As $V_d = 0$, the source voltage during commutation is dropped solely across inductance L_s

$$V_s\sqrt{2}\sin\omega t = \omega L_s \frac{di_s}{d(\omega t)} \qquad (3.21)$$

Through integration for the commutation interval (0,u)

$$\int_0^u V_s\sqrt{2}\sin\omega t \cdot d(\omega t) = \omega L_s \cdot \int_{-I_d}^{I_d} di_s = 2\omega L_s \cdot I_d \qquad (3.22)$$

We find

$$\cos u = 1 - \frac{2\omega L_s}{V_s\sqrt{2}} \cdot I_d \qquad (3.23)$$

Now the average d.c. voltage V_d is

$$V_d = V_{d0} - \frac{2\omega L_s}{\pi} \cdot I_d \qquad (3.24)$$

where $$V_{d0} = \frac{2}{2\pi} \cdot \int_0^\pi V_s\sqrt{2}\sin\omega t \cdot d(\omega t) = \frac{2\sqrt{2}}{\pi}V_s = 0.9 V_s \qquad (3.25)$$

is the ideal ($L_s = 0$) average d.c. voltage (Figure 3.12c).

So the source inductance L_s produces a reduction in the d.c. output voltage for constant d.c. output current. The current commutation is not instantaneous and during the overlapping period angle u all four diodes are conducting.

3.5. THREE-PHASE DIODE RECTIFIERS

In industrial applications three phase a.c. sources are available, so three-phase rectifiers seem the obvious choice (Figure 3.13).

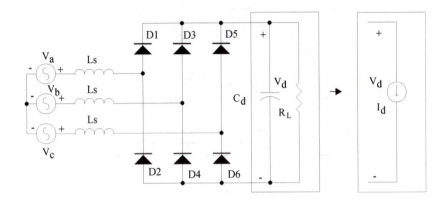

Figure 3.13. Three-phase diode rectifier

The load resistance R_L with a filtering capacitor C_d may be replaced by a constant d.c. current source I_d. Using the same rationale as in the previous paragraph we obtain

$$V_d = V_{d0} - \frac{3\omega L_s}{\pi} \cdot I_d \qquad (3.26)$$

$$\cos u = 1 - \frac{2\omega L_s}{V_{LL}\sqrt{2}} \cdot I_d \qquad (3.27)$$

with $V_{d0} = \dfrac{3\sqrt{2}}{\pi} V_{LL}$ where V_{LL} is the line voltage (rms).

The corresponding waveforms for $L_s = 0$ are shown in Figure 3.14 and for $Ls \neq 0$ in Figure 3.15.

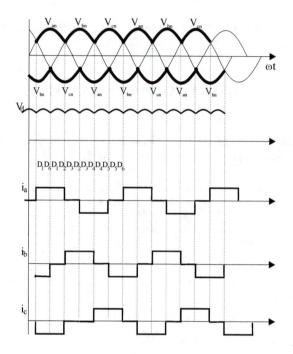

Figure 3.14. Three-phase ideal waveforms for $L_s = 0$

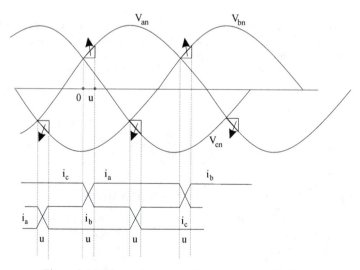

Figure 3.15. Three-phase current commutation with $L_s \neq 0$

For nonzero L_s a reduction of output d.c. voltage (3.26) is accompanied by all three phases conducting during the commutation angle u (Figure 3.15).

On the other hand, for constant d.c. voltage (infinite capacitance C_d), as for the single-phase rectifier, the source current waveform is as in Figure 3.16.

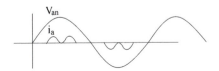

Figure 3.16. Three-phase rectifier with finite L_s and infinite C_d (V_d = ct.)

- the source current and voltage

We should note that in both cases the source currents are distorted and thus current harmonics are present in the power source.

Both current source and d.c. output voltage (current) harmonics in actual single-phase rectifiers are higher than in three-phase rectifiers.

Example 3.3. Commutation overlapping angle u.

For a single-phase or three-phase a.c. system (star connection) with phase voltage $V_s(t) = 120\sqrt{2}\sin 367t$, calculate the commutation angles u, ideal no-load voltage and load voltage of a single or three-phase diode rectifier delivering a constant d.c. current $I_d = 10A$ for the source inductance $L_s = 5mH$.

Solution:

For the single-phase diode rectifier, using (3.23)

$$\cos u = 1 - \frac{2\omega L_s}{V_s\sqrt{2}} \cdot I_d = 1 - \frac{2 \cdot 367 \cdot 5 \cdot 10^{-3}}{120\sqrt{2}} \cdot 10 = 0.783 \qquad (3.28)$$

$u = 22.727^0$.

The ideal no-load voltage V_{d0} (3.25) is

$$V_{d0} = \frac{2\sqrt{2}}{\pi}V_s = \frac{2\sqrt{2}}{\pi}120 = 108V \qquad (3.29)$$

$$V_d = V_{d0} - \frac{2\omega L_s}{\pi} \cdot I_d = 108 - \frac{2 \cdot 367 \cdot 5 \cdot 10^{-3}}{\pi} \cdot 10 = 96.312V \qquad (3.30)$$

For the three-phase diode rectifier (3.27) u is

$$\cos u = 1 - \frac{2 \cdot 367 \cdot 5 \cdot 10^{-3}}{120\sqrt{2} \cdot \sqrt{3}} \cdot 10 = 0.8746 \qquad (3.31)$$

$u = 12.22^0$.

V_{d0} and V_d (from 3.26) are

$$V_{d0} = \frac{3\sqrt{2}}{\pi} V_{LL} = \frac{3\sqrt{2}}{\pi} 120\sqrt{3} = 279.66V \qquad (3.32)$$

$$V_d = V_{d0} - \frac{3\omega L_s}{\pi} \cdot I_d = 279.66 - \frac{3 \cdot 367 \cdot 5 \cdot 10^{-3}}{\pi} \cdot 10 = 262.128V \quad (3.33)$$

Thus the filtering capacitor C_d is notably smaller in three-phase than in single-phase diode rectifiers.

Note: This rather detailed introduction to diode rectifiers has been given because they are used frequently as source-side PECs in most electrical drives while they show a strong indication of line current harmonics and commutation aspects, to be met with in all other PECs.

3.6. PHASE-CONTROLLED RECTIFIERS (A.C.-D.C. CONVERTERS)

Phase-controlled rectifiers — a.c.-d.c. converters — are used to provide controlled d.c. output to directly supply d.c. brush motors or as line-source (first-stage) PECs into two stage a.c.-d.c.-a.c. converters for a.c. drives.

In principle phase-controlled rectifiers might be fully controlled or semicontrolled. A rather complete survey of various phase-controlled rectifiers is given in Table 3.1.

It is important to note the power range and quadrant operation of various configurations. We also should note that besides thyristors, GTOs or IGBTs may be used. For high powers (MW and tens of MW), and as a.c. side converters in a.c.-d.c.-a.c. converters for a.c. motors, special configurations are used. They will be treated in their respective chapters. Also, unity input power factor configurations are available.

Principal details, numerical examples or digital simulation results on various rectifier configurations will occur in pertinent chapters on d.c. brush or large power a.c. motors.

Table 3.1. Phase-controlled rectifier circuits

Circuit type		Power range	Ripple frequency	Quadrant operation
	half wave single-phase	below 0.5KW	f_s	V_{an} i_{an} one quadrant

Table 3.1. (cont'd.)

	half wave three phase	up to 50KW	$3f_s$	two quadrant
	semi-converter single-phase	up to 75KW	$2f_s$	one quadrant
	semi-converter three-phase	up to 100KW	$3f_s$	one quadrant
	full converter single-phase	up to 75KW	$2f_s$	two quadrant
	full converter three-phase	up to 150KW	$6f_s$	two quadrant
	Dual converter single-phase	up to 15KW	$2f_s$	four quadrant
	Dual converter three-phase	up to 1500KW	$6f_s$	four quadrant

3.7. D.C.-D.C. CONVERTERS (CHOPPERS)

Choppers are d.c.-d.c. switch-mode converters with unipolar or bipolar current output capability. They are widely used in d.c. brush motor drives in single-phase output unipolar current configurations (Table 3.2) and for switched reluctance motors (SRM) in multiphase configurations (Figure 3.17).

Multiphase choppers for SRMs of various configurations have been proposed and they will be treated in more detail in Chapter 12 on SRM drives.

Table 3.2. Single-phase chopper configurations for d.c. brush motors

Type	Chopper configuration	e_a-I_a characteristics	Function
First quadrant (step-down) choppers			$V_a = V_0$ for S_1 on $V_a = 0$ for S_1 off and D_1 on
Second quadrant, regeneration (step-up) chopper			$V_a = 0$ for S_2 on $V_a = V_0$ for S_2 off and D_2 on
Two-quadrant chopper			$e_a = V_0$ for S_1 or D_2 on $e_a = V_0$ for S_2 or D_1 on $i_a > 0$ for S_1 or D_1 on $i_a < 0$ for S_2 or D_2 on
Two-quadrant chopper			$V_a = +V_0$ for S_1&S_2 on $V_a = -V_0$ for S_1&S_2 off and D_1&D_2 on
Four-quadrant chopper			S_4 on &S_3 off S_1&S_2 operated $V_a > 0$ i_a - reversible S_2 on &S_1 off S_3&S_4 operated $V_a < 0$ i_a - reversible

If an a.c. source is available, a diode rectifier and filter are used in front of all choppers (Figure 3.17).

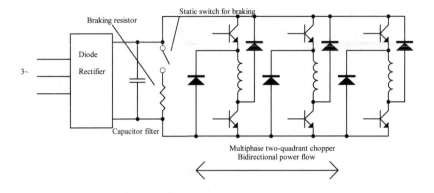

Figure 3.17. Multiphase d.c.-d.c. converters for switched reluctance motors

3.8. D.C.-A.C. CONVERTERS (INVERTERS)

D.c.-a.c. converters (inverters) are of two main types: voltage-source inverters (Figure 3.18) and current-source inverters (Figure 3.19) in the sense that the input source is of constant voltage or constant current character.

Voltage-source inverters are built in general with IGBTs or GTOs while current-source inverters are, in general, built with thyristors and GTOs and refer to high power levels (MW level and higher).

Inverters may be single phase or multiphase and they deliver bipolar current waveforms and allow for bi-directional power flow. They are used for a.c. motor drives.

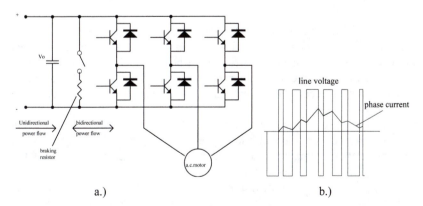

Figure 3.18. Voltage source PWM inverter:

a.) basic configuration, b.) output waveforms

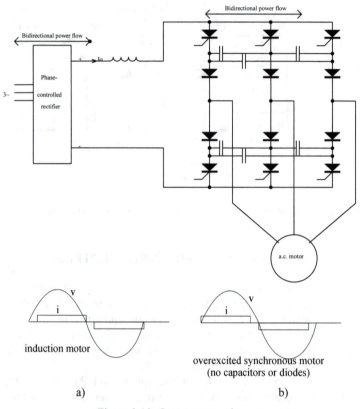

Figure 3.19. Current source inverter

a.) basic configuration; b.) ideal output waveforms (i – current, u – voltage)

When the a.c. motor exhibits a lagging power factor (induction motors), capacitors and diodes are required for successful commutation of thyristors in the current source inverter. In contrast, when the motor exhibits a leading power factor (overexcited synchronous motors), induced voltage (load) commutation is accomplished and thus the capacitors and the diodes are eliminated.

Though current source inverters allow for bidirectional power flow, the energy retrieved from the motor has to be either returned to the power source (Figure 3.19) or dumped into a braking resistor (Figure 3.18) when the source-side converter does not allow for bidirectional power flow.

The phase-controlled rectifier allows for bidirectional power flow that is so necessary during fast braking of high inertia loads. So, from this point of view, the configuration in Figure 3.19 is superior to that of Figure 3.18.

However, switching frequency is smaller with current source inverters and thus the motor current waveform is distorted, leading to a larger derating of the motor to avoid overheating.

On the other hand, with both a.c.-d.c.-a.c. converters unity input power factor and full bidirectional power flow are possible only with configurations such as those in Figure 3.20.

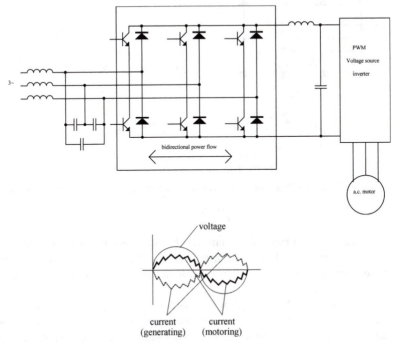

Figure 3.20. Bidirectional power flow (dual) a.c.-d.c. converter

with unity power factor and sinusoidal inputs

The diode rectifier works during motoring while the IGBTs work during generating when the source-side converter will work as a PWM voltage source inverter. Unity power factor and quasisinusoidal source currents and voltages may be obtained.

Though the phase-controlled rectifier-current source inverter configuration (Figure 3.19) provides for bidirectional power flow, the input power factor decreases with rectifier d.c. voltage (motor speed) reduction. To provide for unity power factor and sinusoidal input and output waveforms, the structure in Figure 3.21 with GTOs has been introduced.

The capacitors on the a.c. source-side converter are commutated properly by an additional leg of GTOs. The same is done on the current source side. Thus the voltages and currents are ideally sinusoidal both on the source side and on the motor side through PWM. Also, the unity power factor on the source side is performed with bidirectional power flow. Note, however, that rather large capacitors and filters are required while the number of PESs is 12 in Figure 3.20 and 16 in Figure 3.21.

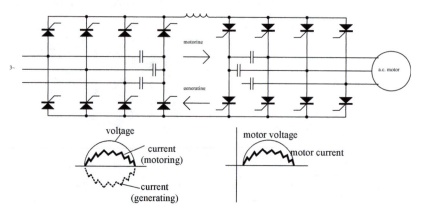

Figure 3.21. A.c.-d.c.-a.c. converter with bidirectional power flow and unity input

power factor

All the PECs introduced so far are characterized by hard switching of PESs – at nonzero voltage or current. As in PECs both conducting and commutation losses count, when switching frequency increases so does commutation loss.

A new breed of PECs, with a very rich literature, called soft-switching (or resonant) PECs, that provide commutation at zero voltage (for IBGTs) or at zero current (for GTOs) have been proposed. They boost the switching frequency one order of magnitude [1].

As soft-switching PECs are not yet commercial for drives we do not pursue them further here. Besides a.c.-d.c.-a.c. converters, there are also direct a.c.-a.c. converters.

3.9. DIRECT A.C.-A.C. CONVERTERS

Direct a.c.-a.c. converters are used in industry for high power synchronous and induction motor drives (for frequencies lower than that of the power source [2]) as cycloconverters).

Cycloconverters (Figure 3.22) represent, in the six pulse configuration, a group of 2 back-to-back fully controlled rectifiers for each phase.

The source voltage and current waveforms are both ideally sinusoidal. The commutation of PESs is provided by the power source and the maximum output frequency is about one third of the input frequency. More involved configurations allow higher output frequencies [2].

A general concept of direct a.c.-a.c. converters called matrix converter has been proposed in 1980 [3-6]. Matrix converters have just reached the markets.

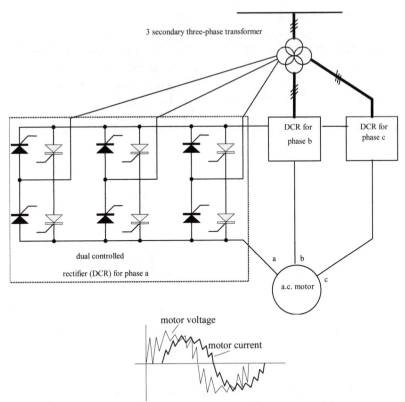

Figure 3.22. Six-pulse cycloconvertor for a.c. motor drives

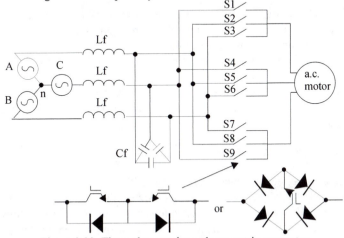

Figure 3.23. Three phase to three phase matrix converter

The matrix converter (Figure 3.23.) is a matrix of switches which allows for the connection of any input to any output phase.

The PWM techniques for matrix converters do in essence provide for a fictitious diode bridge rectifier d.c. link (Figure 3.24b) and then recompose the output line voltage (v_{ab}, Figure 3.24c) by pulse width modulating (PWM) this fictitious d.c. link voltage. One switch in each of the three banks conducts at any instant, so there are $3^3 = 27$ switching states; only lateral line switches are not closed in order to avoid line shorting.

The nine bidirectional power switches may be built either with two IGBTs and two fast switching diodes or from one IGBT and four switching diodes.

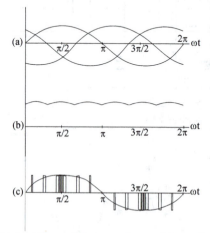

Figure 3.24. Matrix converter waveforms: a) input line voltage;
b) fictitious diode rectifier bridge like d.c. voltage link; c) output line voltage (v_{ab})

The $L_F C_F$ filter is essential for the commutation of a.c. switches when the inductive load current has to be transfered from one line to another; it also serves to filter the line current harmonics.

The matrix converter is bidirectional in principle, with the line current almost sinusoidal.

There are no constraints on the output frequency but the maximum voltage gain is 0.86 and thus the motor has to be sized at lower voltage and higher current.

3.9.1. Low cost PWM converters

While it has been proved that the 6-leg PWM voltage source converter (Figure 3.18) is the best choice for a.c. sinusoidal current output in three phase loads [9], for single phase loads lower power switch count converters have been analyzed [9]. As none of them is without notable demerits we represent here just one of them (Figure 3.25), of low cost, eventually suitable for split-phase capacitor-run a.c. (induction and permanent magnet synchronous) small motors for home appliance (HVAC: heating, ventilating, air conditioning).

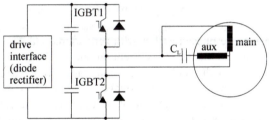

Figure 3.25. B2 inverter supplying a single phase load.

Only half of the d.c. link voltage is applied to the load, so the voltage rating of IGBTs is reduced, but, for the same load power, their current rating is increased. As the capacitor C_L in the load is fixed, the lowest frequency of the converter is also limited to perhaps about 6 to 10 Hz. Fortunately in HVAC applications this is acceptable.

3.10. SUMMARY

- A modern electric drive, capable of variable speeds, comprises, in general, a motor, a multistage power electronic converter, and a controller.

- Power electronic converters (PECs) process the power supplied to the motor and are classified in many ways.

- Basically the input-output waveforms are most important. PECs may be single stage or two stage. Single-stage converters are a.c.-d.c. or d.c.-a.c. or direct a.c.-a.c.

- Two-stage PECs are a.c.-d.c.-a.c. with an intermediate d.c. voltage type or current type link. Most commercial PECs are hard switched, that is, the PESs are turned on (or off) when the voltage (current) is nonzero.

- Soft-switching (resonant) PECs make use of soft switching (under zero voltage or current) and thus are characterized by reduced commutation losses for given switching frequency. Conversely, they allow switching frequencies an order of magnitude higher than for hard switching.

- The various power electronic switches (PESs) are characterized by voltage, current, dv/dt and di/dt limitations and a certain switching frequency limit.

- The line-frequency diode rectifiers presented in some detail in this chapter show the strong influence of source inductance on d.c. output voltage reduction with rectified current (and highly distorted source current) though the displacement power factor (DPF) is above 0.9 in all cases.

- Filters on the source side are required to attenuate the current harmonics in the a.c. power source. These aspects are common to all PECs, as will be seen in subsequent chapters.

3.11. PROBLEMS

3.1.   An ideal single-phase rectifier with zero source inductance L_s (Figure 3.23) produces constant d.c. current output $I_d = 50A$.

      a.   show the source current and load current waveforms;

      b.   with a sinusoidal voltage $V_s = 120V$ (rms) at 60 Hz, calculate the d.c. output voltage V_d and the d.c. power;

      c.   for case b, calculate the source current fundamental, the displacement power factor (DPF) and the power factor (PF).

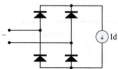

Figure 3.23. Ideal single-phase diode rectifier with zero source inductance
($L_s = 0$) and constant d.c. current I_d.

3.2.   A single-phase diode rectifier with constant output voltage $V_d = 120V$ is fed from an a.c. supply whose voltage is $V_s(t) = 120\sqrt{2} \cdot \sin 376t$. For the source inductance $L_s = 1mH$; calculate the time dependence of the d.c. output current.

3.3.   A three-phase diode rectifier has $V_s(t) = 120\sqrt{2} \cdot \sin 376t$ and the source inductance $L_s = 5mH$. The maximum commutation overlapping angle is $u = 30^0$. For this situation calculate the rectifier current I_d and the ideal and actual d.c. output voltage V_d.

3.12. SELECTED REFERENCES

1.   **N. Mohan, T.M. Underlund, W.P. Robbins**, Power electronics, second edition, John Wiley & Sons Inc., 1995.

2.   **L. Gyugyi, B.R. Pelly**, Static power frequency converters, John Wiley & Sons Inc., 1976.

3.   **A. Alessia, M.G.B. Venturini**, "Analysis and design of optimum-amplitude nine-switch direct a.c.-a.c. converter", IEEE Trans.vol.PE - 4, no.1, 1989, pp.101-112.

4.   **M. Milenovic, F. Mihalic**, "A switching matrix analysis of three phase boost rectifier", Record of IEEE-IAS-1996 Annual Meeting, vol.2, pp.1218-1224.

5.   **S. Bernet, K. Bernet, T.A. Lipo**, "The auxiliary resonant pole matrix converter - a new topology for high power applications", IBID, pp.1242-1249.

6.   **L. Huber, D. Borojevic**, "Space vector modulated three phase to three phase matrix converter with input power factor correction", IEEE Trans.vol.IA-31, no.6., 1995, pp.1234-1246.

7.   **T.Sawa, T. Kuine**, "Motor drive technology – history and vision for the future", keynote address at IEEE – PESC – 2004, Aachen, Germany.

8.   **C. Klumpner, F. Blaabjerg**, "Fundamental of the matrix converter technology", Chapter 3 in the book "Control in Power Electronics", Academic Press 2002, editors: M. Kazmierkowski, R. Khrishnan, F. Blaabjerg.

9.   **H. Kragh**, "Modeling, analysis and optimization of power electronic circuits for low cost drives", PhD Thesis, IET, Aalborg University, Denmark, 2000.

Chapter 4

D.C. BRUSH MOTORS FOR DRIVES

D.c. brush or d.c. commutator machines have traditionally been used in variable speed drives in the range of low (a few watts) to medium, 10MW, power ratings at low speeds. The popularity of d.c. brush motors in variable speed drives is primarily due to the lower cost of a single-stage (rectifier) power electronic converter (PEC) required for two-quadrant operation (one direction of motion). Four-quadrant operation, however, implies rather involved PEC configurations and controls.

In this chapter we will discuss d.c. brush motor basic topologies, state space equations, steady-state curves, losses, and transfer functions, as they will be used in subsequent chapters on d.c. brush motor drives.

4.1 BASIC TOPOLOGIES

As in any electric motor, the d.c. brush motor has two main parts: the stator (fixed) part and the rotor (movable) part. The rotor may be cylindrical (Figure 4.1a) or disk-shaped and contains a symmetric winding made of identical coils connected in series to the insulated copper sectors of the so-called mechanical commutator. The unipolar current injected through the brushes is converted into bipolar current in the rotor coils through the commutator copper sectors, in pace with the rotor position. The mechanical commutator is, in fact, an inverter (d.c.-a.c. converter) which changes the frequency from zero to $f_n = p.n$ where 2p is the number of stator and rotor poles (semi-periods) and n is the rotor speed.

Electromagnetic d.c. excitation (EE) may be replaced by permanent magnets (PMs) – Figure 4.1b and 4.2. High energy PMs may be replaced by a constant excitation (field) current fictitious (or superconducting) coil. The PM stator is ideally lossless as the PM magnetization (or demagnetization) losses are fairly small.

The disk rotor does not have an iron core and thus the rotor winding inductance is fairly small. Moreover, the design current density is higher than for coils in iron core slots, due to direct air exposure. Consequently, the electrical time constant of the disk d.c. brush motor is the lowest known (approximately 1ms for 1kW, 3000 rpm motors).

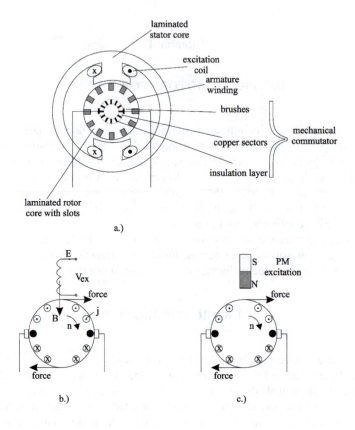

Figure 4.1. Cylindrical d.c. brush motor a.) topology, b.) schematics with electromagnetic excitation, c.) schematics with permanent magnets (PMs).

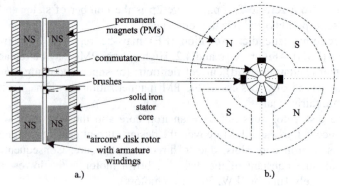

Figure 4.2. Disk d.c. brush motor (2p = 4poles)

a.) cross section; b.) axial view

Unfortunately, the power per unit mass of a disk d.c. brush motor is limited by the mechanical fragility of the "aircore" rotor, to 2-3kW and less than 6000 rpm. On the other hand, the cylindrical d.c. brush motor is power limited by the commutator to 10MW at low speeds for rotors with coils in slots of laminated iron cores. "Air core" cylindrical rotor windings used for small powers are now proposed to improve commutation (due to inductance reduction) and thus increase the power above 10 MW at very low speeds.

For details on d.c. brush motors, refer to [1].

4.2. THE MOTION-INDUCED VOLTAGE (e.m.f.)

As the rotor conductors (Figure 4.1) move in the field produced by the field current (or the PM), an e.m.f. is induced in each rotor coil. The rotor coils are connected in series between neighboring brushes through the copper sectors of the mechanical commutator.

As seen from Figure 4.1, there are at least two current paths in parallel. In general, there are 2a paths. The number of current paths depends on the number of poles, 2p, and the type of the armature winding (lap or wave type). The total number of conductors per rotor periphery is N and the flux linkage per pole is (half-period of excitation field in the airgap) λ_p.

The motion-induced voltage in the rotor, measured at the brushes, E, is also proportional to rotor speed n.

Finally:

$$E = K_e \cdot n \cdot \lambda_p; \quad K_e = \frac{p}{a} \cdot N \qquad (4.1)$$

The pole flux linkage λ_p is proportional to the average flux density in the airgap per pole B_{gav}, to the pole pitch τ and to the core stack length L

$$\lambda_p = B_{gav} \cdot \tau \cdot L \qquad (4.2)$$

As seen from Figure 4.1 the motion induced voltage (e.m.f.) in each coil does not change sign under one pole as the fixed brushes collect the voltage of the coils temporarily located (though in continuous motion) under one pole (field) polarity. The armature current changes polarity when the respective coil is short-circuited by the brushes.

This phenomenon is called mechanical commutation of currents. For more details on the d.c. brush motor topology, principle, performance and design, see [1].

4.3. PERFORMANCE EQUATIONS: d-q MODEL

The d.c. brush motor schematics in Figure 4.1 has ideally an electrical 90° spatial phase difference between the stator (excitation, or PM) magnetic field and rotor (armature) magnetic field (brush axis). Consequently, in the absence of magnetic saturation, there is no interaction (transformer induced voltage) between excitation and armature windings. The excitation (PM) circuit may be termed the field circuit while the rotor (armature) winding may be termed the torque circuit. *Thus the d.c. brush motor allows for separate (decoupled) control of field and torque currents (or torque)*, which is an extraordinary built-in property of d.c. brush motor. Though the stator excitation winding may also be series-connected to the brushes, separate excitation is considered in what follows.

There is an interaction between the stator and rotor windings only through motion voltage, E, induced in the rotor by the stator excitation current. The state-space equations of a d.c. brush motor are

$$V_{ex} = R_e \cdot i_e + L_e \cdot \frac{di_e}{dt}$$
$$V_a = R_a \cdot i_a + E + L_a \cdot \frac{di_a}{dt} \tag{4.3}$$

The pole flux, λ_p, in the rotor windings, is dependent of i_e, in the absence of magnetic saturation, and also of i_a at high rotor currents if magnetic saturation occurs.

So, in general

$$\lambda_p = G \cdot i_e \tag{4.4}$$

where G is the "motion-produced" inductance between the stator and rotor orthogonal windings. The electromagnetic torque, T_e, expression is obtained from the electromagnetic power P_e

$$T_e = \frac{P_e}{2\pi n}; \quad P_e = E \cdot i_a \tag{4.5}$$

Thus

$$T_e = \frac{E \cdot i_a}{2\pi n} = \frac{K_e}{2\pi} \cdot \lambda_p \cdot i_a \tag{4.6}$$

The d.c. brush motor parameters are R_e, R_a, as resistances and L_e, G, L_a as inductances. We should add the inertia, J, as defined through motion equation

$$J 2\pi \cdot \frac{dn}{dt} = T_e - T_{load} - B \cdot n; \quad \frac{d\theta_r}{dt} = 2\pi n \tag{4.7}$$

where T_{load} is the load torque and B is the friction torque coefficient.

4.4. STEADY-STATE MOTOR CHARACTERISTICS

Steady-state means constant speed (dn/dt = 0) and constant currents (i_e = ct, i_a = ct). The steady-state voltage equations, obtained from (4.1)-(4.3) are

$$V_{ex} = R_e \cdot i_e \tag{4.8}$$

$$V_a = R_a \cdot i_a + K_e \cdot n \cdot \lambda_p \qquad (4.9)$$

$$T_e = \frac{K_e}{2\pi} \cdot \lambda_p \cdot i_a = T_{load} + B \cdot n \qquad (4.10)$$

The main characteristic of a motor is the torque (T_e) versus speed (n, or Ω_r) curve called the mechanical characteristics.

From (4.8)-(4.10)

$$V_a = K_e \cdot n \cdot \lambda_p + R_a \cdot \frac{T_e \cdot 2\pi}{K_e \cdot \lambda_p} \qquad (4.11)$$

Modifying the speed, for a given electromagnetic torque (T_e) may be done through:
- voltage, V_a, control;
- flux, λ_p, control.

As apparent from (4.11), it is also possible to add a resistance in series with R_a, to modify speed. This is, however, an energy consuming method which is to be avoided in modern electric drives.

The torque/speed curves obtained through voltage and flux control, all straight lines, are shown on Figure 4.3 in relative units.

In Figure 4.3, n_b, T_{eb} and λ_{pb} are base-speed, base-torque and base-flux values (continuous duty, rated (maximum) voltage).

Above the base-speed flux-weakening at constant armature voltage V_a (limited by the PEC capabilities) is performed, in general, at constant power P_e up to $n_{max} / n_b = 2$-3. Obviously, in PM brush motors flux weakening is not possible.

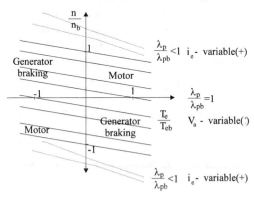

Figure 4.3. The torque/speed curves for variable speed

Also it is evident that both speed control methods illustrated in Figure 4.3, are of the high energy conversion ratio type. So far we have not discussed motor losses though they are crucial in defining the capability of an electric drive.

4.5. D.C. BRUSH MOTOR LOSSES

The performance equations already show two kinds of losses in the motor: the armature (copper) losses represented by R_a and the friction (mechanical) losses P_{mec}

$$P_{mec} \approx B \cdot \omega_r \qquad (4.12)$$

In reality, the friction (mechanical) losses are a complex function of speed, depending on application. For example, in an electric train, the wheel-track and head and lateral surface air-drag losses all constitute load at constant speed and have complex mathematical expressions.

The N-S-N-S pole sequence in the stator poles field produces hysteresis and eddy current losses in the rotor laminated iron core. They are called core losses P_{iron}

$$P_{iron} \approx \left[C_h (p\,n) + C_e (p\,n)^2 \right] \cdot B_{iron}^{\,2} \cdot Gc \qquad (4.13)$$

The first term represents the hysteresis losses while the second takes care of the eddy current losses. B_{iron} is the flux density in the rotor core and Gc the respective core weight.

The slotted rotor core has two main zones: the teeth and back core (or yoke). Further, the slot opening presence leads to stator-pole shoe core loss due to rotor (armature) winding m.m.f. These are called additional (surface) core losses, P_{add}.

Now the armature resistance R_a includes, in principle, the brush, brush-commutator contact and the equivalent resistance of commutator sectors. As the brush-commutator contact surface electrical resistance depends on many factors such as speed n, current i_a, brush-spring tension, especially in low voltage motors, the commutator losses, P_{com}, are to be calculated separately

$$P_{com} = \Delta V_b (i_a, n) \cdot i_a \qquad (4.14)$$

Finally, the excitation winding losses, P_{ex}, have to be considered

$$P_{ex} = R_e \cdot i_e^{\,2} \qquad (4.15)$$

Figure 4.4 summarizes the breakdown of motor losses as developed above.

The efficiency η is: $\qquad\qquad \eta = P_{output} / P_{input} \qquad (4.16)$

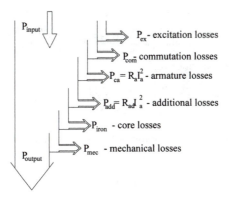

- P_{ex} - excitation losses
- P_{com} - commutation losses
- $P_{ca} = R_a I_a^2$ - armature losses
- $P_{add} = R_{ad} I_a^2$ - additional losses
- P_{iron} - core losses
- P_{mec} - mechanical losses

Figure 4.4. Loss breakdown in d.c. brush motors

<u>Example 4.1. Steady state performance</u>
A d.c. motor with separate excitation has the following data: rated power P_n = 3kW, rated (base) speed n = 1200 rpm, P_{com} = 0.5%P_n, P_{add} = 0.5%P_n, P_{iron} = P_{mec} = 1%P_n and P_{co} = 4%P_n, at rated (maximum) voltage V_{an} = 110V. The excitation losses are neglected.
Calculate:
a. total losses and rated efficiency, η_n;
b. rated current, I_n, and armature resistance, R_a;
c. brush voltage drop, ΔV_b;
d. motion-induced voltage, E_n;
e. rated (base) electromagnetic torque, T_{eb};
f. shaft torque, T_{load};
g. ideal no-load speed, n_0;
h. the armature voltage, V_{ag}, required to produce rated generator braking torque at rated speed.
<u>Solution</u>:
a. The total losses $\sum P$ are

$$\sum P = P_{co} + P_{com} + P_{add} + P_{mec} + P_{iron} =$$
$$\left(4.0 + 0.5 + 0.5 + 1 + 1\right) \cdot \frac{3000}{100} = 210 \text{ W} \tag{4.17}$$

The efficiency:

$$\eta_n = \frac{P_n}{P_n + \sum P} = \frac{3000}{3000 + 210} = 0.9345 \tag{4.18}$$

b. The rated current I_n is

$$I_n = \frac{P_n}{\eta_n \cdot V_n} = \frac{3000}{0.9345 \cdot 110} = 29.18 \text{ A} \qquad (4.19)$$

The armature resistance R_a is

$$R_a = \frac{P_{co}}{I_n^2} = \frac{0.04 \cdot 3000}{29.18^2} = 0.140 \ \Omega \qquad (4.20)$$

c, d. From the voltage equation (4.9), adding the brush voltage drop ΔV_b, we obtain

$$E_n = V_n - R_a \cdot I_n - \Delta V_b \qquad (4.21)$$

with

$$\Delta V_b = \frac{P_{com}}{I_n} = \frac{0.005 \cdot 3000}{29.18} = 0.514 \text{ V} \qquad (4.22)$$

From (4.9)

$$E_n = 110 - 0.14 \cdot 29.18 - 0.514 = 105.4 \text{ V} \qquad (4.23)$$

e. Rated electromagnetic torque, T_{en}, is calculated from (4.6)

$$T_{en} = \frac{E_n \cdot I_n}{2\pi n} = \frac{105.4 \cdot 29.18}{2\pi \cdot \frac{1200}{60}} = 24.48 \text{ Nm} \qquad (4.24)$$

f. The shaft torque, T_{load}, comes directly from rated power P_n and speed n_n

$$T_{load} = \frac{P_n}{2\pi n_n} = \frac{3000}{2\pi \cdot \frac{1200}{60}} = 23.88 \text{ Nm} \qquad (4.25)$$

g. The ideal no-load speed, n_0, corresponds to zero armature current in the voltage equation (4.9)

$$V_{an} = K_e \cdot n_0 \cdot \lambda_p \qquad (4.26)$$

but $K_e.\lambda_p$ is constant

$$K_e \cdot \lambda_p = \frac{E_n}{n_n} = \frac{105.4}{\frac{1200}{60}} = 5.27 \text{ Wb} \qquad (4.27)$$

Consequently

$$n_0 = \frac{V_{an}}{K_e \cdot \lambda_p} = \frac{110}{5.27} = 20.8728 \text{ rps} = 1252.37 \text{ rpm} \qquad (4.28)$$

h. For regenerative braking the armature current becomes negative $I_{gn} = -I_n$ also $\Delta V_b = -\Delta V_b$, E_n remains the same and thus, from (4.9),

$$V_{ag} = E_n - R_a \cdot i_a - \Delta V_b = 105.4 - 0.14 \cdot 29.18 - 0.514 = 100.8 \text{ V} \quad (4.29)$$

So the voltage produced by the PEC should be simply reduced below the e.m.f. E_n level to produce regenerative braking for a given speed. The operating point moves from A in the first quadrant to A' in the second quadrant (Figure 4.5).

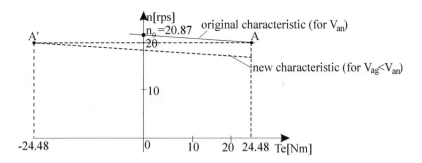

Figure 4.5. Torque/speed curves when switching from motoring to generating

4.6. VARYING THE SPEED

Example 4.2.
For the motor in example 4.1 and assuming that the mechanical losses are proportional to speed squared, while core losses depend, in addition, on flux squared, calculate:
a. the voltage V_a required and efficiency at $n_n / 2 = 600$ rpm for motoring at rated current;
b. for rated power and rated current at $2n_n = 2400$ rpm, calculate the flux weakening ratio and efficiency.
Solution:
a. From (4.21)

$$V_a = R_a \cdot i_a + E_n \cdot \frac{n}{n_n} + \Delta V_b =$$

$$0.14 \cdot 29.18 + 105.4 \cdot \frac{60}{1200} + 0.514 = 57.2992 \text{ V}$$

$$(4.30)$$

The mechanical and core losses are reduced 4 times (as the speed is halved) while the commutator losses remain the same as the current remains constant, I_n. So the total loss ΣP is

$$\sum P = 0.04 P_n + (0.01 + 0.01) \cdot P_n \cdot \left(\frac{1}{2}\right)^2 + (0.005 + 0.005) \cdot P_n \tag{4.31}$$

$$= (0.04 + 0.005 + 0.01) \cdot 3000 = 165 \text{ W}$$

The input power, P_{input}, is

$$P_{input} = V_a \cdot I_n = 57.2992 \cdot 29.18 = 1672 \text{ W} \tag{4.32}$$

The efficiency η is

$$\eta = \frac{P_{input} - \sum P}{P_{input}} = \frac{1672 - 165}{1672} = 0.90 \tag{4.33}$$

It should be noted that even for half the rated speed, at full torque, the efficiency remains high. Consequently, varying speed through varying armature voltage is a high efficiency method.

b. When raising the speed above the base speed, at rated current and voltage, the e.m.f. remains the same as in (4.22): $E = E_n = 105.4$ V.

As the speed is doubled $n = 2400$ rpm the new flux level λ_p' is

$$\frac{\lambda_p'}{\lambda_p} = \frac{n_n}{n} = \frac{1200}{2400} = \frac{1}{2} \tag{4.34}$$

Thus the flux is halved and so is the electromagnetic torque T_e

$$\frac{T_e}{T_{en}} = \frac{\lambda_p'}{\lambda_p} = \frac{1}{2} \tag{4.35}$$

Now the losses, $\sum P'$, are

$$\sum P' = R_a \cdot I_n^2 + p_{add} + p_{iron} + p_{mec} + p_{com} =$$

$$= 0.14 \cdot 29.18^2 + \left(0.005 + (0.01 + 0.01) \cdot \left(\frac{2400}{1200}\right)^2 + 0.005\right) \cdot 3000 = 300 \text{W}$$

$$\tag{4.36}$$

As the input power, P_{input} is still

$$P_{input} = V_n \cdot I_n = 110 \cdot 29.18 = 3209.8 \text{ W} \tag{4.37}$$

The efficiency η is

$$\eta = \frac{P_{input} - \sum P'}{P_{input}} = \frac{3209.8 - 300}{3209.8} = 0.9065 \tag{4.38}$$

Again, the efficiency is high.

The above results are summarized in Figure 4.6.

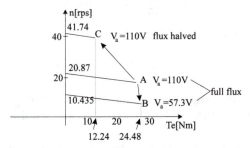

Figure 4.6. Torque/speed curves for varying speed from 1200 rpm (A)

to 600 rpm (B) and, respectively, to 2400 rpm (C) for constant current.

Note that the electromagnetic power for points A and C is the same, while the electromagnetic torque for points A and B is the same. A 3 to 1 speed range (n_{max} / n_b) for constant power and current is quite feasible.

We may infer from here that below rated (base) speed the torque may be maintained constant for variable armature voltage while above rated (base) speed the electromagnetic power may be maintained constant through flux weakening at constant armature voltage as shown schematically in Figure 4.7.

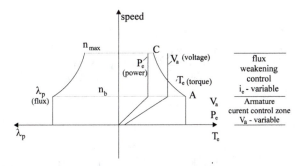

Figure 4.7. Speed/torque, power, voltage envelopes for constant current

Note: PM d.c. brush motors do not allow flux weakening and thus n_b = n_{max}.

There are applications with a constant power-wide speed zone (from n_b to n_{max}: 2-4n_b) such as in spindle drives where only electromagnetic excitation is adequate.

4.7. TRANSIENT OPERATION FOR CONSTANT FLUX

Constant flux means constant excitation (field) current or PM excitation. Consequently, in the performance equations (4.1)-(4.7), $di_e/dt = 0$ (λ_p = ct). Torque equation (4.6) becomes

$$T_e = \frac{K_e}{2\pi} \cdot \lambda_p \cdot I_a = K_T \cdot I_a \qquad (4.39)$$

Similarly, the e.m.f. equation (4.3) is

$$E = K_e \cdot n; \quad K_T = \frac{K_e}{2\pi} \qquad (4.40)$$

K_T is called the torque/current constant and in reality is constant only when the magnetic saturation level is constant or in the absence of magnetic saturation.

The variables, in performance equations, are now the armature current, i_a, and the speed, n. Equations (4.1)-(4.7) reduce to

$$V_a = R_a \cdot i_a + L_a \cdot \frac{di_a}{dt} + K_e n$$

$$2\pi J \cdot \frac{dn}{dt} = \frac{K_e}{2\pi} \cdot i_a - T_{load} - B \cdot n \qquad (4.41)$$

$$\frac{d\theta_r}{dt} = 2\pi n$$

As (4.41) constitutes a system of linear differential equations (no product of variables), Laplace transform may be used. For zero initial conditions we obtain

$$\tilde{V}_a = (R_a + sL_a) \cdot \tilde{i}_a + K_e \tilde{n}$$

$$s\tilde{n} = \frac{1}{2\pi J} \cdot \left(\frac{K_e}{2\pi} \cdot \tilde{i}_a - \tilde{T}_{load} - B \cdot \tilde{n} \right) \qquad (4.42)$$

$$s\tilde{\theta}_r = 2\pi \tilde{n}$$

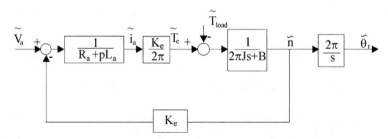

Figure 4.8. Constant flux block diagram of d.c. brush motor

Equation (4.42) suggests the block diagram of Figure 4.8, valid for small or large transients. So under constant flux (field current), the d.c. brush motor has two input variables $\tilde{V}_a$ and $\tilde{T}_{load}$ and three state variables $\tilde{i}_a$, n and $\tilde{\theta}_r$. Only with a single input – either voltage $\tilde{V}_a$ for speed control ($\tilde{T}_{load}$ = 0) or load torque $\tilde{T}_{load}$ for torque control ($\tilde{V}_a$ = 0) – cascaded transfer functions can be obtained

$$\tilde{i}_a = \frac{\tilde{V}_a}{\left(R_a + sL_a\right) + K_e \cdot \dfrac{K_e}{2\pi} \cdot \dfrac{1}{2\pi Js + B}} \; ; \tilde{n} = \frac{K_e \cdot \tilde{i}_a}{2\pi \cdot \left(2\pi Js + B\right)} ; \tilde{T}_{load} = 0 \qquad (4.43)$$

or

$$\tilde{i}_a = \frac{\tilde{T}_{load}}{\dfrac{K_e}{2\pi} + \dfrac{\left(2\pi Js + B\right) \cdot \left(R_a + sL_a\right)}{K_e}} ; \tilde{n} = \frac{-\left(R_a + sL_a\right)\tilde{i}_a}{K_e} ; \tilde{V}_a = 0 \quad (4.44)$$

The cascaded block diagrams described by (4.43)-(4.44) are shown in Figure 4.9.

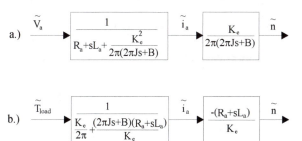

Figure 4.9. Cascaded transfer function for constant flux

a.) for speed control; b.) for torque control

The cascaded block diagrams are instrumental in designing the torque, speed or position control closed loops as shown in Chapter 7. The $\tilde{V}_a / \tilde{i}_a$ transfer functions have in both cases, Figure 4.9, the same poles (denominator's zeros) as a motor property

$$2\pi \cdot \left(2\pi Js + B\right) \cdot \left(R_a + sL_a\right) + K_e^{\,2} = 0 \qquad (4.45)$$

Both poles $s_{1,2}$ have a negative real part, so the response is always stable. For B = 0 $s_{1,2}$ we have the simplified form

$$s_{1,2} = \frac{-1 \pm \sqrt{1 - 4\tau_e / \tau_{em}}}{2\tau_e} \qquad (4.46)$$

with $$\tau_e = \frac{L_a}{R_a}; \quad \tau_{em} = \frac{4\pi^2 JR_a}{K_e^2} \quad\quad (4.47)$$

where τ_e is the electrical time constant while τ_{em} is the electromechanical time constant. As seen from (4.46) for $4\tau_e \le \tau_{em}$, the response is aperiodically stable, while for $4\tau_e > \tau_{em}$ the response of the motor is periodically stable.

Low inertia fast response drives qualify for $4\tau_e > \tau_{em}$ and thus open-loop motor response is periodically stable.

4.8. PM D.C. BRUSH MOTOR TRANSIENTS

Example 4.3.
A PM d.c. brush motor has the following data: V_n = 110V, I_n = 10A, R_a = 0.5Ω, n_n = 1200rpm, τ_e = 2ms. For a 10V step voltage increase, calculate the speed response for J = 0.005kgm^2 and J' = 0.05kgm^2, for constant load torque ($\tilde{T}_{load}$ = ct.) and B = 0 (no friction torque).

Solution:
From the voltage equation under steady state (4.20) we may calculate the e.m.f. E_n

$$E_n = V_n - R_a \cdot I_n = 110 - 0.5 \cdot 10 = 105 \text{ V} \quad\quad (4.48)$$

From (4.40)

$$K_e = \frac{E_n}{n_n} = \frac{105}{\dfrac{1200}{60}} = 5.25 \text{ Wb} \quad\quad (4.49)$$

Also from (4.40)

$$K_T = \frac{K_e}{2\pi} \quad\quad (4.50)$$

The electromagnetic torque T_{em} (4.39) is

$$T_{em} = K_T \cdot I_n = 0.836 \cdot 10 = 8.36 \text{ Nm} \quad\quad (4.51)$$

Under steady state, the load torque T_{load} is equal to the motor torque T_{em}.
In our case, it means that the initial and final current values are the same. Eliminating the current i_a in (4.41) we obtain

$$\tau_e \tau_{em} \frac{d^2 n}{dt^2} + \tau_{em} \frac{dn}{dt} + n = \frac{V_a}{K_e} - \frac{T_{load} \cdot R_a}{K_T \cdot K_e} \quad\quad (4.52)$$

The roots of the characteristic equation of (4.52) are evidently equal to $s_{1,2}$ of (4.46). With $\tau_e = 2\text{ms}$

$$\tau_{em} = \frac{4\pi^2 JR_a}{K_e^2} = \begin{cases} \dfrac{4\pi^2 0.005 \cdot 0.5}{5.25^2} = 3.577 \cdot 10^{-3}\text{s} \\[4mm] \dfrac{4\pi^2 0.05 \cdot 0.5}{5.25^2} = 35.77 \cdot 10^{-3}\text{s} \end{cases} \tag{4.53}$$

Consequently (from 4.46)

$$s_{1,2} = \frac{-1 \pm \sqrt{1 - 4 \cdot 2 / 3.577}}{4 \cdot 10^{-3}} = -250 \pm j280; \quad \text{for } \tau_{em} = 3.577\text{ms}$$

and $\hspace{10cm}$ (4.54)

$$s_{1,2} = \frac{-1 \pm \sqrt{1 - 4 \cdot 2 / 35.77}}{4 \cdot 10^{-3}} = -250 \pm 218.8; \quad \text{for } \tau_{em} = 35.77\text{ms}$$

The final value of speed is obtained from (4.52) with zero derivatives

$$\begin{aligned}(n)_{t=\infty} &= \frac{(V_a)_{t=\infty}}{K_e} - \frac{T_{load} \cdot R_a}{K_T K_e} = \frac{(110+10)}{5.25} - \frac{8.36 \cdot 0.5}{0.836 \cdot 5.25} = \\ &= 21.904\text{rps} = 1314.27 \text{ rpm}\end{aligned} \tag{4.55}$$

The solution of (4.52) for the two cases is

$$n(t) = (n)_{t=\infty} + A \cdot e^{-250t} \cdot \cos(280t + \gamma); \quad \text{for } \tau_{em} = 3.577 \text{ ms}$$

and $\hspace{10cm}$ (4.56)

$$n(t) = (n)_{t=\infty} + A_1 \cdot e^{-468.8t} + A_2 \cdot e^{-31.2t}; \quad \text{for } \tau_{em} = 35.77 \text{ ms}$$

The initial conditions refer to the fact that

$$(n)_{t=0} = 20\text{rps}; \quad \left(\frac{dn}{dt}\right)_{t=0} = 0 \tag{4.57}$$

Finally,

$$\begin{aligned} A &= 2.54; \quad \gamma = -41.76^0 \\ A_1 &= 0.278; \quad A_2 = -2.046 \end{aligned} \tag{4.58}$$

The two speed responses are drawn in Figure 4.10.

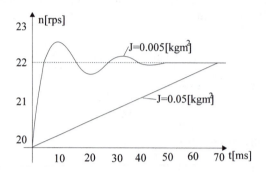

Figure 4.10. Speed responses to a step voltage increase from 110V to 120V at

constant load torque

4.9. TRANSIENT OPERATION FOR VARIABLE FLUX

Variable flux means variable excitation current. This time the complete set of performance equations (4.1)-(4.7) has to be used. Even in the absence of magnetic saturation, when $\lambda_p = G \cdot i_e$, there are products of variables ($i_e * i_a$ or $i_e * n$) which render the system nonlinear

$$
\frac{di_e}{dt} = \frac{V_{ex} - R_e \cdot i_e}{L_e}
$$

$$
\frac{di_a}{dt} = \frac{V_a - R_a \cdot i_a - n \cdot G \cdot i_e}{L_a} \tag{4.59}
$$

$$
\frac{dn}{dt} = \frac{1}{2\pi J}\left(\frac{G \cdot i_e}{2\pi}i_a - T_{load} - B \cdot n\right)
$$

The position θ_r is left out as position control is primarily performed with PM d.c. brush motor drives.

Through numerical methods, such as Runge Kutta-Gill, etc. the system (4.59) may be solved for large signal variables V_{ex}, V_a and T_{load}.

However, for the closed-loop control, design linearization around a steady-state point is standard

$$
\begin{aligned}
&V_{ex} = V_{ex0} + \Delta V_{ex}; \quad V_a = V_{a0} + \Delta V_a; \quad i_a = i_{a0} + \Delta i_a; \\
&i_{ex} = i_{ex0} + \Delta i_{ex}; \quad T_{load} = T_{load0} + \Delta T_l; \quad n = n_0 + \Delta n;
\end{aligned} \tag{4.60}
$$

For the initial steady-state point, $d/dt = 0$ in (4.59)

$$V_{ex0} = R_e \cdot i_{e0}$$
$$V_{a0} = R_a \cdot i_{a0} + n_0 \cdot G \cdot i_{e0} \tag{4.61}$$
$$\frac{G \cdot i_{e0}}{2\pi} i_{a0} = T_{load0} + B \cdot n_0$$

From (4.59) with (4.60)-(4.61) we get the matrix

$$
\begin{vmatrix} \Delta V_{ex} \\ \Delta V_a \\ \Delta T_1 \end{vmatrix} =
\begin{vmatrix}
R_e + sL_e & 0 & 0 \\
n_0 \cdot G & R_a + sL_a & G \cdot i_{e0} \\
\dfrac{G \cdot i_{a0}}{2\pi} & \dfrac{G \cdot i_{e0}}{2\pi} & -(J2\pi s + B)
\end{vmatrix} \cdot
\begin{vmatrix} \Delta i_e \\ \Delta i_a \\ \Delta n \end{vmatrix} \tag{4.62}
$$

It is now evident, again, that the excitation circuit is decoupled from the armature circuit. The eigenvalues of matrix (4.62) are obtained by solving its determinant equation

$$\left(R_e + sL_e \right) \cdot \left(\left(R_a + sL_a \right)(J2\pi s + B) + \frac{G^2 \cdot i_{e0}^2}{2\pi} \right) = 0 \tag{4.63}$$

The first root of (4.63) s_0 is related to excitation

$$s_0 = -\frac{L_e}{R_e} \tag{4.64}$$

The other two roots $s_{1,2}$ are identical to those obtained for constant flux (4.43)-(4.44). The field current i_e, when varied through ΔV_{ex}, produces a notable delay in the current Δi_a and speed Δn response and thus should be avoided when fast transients are required. On the other hand, variable flux is useful for extending the speed/torque envelope for a given armature voltage V_{an}.

4.10 SPEED/EXCITATION VOLTAGE TRANSFER FUNCTION

Example 4.4.
For a d.c. brush motor with separate excitation, find the speed to excitation voltage transfer function based on the following numerical data: i_{e0} = 5A, R_e = 1Ω, L_e = 1H, R_a = 0.1Ω, L_a = 5mH, I_{a0} = 100A, J = 1Kgm^2, n_0 = 1200rpm, B = 0, V_{a0} = 210V.
Solution:
The required transfer function may be obtained from (4.62), with ΔV_a = 0 and $\Delta T_1 = 0$, by eliminating Δi_e and Δi_a

$$\frac{dn}{dv_{ex}} = \frac{G \cdot i_{a0} \cdot \left(R_a + sL_a \right) - n_0 \cdot G^2 \cdot i_{e0}}{\left(R_e + sL_e \right) \cdot \left[G^2 \cdot i_{e0}^2 + 4\pi^2 Js \cdot \left(R_a + sL_a \right) \right]} \tag{4.65}$$

In (4.65) we have all the parameters with the exception of G

$$E_0 = V_{a0} - R_a \cdot i_{a0} = 210 - 0.1 \cdot 100 = 200 \text{ V} \qquad (4.66)$$

$$G = \frac{E_0}{n \cdot i_{e0}} = \frac{200}{\dfrac{1200}{60} \cdot 5} = 2 \text{ H} \qquad (4.67)$$

Finally,

$$\frac{dn}{dV_{ex}} = \frac{2 \cdot 100 \cdot (0.1 + 0.005s) - 20 \cdot 2^2 \cdot 5}{(1 + 1 \cdot s)(2^2 \cdot 5^2 + 4\pi^2 \cdot 1 \cdot s \cdot (0.1 + 0.005s))} =$$
$$= \frac{s - 380}{(1 + s)(100 + 3.4438s + 0.1972s^2)} \qquad (4.68)$$

4.11. THE D.C. BRUSH SERIES MOTOR

The d.c. brush series motor schematics is shown in Figure 4.11. The excitation and armature currents are equal to each other unless an additional resistance R_{ead} is connected in parallel to the excitation circuit to produce flux weakening.

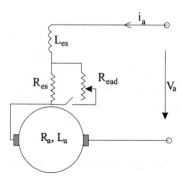

Figure 4.11. The d.c. brush series motor

With $R_{ead} = \infty$, the governing equations in terms of symbols shown in Figure 4.11 are

$$V_a = (R_a + R_{es}) \cdot i_a + (L_a + L_{es}) \frac{di_a}{dt} + nGi_a$$
$$T_e = \frac{Gi_a^2}{2\pi} = J2\pi \frac{dn}{dt} + T_{load} + B \cdot n \qquad (4.69)$$

As it contains products of variables, the system (4.69) is nonlinear and thus solving it directly may be performed only through numerical methods.

For small perturbations, we may, however, linearize the equations

$$V_a = V_{a0} + \Delta V_a; \quad T_{load} = T_{load0} + \Delta T_l;$$
$$n = n_0 + \Delta n; \quad i_a = i_{a0} + \Delta i_a$$

(4.70)

to find

$$\left| \begin{array}{c} \Delta V_a \\ \Delta T_l \end{array} \right| = \left| \begin{array}{cc} R_a + R_{es} + n_0 \cdot G + s(L_a + L_{es}) & G i_{a0} \\ 2\dfrac{G}{2\pi} i_{a0} & -(2\pi Js + B) \end{array} \right| \cdot \left| \begin{array}{c} \Delta i_a \\ \Delta n \end{array} \right|$$

(4.71)

The determinant of (4.71) leads to the eigenvalues of the system

$$\left(R_a + R_{es} + n_0 \cdot G + s \cdot (L_a + L_{es})\right) \cdot (2\pi Js + B) + 2\dfrac{G^2}{2\pi} \cdot i_{a0}^{\ 2} = 0 \quad (4.72)$$

The apparent electrical time constant in (4.72), if compared to the constant flux case (4.43), depends on speed through the term $n_0 G$. At zero speed as $L_{es} > L_a$, and $R_{es} < R_a$, the electrical time constant is again larger than for the constant flux case when $\tau_e = L_a / R_a$.

Fast torque response is thus not easily expected from the d.c. brush series motor. However, it does not require a separate source to supply the field current, and flux weakening is simply possible through the shunt resistor R_{ead} (Figure 4.11).

For steady state, the speed/torque curve may be obtained from (4.67) with d/dt = 0.

$$i_a = \dfrac{V_a}{(R_a + R_{es}) + n \cdot G} = \sqrt{\dfrac{2\pi}{G} T_e}$$

(4.73)

At zero current (torque) the speed is infinite and thus the speed/torque curve is called mild (Figure 4.12).

Reducing the speed may be accomplished through V_a reduction, as done with d.c. brush motors with separate excitation.

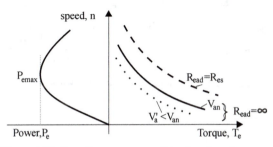

Figure 4.12. Speed/torque curve of d.c. brush series motor

Flux weakening may be performed through R_{ead} to obtain speeds above the rated (base) value for full armature voltage V_{an}. The electromagnetic power P_e is

$$P_e = T_e \cdot 2\pi n = \frac{V_a^2 \cdot G \cdot n}{\left(R_a + R_{es} + nG\right)^2} \qquad (4.74)$$

Over a certain speed range electromagnetic power does not vary much and is, anyway, limited. This is a unique characteristic of the d.c. brush series motor that is so beneficial in transportation applications where the on-board installed power level is limited.

4.12. THE A.C. BRUSH SERIES MOTOR

Known also as the universal motor, the a.c. brush series motor is still used extensively in some home appliances such as washing machines, kitchen robots and vacuum cleaners. It is also predominant in hand-held (portable) tools and is fabricated up to 30.000 rpm at 1kW, in general. Despite the brush-burden the universal motor survived due to its low cost/performance.

Topologically is very similar to the d.c. brush series motor (Fig. 4.13) but the stator is made of stamped laminations because the stator excitation coils are a.c. fed. The commutation of the a.c. current at brushes is more difficult in comparison with d.c. brush motors, due to the a.c. transformer–type (speed independent)–additional–e.m.f. induced in the commutating coil in the rotor by the stator excitation a.c. current.

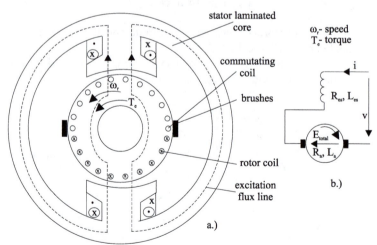

Figure 4.13. The two pole a.c. brush series motor

a) cross section; b) equivalent scheme

Because the stator excitation coils are series connected to the brushes, the field current and rotor current are equal to each other.

Also, as at brushes and in the stator, there is a.c. current; phasors may be used to investigate the machine steady state, despite the fact that, in the rotor, the current has two frequencies $\omega_r \pm \omega_1$ (ω_1 – stator frequency, $\omega_r = 2\pi np$ – electric rotor angular speed).

The voltage equation, valid also for transients, is straightforward:

$$(R_a + R_{es}) \cdot i - v = E_{pulse} - E_{rot} \tag{4.75}$$

E_{pulse} is the transformer (self induced) voltage in the stator and rotor inductances L_{es} and L_a:

$$E_{pulse} = -(L_{es} + L_a)\frac{di}{dt} \tag{4.76}$$

The rotational induced voltage (the motion e.m.f.) E_{rot} is, as for the d.c. brush machine, in phase with the stator current:

$$E_{rot} = k_e k_\Phi ni \tag{4.77}$$

It is evident that only the motion e.m.f. produces electromagnetic power P_{elm}:

$$P_{elm} = E_{rot} i = T_e 2\pi n \tag{4.78}$$

So the instantaneous torque T_e (from 4.78 with 4.77) is:

$$T_e = k_e k_\Phi \frac{i^2}{2\pi} \tag{4.79}$$

Now for steady state the stator voltage and current are sinusoidal:

$$v = V\sqrt{2}\cos\omega_1 t; i = I\sqrt{2}\cos(\omega_1 t - \varphi_1); \tag{4.80}$$

Consequently with I from (4.80), the instantaneous torque T_e is:

$$T_e = \frac{k_e k_\Phi I^2}{2\pi}[1 - \cos(2\omega_1 t - \varphi_1)] \tag{4.81}$$

So the instantaneous torque, for steady state, has a constant component which is the average torque and an a.c. (pulsating) component at $2\omega_1$ (Fig. 4.14a). The $2\omega_1$ torque pulsations mean vibrations, noise, and also additional losses.

For steady state from now on, we make use of phasors concept:

$$\underline{V} = V\sqrt{2} \cdot e^{j\omega_1 t}; \underline{I} = I\sqrt{2} \cdot e^{j(\omega_1 t - \varphi_1)} \tag{4.82}$$

With this denomination, the voltage equation (4.75) turns into:

$$\underline{V} = (R_{es} + R_a)\underline{I} + j\omega_1(L_{es} + L_a)\underline{I} + k_e k_\Phi n\underline{I} \tag{4.83}$$

$$R_{ae} = R_a + R_{es}; \omega_1(L_a + L_{es}) = X_{ae}$$

We may also include the stator and rotor core losses by replacing the total reactance X_{ae} to a series impedance Z_{ae}:

$$\underline{Z}_{ae} \approx R_{core} + jX_{ae} \qquad (4.84)$$

Consequently, with core losses included, voltage eqn. (4.83) becomes:

$$\underline{V} = (R_{ae} + \underline{Z}_{ae} + k_e k_\Phi n)\underline{I} \qquad (4.85)$$

The average torque is now evident:

$$T_{eav} = \frac{(k_e k_\Phi n \cdot I) \cdot I^*}{2\pi n} = \frac{k_e k_\Phi I^2}{2\pi} \qquad (4.86)$$

The dependence of torque on speed n is very similar to the case of the d.c. brush series motor (Fig. 4.14b) but with the torque reduced by the presence of total reactance X_{ae}.

Also the power factor $\cos\varphi_1$ is:

$$\cos\varphi_1 = \frac{1}{\sqrt{\left(\dfrac{X_{ae}}{R_{ae} + k_e k_\Phi n + R_{core}}\right)^2 + 1}} \qquad (4.87)$$

The larger the speed n (or the np/f_1 ratio) the larger the power factor (Fig. 4.14b). A ratio $np/f_1 > 6 / 1$ is common to vacuum cleaners and for kitchen robots with only $3 / 1$ in washing machines.

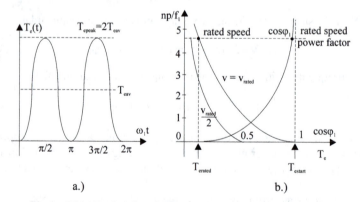

Figure 4.14. A.c. brush series motor steady state characteristics:

a) instantaneous torque, b) torque and power factor versus relative speed np / f_1

In general the universal motor is designed for a power factor above 0.9 for rated speed, which, in low power drives is a good asset in the absence of capacitors (required for the a.c. brushless motors). For a comprehensive analysis see Ref. 5.

Speed control may be approached simply by an a.c.-a.c. voltage changer that makes use of a single bidirectional thyristor module (called Triac). Torque equation (4.86) illustrates its quadratic dependence on the voltage amplitude.

The Triac is a very low cost device with simple controllability but it produces notable line current harmonics which have to be filtered. Also, when the voltage is decreased, the displacement power factor (DPF) is reduced.

Example 4.5.

Let us consider a universal motor for a home appliance supplied at 120 V a.c. (RMS), 60 Hz, that produces P_n = 600W at 18000rpm. It has two poles. The core loss is equal to half the copper loss and the mechanical loss is 2% of rated power. The efficiency is $\eta = 0.9$ and the power factor $\cos\varphi_u$ = 0.97. Calculate:

a) The rated current;
b) The total winding resistance R_{ae} and the core loss resistance R_{core};
c) The motion induced e.m.f. E_{rot};
d) The total inductance L_{ae};
e) The electromagnetic average torque;
f) The shaft torque;
g) The ratio of speed to frequency $n_n p/f_1$;
h) The starting current and torque;
i) The starting current when a.c. fed at 120 V d.c.

Solution:

a) From the definition of efficiency as output to input power:

$$\eta_n = \frac{P_n}{V_n I_n \cos\varphi_n}$$

The rated current I_n is:

$$I_n = \frac{600}{120 \cdot 0.97 \cdot 0.9} = 5.7273A$$

b) The total losses in the machine Σp are:

$$\Sigma p = \frac{P_n}{\eta_n} - P_n = 66.66W$$

With mechanical losses:

$$p_{mec} = 0.02\, P_n = 12W,$$

The winding plus iron losses are:

$$P_{copper} + P_{core} = R_{ae} I_n^2 + R_{core} I_n^2 = \Sigma p - p_{mec} = 66.66 - 12 = 54.66W$$

With the core losses as half the winding losses:

$$R_{ae}I_n^2 = 54.66; R_{ae} = \frac{54.66 \cdot 2}{3} \frac{1}{5.7273^2} = 1.111\Omega$$

$$R_{core} = \frac{1}{2}R_{ae} \approx 0.555\Omega$$

From (4.87)

$$I_n = \frac{U_n \cos\varphi_n}{R_{ae} + R_{core} + k_e k_\Phi n}$$

$$\tan\varphi_n = \frac{X_{ae}}{R_{ae} + R_{core} + k_e k_\Phi n}$$

So:

$$R_{ae} + R_{core} + k_e k_\Phi n = \frac{120 \cdot 0.97}{5.7273} = 20.3237\Omega$$

The motion induced e.m.f.:

$$E_{rot} = k_e k_\Phi n I_n = (20.3237 - 1.111 - 0.555) \cdot 5.7273 = 106.858V$$

Finally the machine reactance X_{ae}, from the impedance definition is:

$$X_{ae} = \frac{V_n}{I_n}\sin\varphi_n = \frac{120}{5.7273}0.243 = 5.0936\ \Omega$$

So the machine inductance:

$$L_{ae} = \frac{X_{ae}}{\omega_1} = \frac{5.0936}{2\pi 60} = 0.0135H$$

e) The average torque T_{eav} comes from (4.79):

$$T_{eav} = \frac{k_e k_\Phi I_n^2}{2\pi} = \frac{E_{rot}I_n}{2\pi \cdot n_n} = 0.3248\ Nm$$

f) The shaft torque T_{shaft} is approximately:

$$T_{shaft} \approx T_{eav} - \frac{p_{mec}}{2\pi n_n} = 0.3248 - \frac{12}{2\pi \cdot 300} = 0.31847\ Nm$$

g) The ratio $n_n p/f_1 = (18000/60)/60 = 5/1$ explains the good power factor of the machine.

h) The starting current is obtained again from (4.85) but with n=0:

$$I_{start} = \frac{V_1}{\sqrt{(R_{ae} + R_{core})^2 + X_{ae}^2}} = 22.39A$$

The average starting torque $(T_{eav})_{start}$ is (4.79):

$$\left(T_{eav}\right)_{start} = T_{eav}\left(\frac{I_{start}}{I_n}\right)^2 = 4.9646 \text{ Nm}$$

Note: In reality for this high current the magnetic flux saturates heavily and thus the ideal starting torque above is reduced considerably.

i) For d.c. at start, there is no a.c. (pulse) voltage across the machine inductances (X_{ae}=0) and R_{core}=0 (no core losses in d.c.).

$$I_{startd.c.} = \frac{V_{dc}}{R_{ae}} = \frac{120}{1.111} = 108 \text{Ad.c.}$$

So, the brush series motor is to be fed from a much smaller d.c. voltage if the starting current is to be limited to reasonable values (3 – 5 times the rated value).

Note: In conjunction with PECs, the d.c. brush series motor is still widely used in standard electric propulsion systems for urban, interurban or water transportation and some heavy duty off-highway vehicles. Though the a.c. drives are taking over the electric propulsion technologies, we felt it was useful to devote two pages to the d.c. brush series motor, the workhorse of electric propulsion in the 20th century; not in the 21st century, however. A similar fate faces the a.c brush series motor.

4.13. SUMMARY

- The d.c. brush motor may be electromagnetically or PM excited in the stator. For variable speed drives, due to current pulse additional losses for PEC control, both the stator and rotor cores are laminated.
- For the axial airgap disk-rotor PM d.c brush machine, the rotor windings are in air and thus the core losses in the machine are very small and the electrical time constant is small (≤1ms in the kW power range), allowing for fast current (torque) control.
- Separately excited or PM machines, for constant field current (or PM), are second-order linear systems and, due to the feedback of e.m.f., they provide stable responses in speed or current; either periodic or aperiodic.
- The variation of field current introduces an additional, large delay in the response, corresponding to the excitation circuit time constant. Still the responses in speed and current are stable.
- The inherent decoupling between excitation and armature windings allows for quick armature current (torque) control for constant field current.
- The d.c. brush series motor has a mild speed/torque curve while the one with separate excitation has a rigid, linear, speed/torque curve. The former also proved to be very useful in the electrical propulsion of various vehicles.

- The a.c. brush series (universal) motor still enjoys, sizeable markets in home appliance and hand-held tools at high speeds, due to low cost with simple speed control.

4.14. PROBLEMS

4.1.   For the d.c. brush motor in example 4.1 add the fact that the rated field current is $I_n = 1A$ and $V_{excn} = 110$ V, and determine:
        a.   the voltage V_a at standstill for rated current;
        b.   the field current at 3600 rpm and rated current, and the corresponding torque, and total input power.

4.2.   For the PM d.c. brush motor in example 4.3, determine the time variation of armature current for the 10 V step in the armature voltage for constant load torque.

4.3.   A d.c. brush series motor for a light rail urban transportation system has the data: $V_{an} = 800$ V (d.c.), $P_n = 100$ KW, $n_n = 1200$ rpm, rated efficiency $\eta_n = 0.92$, $p_{mec} = p_{iron} = 0.015\ P_n$, $R_{es} = R_a$. The commutator and additional losses are neglected. Calculate:
        a.   the rated current, I_n;
        b.   armature and excitation resistances, R_a and R_{es};
        c.   the rated e.m.f., E_n, and electromagnetic torque, T_{en}.

4.15. SELECTED REFERENCES

1.   **S.A. Nasar (editor),** Handbook of electric machines, Chapter 5, by M.G.Say, McGraw-Hill Book Company, 1987.
2.   **W. Leonhard,** Control of electric drives, first, second, third editions, Springer Verlag, 1985, 1996, 2001.
3.   **P.C. Sen,** Thyristor d.c. drives, John Wiley, 1981.
4.   **S.A. Nasar, I. Boldea, L. Unnewehr,** Permanent magnet, reluctance and self-synchronous motors, book, Chapter 3, CRC Press, 1993.
5.   **A. Di Gerlando, R. Perini, G. Rapi,** "Equivalent circuit for the performance analysis of universal motors", IEEE Trans. Vol EC – 19, no. 1, 2004.

Chapter 5

CONTROLLED RECTIFIER D.C. BRUSH MOTOR DRIVES

5.1. INTRODUCTION

Rectifiers are phase-controlled a.c. to d.c. static power converters. The rectifier provides variable d.c. voltage to the d.c. brush motor. Thyristors, bipolar transistors, IGBTs or MOSFETs may be used as power electronic switches (PESs) in the converter.

In general, the commutation process is natural, from the d.c. source, without any additional circuitry. However, for improved power factor forced commutation is used frequently. Phase-controlled power electronic converters are broadly classified as a.c.-d.c. single-phase or three phase converters [1]. Their main configurations are shown again in Table 5.1 for convenience (after their classification in Chapter 3).

Half-wave single-phase and semiconverters have one polarity of output voltage e_{av} and current i_{av}. That is, they work in one quadrant. Full converters work in two quadrants: the output voltage e_{av} may be positive or negative while the output current remains positive. Only dual converters can operate in four quadrants.

When the PESs are blocked, the stored energy dissipates through the free wheeling diodes in half-wave and semiconverters. In a single, phase half-wave converter the motor current is discontinuous unless a high additional inductance is added, while for the other converters, the output current may be either continuous or discontinuous. In three-phase converters the motor current is mostly continuous.

Table 5.1. Phase controlled rectifier circuits

Circuit	type	Power range	Ripple frequency	Quadrant operation
	half wave single-phase	below 0.5KW	f_s	one quadrant
	half wave three-phase	up to 50KW	$3f_s$	two quadrant

Table 5.1. (continued)

	semi-converter single-phase	up to 75KW	$2f_s$	one quadrant
	semi-converter three-phase	up to 100KW	$3f_s$	one quadrant
	full converter single-phase	up to 75KW	$2f_s$	two quadrant
	full converter three-phase	up to 150KW	$6f_s$	two quadrant
	Dual converter single-phase	up to 15KW	$2f_s$	four quadrant
	Dual converter three phase	up to 1500KW	$6f_s$	four quadrant

5.2. PERFORMANCE INDICES

As the d.c. motor current, when fed from phase controlled a.c.-d.c. converters, is not constant and a.c. supply current is not sinusoidal, adequate performance indexes for the motor-converter combination should be defined.

The main performance indices related to the motor are:
- the torque-speed characteristic;
- nature of motor current-continuous or discontinuous;
- average motor current I_a:

$$I_a = \frac{1}{T} \cdot \int_{t_1}^{t_1+T} i_a \, dt$$

(5.1)

where: i_a - instantaneous armature current;

   T - time period of one cycle of i_a variation.

- Rms motor current I_{ar}

$$I_{ar} = \sqrt{\frac{1}{T} \cdot \int_{t_1}^{t_1+T} i_a^2 \, dt}$$

(5.2)

As known the rms current squared is proportional to the heat produced in the armature winding.

- peak motor current i_{ap}; the mechanical commutator stress depends on the peak value of the current.

   The main performance related to the input (a.c.) source are:

- input power factor PF

$$PF = \frac{\text{main input power}}{\text{Rms input volt} \times \text{amperes}}$$

(5.3)

If the supply voltage is a pure sinusoid, only the fundamental input current will produce the mean input power and thus

$$PF = \frac{V \cdot I_1 \cdot \cos \varphi_1}{V \cdot I}$$

(5.4)

where: V - rms supply phase voltage;

   I - rms supply phase current;

   I_1 - rms fundamental component of a.c. supply current;

   φ_1 - angle between supply voltage and current fundamentals.

- input displacement factor DF or the fundamental power factor

$$DF = \cos \varphi_1$$

(5.5)

- harmonic factor HF

$$HF = \frac{\sqrt{\left(I^2 - I_1^2\right)}}{I_1} = \frac{\sqrt{\sum_{n=2}^{\infty} I_n^2}}{I_1} = \frac{I_h}{I_1}$$

(5.6)

I_h - rms value of the net harmonic current.

The above performance indices are somewhat similar to those for diode rectifiers (Chapter 3). They are summarized here only for convenience.

The basic motor equations are

$$V_a = R_a \cdot i_a + L_a \cdot \frac{di_a}{dt} + e_g; \ e_g = K_e \cdot \lambda_p \cdot n$$

$$2\pi J \frac{dn}{dt} = T_e - T_{load}; \ T_e = \frac{e_g \cdot i_a}{2\pi n} = \frac{K_e}{2\pi} \cdot \lambda_p \cdot i_a$$

(5.7)

with v_a, i_a motor input voltage and current; e_g - motion induced voltage; R_a, L_a - armature resistance and inductance; J - inertia; T_e - motor torque, T_{load} - load torque. For the steady state, $dn/dt = 0$, while in general $di_a / dt \neq 0$ as the armature current is not constant in time with phase-controlled converter supplies. To facilitate a quick assimilation of so many controlled rectifier configurations performance we will proceed directly through numerical examples.

5.3. SINGLE PES-CONTROLLED RECTIFIER

A d.c. brush motor with separate excitation with the data: $K_e\lambda_p = 2Wb$, $R_a = 5\Omega$, $L_a = 0.1H$ is fed through a thyristor (Figure 5.1) from a single phase a.c. source whose voltage is $V = V_1 \cdot \sin\omega_1 t = 120\sqrt{2}\sin 376.8t$. The motor speed is considered constant at n = 750 rpm.

a. Calculate the motor current i_a time variation for a delay angle $\alpha = +30^0$;

b. How does the motor voltage V_a vary in time?

c. How does i_a vary in time in the presence of a free wheeling diode in parallel with the motor armature (Figure 5.1)? The thyristor and the diode are considered as ideal switches.

Solution:

a. For $\omega t > \alpha$ (Figure 5.1) with discontinuous current:

$$V_a(t) = R_a i_a + L_a \frac{di_a}{dt} + K_e \lambda_p n$$

(5.8)

with the initial condition $i_a = 0$ for $\omega_1 t = \alpha = \pi/6$.

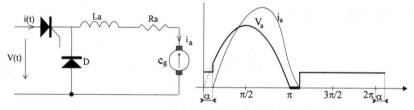

Figure 5.1. A d.c. brush motor supplied through a thyristor

The steady-state solution is

$$i_{ap} = A + B \cdot \cos\omega_1 t + C \cdot \sin\omega_1 t$$

(5.9)

and finally:

$$i_{ap} = -\frac{K_e \cdot \lambda_p \cdot n}{R_a} + \frac{V_1}{\sqrt{R_a^2 + \omega_1^2 \cdot L_a^2}} \cdot \sin(\omega_1 t - \gamma);$$

(5.10)

$$\gamma = \tan^{-1}(\omega_1 L_a / R_a)$$

The complete solution:

$$i_a(t) = i_{ap} + A \cdot e^{-tR_a / L_a} + B$$

(5.11)

For $t = t_1 = \alpha / \omega_1 = \pi/(6.2\pi.60) = 1.3888.10^{-3}$ s $\qquad i_a(t) = 0$
and thus:

$$\gamma = \tan^{-1}\frac{2 \cdot \pi \cdot 60 \cdot 0.1}{5} = 82.44^0;$$

$$\frac{V_1}{\sqrt{R_a^2 + \omega_1^2 \cdot L_a^2}} = \frac{120}{\sqrt{5^2 + (2 \cdot \pi \cdot 60 \cdot 0.1)^2}} = 3.157 \text{ A}$$

(5.12)

and

$$\frac{K_e \cdot \lambda_p \cdot n}{R_a} = \frac{2 \cdot 12.5}{5} = 5 \text{ A}$$

(5.13)

$$A = \left[\frac{K_e \cdot \lambda_p \cdot n}{R_a} - \frac{V_1}{\sqrt{R_a^2 + \omega_1^2 \cdot L_a^2}} \cdot \sin(30^0 - \gamma)\right] \cdot e^{-1.3888 \cdot 10^3 \cdot 5/0.1} =$$

$$= \left[5 - 3.157 \sin(-52.44^0)\right] \cdot 1.07186 = 8.04 \text{ A}$$

(5.14)

Finally, $\qquad i_a(t) = 8.04 \cdot e^{-50t} - 5 + 3.157 \cdot \sin(378.4t - 1.43812)$

(5.15)

Equation (5.15) is valid for $\omega_1 t > \pi/6$ until the current becomes zero soon after $\omega_1 t = \pi$, as shown in Figure 5.1.

b. The motor voltage is equal to the source voltage as long as the thyristor is on and becomes equal to minus the e.m.f.: $-e_a = K_e \cdot \lambda_p \cdot n = 2 \cdot 12.5 = 25$ V, when the motor current is zero (Figure 5.1)

c. In presence of the free wheeling diode D, the latter starts conducting when V(t) becomes negative ($\omega_1 t = \pi$).

From now on, the motor current i_a' flows through the diode until it becomes zero

$$0 = R_a \cdot i_a' + L_a \cdot \frac{di_a'}{dt} + K_e \cdot \lambda_p \cdot n$$

(5.16)

with $\qquad i_a'(0) = i_a(\pi)$

(5.17)

$$i_a' = -\frac{K_e \cdot \lambda_p \cdot n}{R_a} + A' \cdot e^{-t'R_a/L_a} \qquad (5.18)$$

$$i_a'(\pi) = i_a(\pi) = \frac{-2 \cdot 12.5}{5} + A' \cdot e^{-50/120} = 3.42729 \, A \qquad (5.19)$$

$$A' = 12.78319 \, A.$$

Thus $\qquad\qquad\qquad i_a' = -5 + 12.783 \cdot e^{-50t'} \qquad (5.20)$

The current $i_a' = 0$ for $\omega_1 t' = 7.07°$.

The motor voltage V_a becomes zero during the time interval when the free wheeling diode is conducting (180° to 180°+7.07°).

5.4. THE SINGLE-PHASE SEMICONVERTER

A d.c. motor supplied through a single-phase semiconverter has a constant speed and the motor current is constant in time, $I_a = 3A$, $R_a = 5\Omega$, $K_e\Phi = 2Wb$, $v = 120\sqrt{2} \sin 120\pi t$ with $\alpha = \pi/4$.

a. Determine the time waveforms of voltages at motor terminals and along the thyristors and diodes and the corresponding current waveforms.

b. Calculate the rms values of currents through diodes, thyristors and from the a.c. supply.

c. Determine the motor voltage average value dependence and its maximum value.

d. Calculate the rms of a.c. supply current fundamental and its phase shift with respect to the a.c. supply voltage.

Solution:

a. As the filter inductance L_a is large, the motor current i_a is considered constant (Figure 5.2). The thyristors T_1 and T_3 conduct for positive voltage applied to them along 180°–α regions while the diodes D_1, D_3 conduct for zones of 180°+α degrees (Figure 5.2.d). All currents are rectangular (Figure 5.2d, e) while the motor voltage is either positive or zero (Figure 5.2b).

b. The rms value of thyristor current is

$$I_{T_{1,3}} = \sqrt{\frac{1}{2\pi} \int_0^{2\pi} i_{T_1}^2 d(\omega_1 t)} = \sqrt{\frac{1}{2\pi} \int_\alpha^\pi i_a^2 d(\omega_1 t)} =$$

$$= I_a \sqrt{\frac{\pi - \alpha}{2\pi}} = 3 \cdot 0.6123 = 1.837 \, A \qquad (5.21)$$

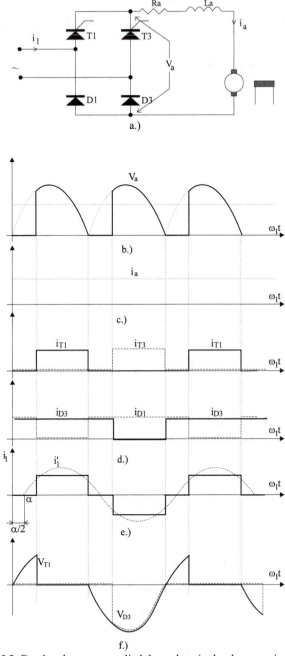

Figure 5.2. D.c. brush motor supplied through a single-phase semiconverter (with large inductance L_a)

The diode current $I_{D_{1,3}}$

$$I_{D_{1,3}} = \sqrt{\frac{1}{2\pi} \int_0^{\pi+\alpha} i_a^2 d(\omega_1 t)} = I_a \sqrt{\frac{\pi+\alpha}{2\pi}} = 3 \cdot 0.79 = 2.37 \text{ A} \qquad (5.22)$$

The rms primary (supply current)

$$I_1 = \sqrt{\frac{1}{\pi} \int_\alpha^\pi i_a^2 d(\omega_1 t)} = I_a \sqrt{\frac{\pi-\alpha}{\pi}} = 3 \cdot 0.866 = 2.598 \text{ A} \qquad (5.23)$$

c. The motor average voltage V_{av} is

$$V_{av} = \frac{1}{\pi} \int_\alpha^\pi v_a d(\omega_1 t) = \frac{1}{\pi} \int_\alpha^\pi V\sqrt{2} \sin \omega_1 t \cdot d(\omega_1 t) = \frac{V_1\sqrt{2}}{\pi}(1+\cos\alpha) \qquad (5.24)$$

For $\alpha = \pi/4$

$$\left(V_{av}\right)_{\alpha=\pi/4} = \frac{120\sqrt{2}}{\pi}(1+0.707) = 91.98 \text{ V} \qquad (5.25)$$

The maximum average voltage is obtained for $\alpha = 0$;

$$\left(V_{av}\right)_{\alpha=0} = \frac{120\sqrt{2}}{\pi}(1+1) = 107.77 \text{ V} \qquad (5.26)$$

and corresponds to the full bridge diode rectifier (Chapter 3).

d. The a.c. supply current harmonics I_{1v} is

$$I_{1v} = \frac{1}{\pi} \int_0^{2\pi} i_1 \sin v\omega_1 t \cdot d(\omega_1 t) = \frac{4}{\pi v} I_a \cos\frac{v\alpha}{2} \qquad (5.27)$$

The fundamental is: $I_{1(1)} = \frac{4}{\pi} 3 \cos\frac{\pi}{2} = 3.3313 \text{ A} \qquad (5.28)$

The input displacement power factor DPF $= \cos\varphi_1$ refers to the cosine of the angle between the a.c. source voltage and current fundamentals. As seen from Figure 5.2b, e $\varphi_1 = \alpha/2 = \pi/8$. Thus

$$\text{DPF} = \cos\varphi_1 = \cos\pi/8 = 0.9238 \qquad (5.29)$$

On the other hand, the power factor PF (5.4) is

$$\text{PF} = \frac{I_{1(1)} \cdot \cos\varphi_1}{\sqrt{2} \cdot I_1} = \frac{3.3313 \cdot 0.9238}{1.41 \cdot 2.598} = 0.84 \qquad (5.30)$$

5.5. THE SINGLE-PHASE FULL CONVERTER

A d.c. motor of 7kW and 1200 rpm rating with separate excitation is supplied through a single-phase full converter as shown in Figure 5.3a. The other data are $R_a = 0.2\Omega$, rated current $I_{ar} = 40A$, $K_e\lambda_p = 10Wb$; the a.c. supply voltage (rms) is 260V.

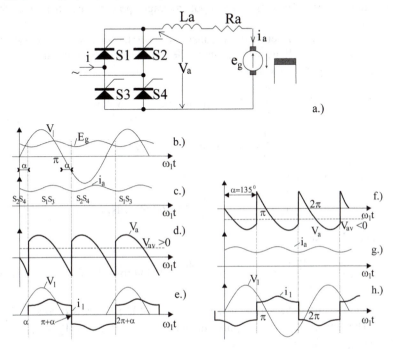

a.)

Figure 5.3. The d.c. brush motor fed through a single-phase full converter:

a.) the converter,

b.), c.), d.), e.) - voltage and current waveforms for motor action ($\alpha=45°$),

f.), g.), h.) - for generator action ($\alpha = 135°$).

a. Draw the voltage and current waveforms for steady state and finite motor armature inductance and $\alpha = 45°$ and $\alpha = 135°$.

b. For a firing angle of $\alpha = 30°$ and rated motor current (rectifier regime), calculate: motor speed, torque and supply power factor, neglecting the motor current ripple.

c. By reversing the field current, the motor back e.m.f. E_g is reversed; for rated current, calculate: converter firing angle α' and the power fed back to the supply.

Solution:

a. The waveforms of currents and voltages are shown on Figure 5.3. It should be noted that because the motor inductance is not very high the armature current, and speed pulsate with time.

Motor action is obtained for $\alpha = 45°$ when both the motor and converter average input powers are positive. For generator action ($\alpha = 135°$) the motor current remains positive but the motor average voltage V_a is negative. Thus both the motor and converter input average powers are negative for generating.

b. As the motor current is considered ripple-free, we use only the average values of voltage and current while the speed is also constant.

The average motor voltage V_{av} is

$$V_{av} = \frac{1}{\pi} \int_{\alpha}^{\pi+\alpha} V\sqrt{2} \sin\omega_1 t \cdot d(\omega_1 t) = \frac{2\sqrt{2}}{\pi} V \cos\alpha =$$

$$= \frac{2\sqrt{2}}{\pi} 260 \cos 30° = 202.44 \text{ V}$$

(5.31)

The motor torque T_e is

$$T_e = \frac{K_e \lambda_p}{2\pi} I_a = \frac{10}{2\pi} 40 = 63.6942 \text{ Nm}$$

(5.32)

The motor speed n is

$$n = \frac{V_{av} - R_a I_a}{K_e \phi} = \frac{202.44 - 0.2 \cdot 40}{10} = 19.444 \text{ rps} = 1166.64 \text{ rpm}$$

(5.33)

As the primary current is now rectangular (40A), the rms current is $I_1 = 40A$. The power from the supply, neglecting the losses, is $P_s = V_{av} \cdot I_a = 202.44 \cdot 40 = 8097.6$ W.

Thus, the supply power factor

$$PF = \frac{P_s}{V_1 \cdot I_1} = \frac{8097.6}{260 \cdot 40} = 0.7786$$

(5.34)

c. For generator action the polarity of induced voltage should be reversed

$$e_g = -K_e \lambda_p n = -10 \cdot 19.44 = -194.44 \text{ V}$$

(5.35)

Thus the motor voltage V_a becomes

$$V_a = e_g + R_a I_a = -194.44 + 0.2 \cdot 40 = -186.44 \text{ V}$$

(5.36)

Finally,
$$\alpha = \cos^{-1}\left[\frac{V_a \pi}{2\sqrt{2}V}\right] = \cos^{-1}\left[\frac{-186.44 \cdot \pi}{2\sqrt{2} \cdot 260}\right] \approx 143°$$

(5.37)

The regenerated power P_{sg} is:

$$P_{sg} = V_a I_a = 186.44 \cdot 40 = 7457.6 \text{ W}$$

(5.38)

As for this regime either the flux λ_p or the speed n changes sign, the motor torque opposes the motion providing generator braking.

In the process, the motor speed and e.m.f. decrease and, to keep the current constant, the firing angle $\alpha>90^0$ in the converter should be modified accordingly.

For the motor as above, having $L_a = 2mH$, calculate the following:

a. For $\alpha = 60°$ and constant speed n = 1200 rpm draw the voltage and current waveforms knowing that the motor current is discontinuous.

b. Calculate the motor current waveform for n = 600 rpm.

Solution:

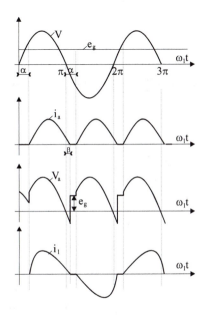

Figure 5.4. D.c brush motor supplied from a full converter –

discontinuous current mode

a. As the motor electrical time constant $\tau_e = L_a / R_a = 2.10^{-3} / 0.2 = 10$ ms is small, when the voltage becomes zero (for $\omega_1 t = \pi$), the motor current decays quickly to zero along the angle β (from π to $\pi+\beta$) with $\beta<\alpha$ and thus the current is discontinuous. As α increases, for a given speed n (and e.m.f. e_g), the average voltage V_{av} increases with respect to the case of continuous current.

The discontinuity of current contributes to a sluggish dynamic response as, during zero current, motor internal torque control is lost. The situation occurs especially at low torques (and currents). Special measures such as flux weakening (reducing λ_p) for low torques leads to reduced α at high speeds, avoiding the discontinuity in current.

b. With the speed considered constant, the voltage equation is

$$V\sqrt{2}\sin\omega_1 t = R_a i_a + L_a \cdot \frac{di_a}{dt} + K_e \lambda_p n \;; \tag{5.39}$$

with $i_a = 0$ for $\omega_1 t_1 = \alpha$, the solution of above equation is

$$i_a = A \cdot e^{-t R_a / L_a} - \frac{K_e \lambda_p n}{R_a} + \frac{V\sqrt{2}\sin(\omega_1 t - \varphi_1)}{\sqrt{R_a^2 + \omega_1^2 L_a^2}}; \tag{5.40}$$

$$\varphi_1 = \tan^{-1}(\omega_1 L_a / R_a)$$

for $t = t_1 = \dfrac{\alpha_0}{\omega_1}$; $(i_a)_{t_1} = 0$ and thus

$$\varphi_1 = \tan^{-1}\left(\frac{2\pi 60 \cdot 2 \cdot 10^{-3}}{0.2}\right) = 75.136° \tag{5.41}$$

$$i_a = 1402 \cdot e^{-100t} - 1000 + 470\sin(\omega_1 t - \varphi_1) \tag{5.42}$$

The current becomes zero again for $\omega_1 t_2 = \pi + \beta$

$$0 = 1402 \cdot e^{-\frac{100(\pi+\beta)}{2\pi 60}} - 1000 + 470\sin(\pi + \beta - 1.308) \tag{5.43}$$

The solution for β is $\beta \approx 12°$.

5.6. THE THREE-PHASE SEMICONVERTER

Consider a three-phase semiconverter (Figure 5.5) supplying a d.c. motor. The armature current is considered ripple free and the commutation is instantaneous.

a. Draw the waveform of output voltage for $\alpha = 30°$.

b. Calculate the average (rms) voltage as a function of α.

c. For a line voltage (rms) $V_L = 220V$, $f = 60Hz$, $R_a = 0.2\Omega$, $K_e\lambda_p = 4Wb$ and an armature current $I_a = 50A$, determine the motor speed for $\alpha = 30°$.

Solution:

a. As seen from Figure 5.5b if the thyristor T_a is turned on with a delay angle $\alpha = 30°$ (point A) from the moment corresponding to $\alpha = 0$, it will conduct together with the diode D_b for $30°$ as long as V_{ab} is positive and higher than the other line voltages. At point B, T_a will conduct together with the diode D_c for $90°$ as $V_{ac} > V_{ab}$. At point C, after $120°$ of conduction, the thyristor T_a is turned off and T_b turned on and so on.

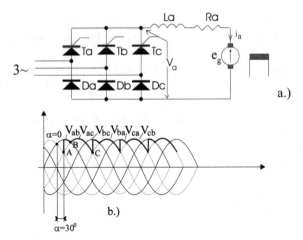

Figure 5.5. D.c. brush motor supplied through a three-phase semiconverter

b. The expression for average voltage is

$$V_{av} = \frac{3}{2\pi}\left[\int_{\alpha}^{\pi/3}\sqrt{2}V_L\sin(\omega_1 t + \pi/3)d(\omega_1 t) + \int_{\pi/3}^{\alpha+2\pi/3}\sqrt{2}V_L\sin(\omega_1 t)d(\omega_1 t)\right] =$$

$$= \frac{3\sqrt{2}V_L}{2\pi}(1 + \cos\alpha) \tag{5.44}$$

For $\alpha > 60°$ the integration interval is from α to π with the same final result. The output average voltage V_{av} may not be negative and thus the converter may not be used as an inverter, confirming the single quadrant operation.

c. The motor voltage equation for ripple free current is

$$\frac{3\sqrt{2}V_L}{2\pi}(1 + \cos\alpha) = R_a I_a + K_e\lambda_p n \tag{5.45}$$

$$\frac{3\cdot\sqrt{2}\cdot 220}{2\pi}\left(1 + \cos\frac{\pi}{6}\right) = 0.2\cdot 50 + 4\cdot n \tag{5.46}$$

$$n = 66.66 \text{ rps} = 4000 \text{ rpm} \tag{5.47}$$

5.7. THE THREE-PHASE FULL CONVERTER – MOTOR SIDE

A three-phase full converter (Figure 5.6) supplying a d.c. motor has the data: line voltage $V_L = 220V$ (rms), $K_e\lambda_p = 10Wb$, $R_a = 0.2\Omega$, and neglect the ripples in the motor current.

a. For L_s equal to zero, calculate the output average voltage as a function of the delay angle α; and for $\alpha = 30°$, determine the motor speed for $I_a = 50A$.

b. Considering the a.c. source inductance $L_s \neq 0$, determine the expression of the output voltage as a function of α and I_a and calculate the motor speed for $I_a = 50A$, $\alpha = 30°$ and $L_s = 11mH$.

c. Calculate the overlapping angle at commutation for case b.

<u>Solution:</u>

a. The principle of operation is quite similar to that of the previous case. However, only a 60° conduction of each pair of thyristors occurs as shown in Figure 5.6.

In the absence of a.c. source inductances L_s ($L_s = 0$) the commutation is instantaneous. The average output voltage V_{av} is

$$V_{av} = \frac{3}{\pi} \int_{\alpha+\pi/3}^{\alpha+2\pi/3} \sqrt{2}V_L \sin\omega_1 t \cdot d(\omega_1 t) = \frac{3\sqrt{2}V_L}{\pi}\cos\alpha \qquad (5.48)$$

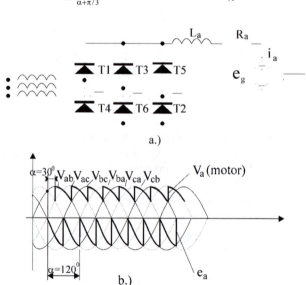

a.)

b.)

Figure 5.6. D.c. brush motor fed from a three-phase full converter

Consequently, $V_{av} < 0$ ($V_{av} > 0$) depending on α. For $\alpha < 90$ the average output voltage V_{av} is positive, that is rectifier mode; for $\alpha > 90$ $V_{av} < 0$ and thus the inverter mode are obtained. The motor current remains positive irrespective of α. A two-quadrant operation is obtained.

The motor voltage equation is

$$\frac{3\sqrt{2}V_L}{\pi}\cos\alpha = R_a I_a + K_e \lambda_p n \qquad (5.49)$$

$$n = \left(\frac{3 \cdot 220\sqrt{2}\cos(\pi/6)}{\pi} - 0.2 \cdot 50\right)/10 = 24.74 \text{ rps} = 1484.58 \text{ rpm} \quad (5.50)$$

b. The effect of a.c. source inductances L_s on current commutation is shown in Figure 5.7 for the case of constant motor current (similar to the diode full converter [chapter 3]). The commutation between T_1 and T_5 is not instantaneous anymore; an overlapping angle u occurs. The effect of actual commutation is, in fact, a reduction of output voltage V_d determined by the area A_u when a kind of short circuit occurs between phases a and c.

$$A_u = \int_{\alpha}^{\alpha+u} V_{Ls}d(\omega_1 t) = \int_{\alpha}^{\alpha+u} L_s \cdot \frac{di_a}{dt}d(\omega_1 t) = \omega L_s I_d \quad (5.51)$$

This is so since the current in phase a is zero for $\omega_1 t = \alpha$ and equal to I_d for $\omega_1 t = \alpha + u$. The average voltage is reduced by $3/\pi A_u$.

Finally, the average voltage is equal to that already found for $L_s = 0$ minus $3/\pi A_u$

$$V_d = \frac{3\sqrt{2}V_L}{\pi}\cos\alpha - \frac{3}{\pi}\omega_1 L_s I_d \quad (5.52)$$

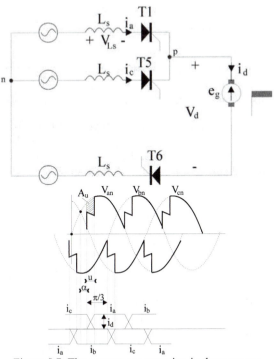

Figure 5.7. The current commutation in the presence of a.c. source side inductances L_s.

As we can see, the angle u is not required to calculate the output voltage V_d. However, it is needed to ensure reliable operation in the inverter mode ($\alpha>90°$). The second Kirchoff law for phase a and c during commutation provides the equation

$$V_{an} - L_s \cdot \frac{di_a}{dt} = V_{cn} - L_s \cdot \frac{di_c}{dt} \tag{5.53}$$

$$\text{with } i_c + i_a = i_d, \text{ that is } \frac{di_a}{dt} = -\frac{di_c}{dt} \tag{5.54}$$

Thus,

$$V_{an} - V_{cn} = V_{ac} = 2L_s \cdot \frac{di_a}{dt} \tag{5.55}$$

$$\text{with } V_{ac} = \sqrt{2}V_L \sin\omega_1 t \tag{5.56}$$

$$\int_\alpha^{\alpha+u} V_{ac} d(\omega_1 t) = 2\omega_1 L_s I_d \tag{5.57}$$

and finally,
$$\cos(\alpha + u) = \cos\alpha - \frac{2\omega_1 L_s I_d}{\sqrt{2}V_L} \tag{5.58}$$

Thus knowing α and I_d we may calculate the overlapping angle u.

In the inverter mode ($\alpha>90°$) the commutation should be finished before $\alpha + u = \pi$ in order to allow the turn-off time t_{off} required for the recombination of charged particles in the thyristors ($\frac{\pi - (\alpha + u)}{\omega_1} > t_{off}$) for negative voltage along the turning-off thyristors.

The rectifier voltage V_d for $\alpha=\pi/6$ and $I_d = 50$ is

$$V_d = \frac{3\sqrt{2}\cdot 220}{\pi}\cos\frac{\pi}{6} - \frac{3}{\pi}2\cdot\pi\cdot 60\cdot 50\cdot 10^{-3} = 239.43 \text{ V} \tag{5.59}$$

$$V_d = R_a I_a + K_e \lambda_p n$$
$$239.43 = 0.2\cdot 50 + 10\cdot n; \quad n = 22.943 \text{ rps} = 1376.58 \text{ rpm} \tag{5.60}$$

The reduction of output voltage for the same α due to commutation with 18V has contributed to a notable reduction in speed, for 50A, from 1484.58 rpm to 1376.58 rpm.

c. To calculate the overlapping angle u we use the expression derived above.

$$\cos(\alpha + u) = \cos\alpha - \frac{3\omega_1 L_s I_d}{\sqrt{2}V_L} = \cos\frac{\pi}{6} - \frac{3 \cdot 2\pi \cdot 60 \cdot 50 \cdot 10^{-3}}{\sqrt{2} \cdot 220} \tag{5.61}$$

$$30° + u = 46.78°; \quad u = 16.78°$$

A considerable value for u has been obtained.

5.8. THE THREE-PHASE FULL CONVERTER – SOURCE-SIDE ASPECTS

For the three-phase full converter and d.c. motor in the previous example with $L_s = 0$ for $\alpha = 0°$, $\alpha = 45°$ with $I_d = 50A$, determine:

a. The waveforms of a.c. source current and its harmonics factor.

b. The rms of fundamental source current and of the total source current.

c. The displacement power factor for $L_s = 0$ and $L_s = 1mH$ for $\alpha = 45°$.

d. Calculate the line voltage and voltage distortion due to $L_s = 1mH$ and $\alpha = 45°$.

Solution:

a. As the motor armature current is considered constant, the a.c. source current is rectangular as shown in Figure 5.8a and b.

The presence of a.c. source inductances $L_s \neq 0$ leads to the overlapping angle $u \neq 0$.

b. The rms value of the current fundamental, I_{a1}, is

$$I_{a1} = \frac{2\sqrt{3}}{\pi} \cdot \frac{I_d}{\sqrt{2}} = \frac{\sqrt{6}}{\pi} \cdot I_d = 0.78 \cdot 50 = 39 \text{ A} \tag{5.62}$$

For the total a.c. source current which is made of rectangular $120°$ wide "blocks",

$$I_a = \sqrt{\frac{2}{3}} \cdot I_d = 0.816 \cdot 50 = 40.8 \text{ A} \tag{5.63}$$

In the absence of L_s, the displacement power factor (DPF) angle is equal to α and thus the DPF is

$$DPF = \cos\varphi_1 = \cos\alpha = \begin{cases} 1 \text{ for } \alpha = 0 \\ 0.709 \text{ for } \alpha = 45° \end{cases} \tag{5.64}$$

In the presence of L_s, the displacement power factor is approximately

$$DPF = \cos\left(\alpha + \frac{u}{2}\right) \tag{5.65}$$

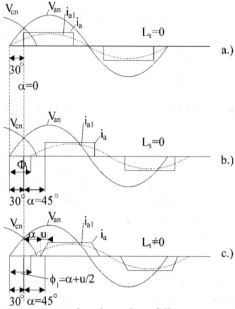

Figure 5.8. A.c. source currents in a three-phase full converter and constant motor current: a.) for $\alpha = 0$, $L_s = 0$; b.) for $\alpha = 45°$, $L_s = 0$; c.) for $\alpha = 45°$, $L_s = 1mH$

To calculate u for $\alpha = 45°$, we use (5.58)

$$\cos(\alpha + u) = \cos\alpha - \frac{3\omega_1 L_s I_d}{\sqrt{2}V_L} = 0.707 - \frac{3 \cdot 2\pi \cdot 60 \cdot 50 \cdot 10^{-3}}{\sqrt{2} \cdot 220} = 0.5248 \qquad (5.66)$$

$$45 + u = 58.34°; \quad u = 13.34°$$

Finally, $$DPF = \cos\left(45 + \frac{13.34}{2}\right) = 0.62$$

The current commutation in the presence of L_s produces a further reduction of the displacement power factor.

In general, the DPF decreases with α increasing, which constitutes a notable disadvantage of phase delay a.c.-d.c. converters. Special measures are required to improve DPF with α increasing.

d. The a.c. source current overlapping during commutation produces notches in the line voltage. From Figure 5.9. we may obtain $V_{ab} = V_{an} - V_{bn}$ of the waveform shown in Figure 5.9.

Considering u as small, the deep notch depth is considered equal to $\sqrt{2}V_L \sin\alpha$ and thus the notch width u is approximately:

$$u = \frac{A_u}{\text{notch depth}} = \frac{2\omega_1 L_s I_d}{\sqrt{2}V_L \sin\alpha} \qquad (5.67)$$

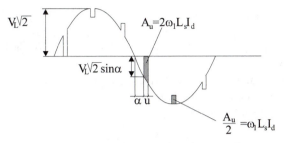

Figure 5.9. A.c. source-line voltage notches due to Ls (commutation)

The depth of shallow notches is considered half of that of deep notches. IEEE standard 519-1981 suggests the limitation of line notches to 250 μs (5.4. electrical degrees) and the deep notch depth to 70% of rated peak line voltage in order to perform satisfactorily.

Special filtering is required to cope with more recent standards. The voltage distortion due to notches depends on the harmonics currents I_v and the a.c. source inductance L_s

$$\text{voltage\%THD} = \frac{\sqrt{\left[\sum_{v>1}\left(I_v \cdot v\omega_1 L_s\right)^2\right]}}{V_{phase\,(fundamental)}} \cdot 100 \qquad (5.68)$$

with $I_v \approx \dfrac{I_{a1}}{v} = \dfrac{\sqrt{6}}{\pi} \cdot \dfrac{I_d}{v}$ due to the almost rectangular form of a.c. source current.

5.9. THE DUAL CONVERTER - FOUR-QUADRANT OPERATION

A high power d.c. motor drive sometimes has to undergo four-quadrant operation. Two full converters are connected back to back for this purpose (Figure 5.10).

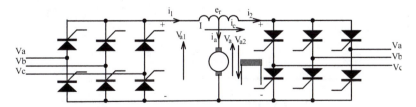

Figure 5.10. Dual converter with circulating current supplying a d.c. brush motor

a. Assuming that the converters are ideal and produce pure d.c. output voltages with one converter as rectifier and the other as inverter, calculate the relationship between the delay angles α_1 and α_2 in the two converters.

b. With $\alpha_1 + \alpha_2$ as above, calculate the circulating current between the two converters and show the voltage and current actual output waveforms. The numerical data are $V_L = 220V$, $\omega_1 = 377.8rad/s$, $L = 10mH$, $\alpha_1 = 60°$.

Solution:

a. In an ideal dual converter the voltages produced by the two full converters should be equal and opposite.

By now we know that

$$V_{a1} = V_{max} \cdot \cos\alpha_1$$
$$V_{a2} = V_{max} \cdot \cos\alpha_2$$

(5.69)

with $V_a = V_{a1} = -V_{a2}$ it follows that $\cos\alpha_1 + \cos\alpha_2 = 0$.

Hence, $\alpha_1 + \alpha_2 = 180°$ (Figure 5.11). In the ideal converter the load voltage is equal to the converter output voltages and thus the current may flow equally through either converter.

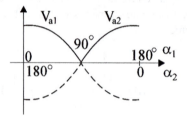

Figure 5.11. Ideal dual converter voltages

b. A real (nonideal) converter produces a voltage with ripples. The ripple voltages of the two converters are out of phase (Figure 5.12). The instantaneous voltage difference produces a circulating current which is limited through a reactor L.

With

$$V_{a,b,c} = \frac{V_L\sqrt{2}}{\sqrt{3}}\sin\left[\omega_1 t - (i-1)\frac{2\pi}{3}\right]$$

(5.70)

during the interval

$$\frac{\pi}{6} + \alpha_1 < \omega_1 t < \frac{\pi}{6} + \alpha_1 + \frac{\pi}{3}$$

(5.71)

$$V_{a1} = V_a - V_b$$
$$V_{a2} = -(V_c - V_b)$$
$$e_r = V_{a1} - V_{a2} = V_a + V_c - 2V_b = -3V_b$$

(5.72)

The circulating current i_c is

$$i_c = \frac{1}{\omega_1 L} \int_{\alpha_1+\frac{\pi}{6}}^{\omega_1 t} e_r dt = \frac{\sqrt{6}V_L}{\omega_1 L}\left[\cos\left(\omega_1 t - \frac{2\pi}{3}\right) - \cos\left(\alpha_1 - \frac{\pi}{2}\right)\right] \quad (5.73)$$

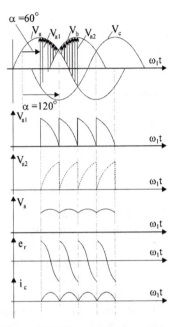

Figure 5.12. The real dual converter

When the motor current is zero, the converter currents are equal to the circulating current $i_1 = i_2 = i_c$ ($i_a = 0$). Consequently, the converters have a continuous current though the load current is zero. However for $\alpha_1 = 60°$ and $\alpha_2 = 120°$ the peak value of circulating current occurs at $\omega_1 t = 2\pi/3$

$$\left(i_c\right)_{peak} = \frac{V_L\sqrt{6}}{\omega_1 L}\left[1-\cos\frac{\pi}{6}\right] = \frac{220\sqrt{6}}{2\cdot\pi\cdot60\cdot10^{-2}}\left[1-0.867\right]=19.02 \text{ A} \quad (5.74)$$

If the load current i_a is constant (no ripples), the first converter ($\alpha_1 = 60°$) carries $i_a + i_c$ while the second converter ($\alpha_1 = 120°$) has the circulating current i_c only. Thus the first converter is "overloaded" with the circulating current. However, for low load current, the discontinuous current mode in the converters is avoided as shown above. This could be an important advantage in terms of control performance.

5.10. A.C. BRUSH SERIES (UNIVERSAL) MOTOR CONTROL

The universal motor is a.c. voltage supplied but it is still a brush (mechanical comutater) series connected motor (Chapter 3), Figure 5.13.

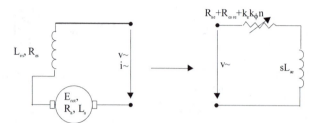

Figure 5.13. The equivalent scheme of universal motor

The motor voltage equation is (Chapter 3)

$$L_{ae}\frac{di}{dt} \approx v - \left(R_{ae} + R_{core} + k_ek_{\Phi}n\right)i; \quad T_e = \frac{k_ek_{\Phi}i^2}{2\pi} \qquad (5.75)$$

The mechanical equation is straightforward

$$J2\pi\frac{dn}{dt} \approx \frac{k_ek_{\Phi}i^2}{2\pi} - Bn - T_{load} \qquad (5.76)$$

All core losses are lumped into the stator while in the motion equation they are left out for simplicity. Magnetic saturation of the magnetic circuit is considered constant or absent (L_{ae} = ct, k_{Φ} = const.).

The investigation of transients with any input voltage waveform and load torque perturbation is rather straightforward for such a second order system.

The motion induced voltage (e.m.f.), E_{rot}, may be considered as an additional, variable resistance voltage drop ($k_ek_{\Phi}ni$), eqn. (5.75).

The torque is essentially proportional to current squared and, for given voltage, it is proportional to voltage squared.

An a.c. voltage changer is required.

At start (n=0) the machine is represented by a small resistance $R_{ae}+R_{core}$ plus the inductance L_{ae}. So the electric circuit of the machine at start is strongly inductive.

On the other end, at high speed, due to the large e.m.f. (E_{rot}) − eqn. (5.75) − the machine equivalent circuit is notably resistive.

The equivalent circuit is composed simply of an equivalent inductance L_{ae} − relatively constant − and a resistance R_{en} = ($R_{ae}+R_{core}+k_ek_{\Phi}n$) which strongly increase with speed (Fig. 5.13).

A typical a.c. voltage charger − which may be used also as a power switch − may be obtained by using two antiparalleled thyristors which may be assembled into a single bidirectional power switch (Fig. 5.14) − the Triac.

The Triac is turned on by applying a short resistive current pulse on the thyristor gate. The thyristor will turn off when the current decays naturally to zero. Then the thyristor for the negative voltage polarity the antiparalleled thyristor is turned on. As the turn-on angle α − with respect to zero crossing of the voltage waveform − increases, the average voltage decreses. But the average voltage depends also on the machine equivalent resistance R_e which increases with speed.

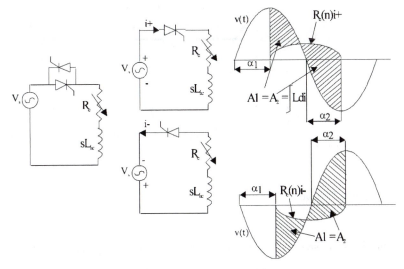

Figure 5.14. The Triac variac and the current

Let us consider separately the two voltage polarities (Figure 5.14).

The difference between supply voltage v(t) and the resistance voltage $R_e(n)i$ is equal to the voltage drop along the machine inductance L_{ae}:

$$V_{Lae}(t) = L_{ae}\frac{di}{dt} = v_s(t) - R_e(n)i \qquad (5.77)$$

$$\int_{\alpha 1}^{\pi+\alpha 2} V_{Lae}(t)dt = 0 \qquad (5.78)$$

The current goes to zero when areas A1 and A2 on Figure 5.14 become equal to each other.

It is evident that for the universal motor the current goes to zero after the voltage changes polarity by the angle α_2. However this delay angle decreases with speed as the circuit becomes more and more resistive as the e.m.f. increases.

There may be situations with $\alpha_2 > \alpha_1$ and $\alpha_2 < \alpha_1$. In the first case the current will be continuous while in the second case it will be discontinuous.

In general α_1 has to be larger than the displacement power factor angle φ1 of the equivalent circuit, to secure the fundamental output voltage control by the Triac.

So, at low speed α_1 should be large to reduce the voltage fundamental and will decrease with speed. But the output voltage is full of harmonics as only parts of the sinusoid are active. Also there are harmonics in the machine (input) current (Fig. 5.15). An input power filter is required.

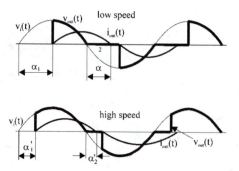

Figure 5.15. Output voltage and current waveforms at low and high speed

For constant speed and given delay angle α_1, eqn (5.77) has an analytical solution:

$$i = Ae^{-\frac{Re(n)}{Lae}t} + B\sin(\omega_1 t - \varphi_1); \quad B = \frac{V\sqrt{2}}{\sqrt{R_e^2(n) + \omega_1^2 L_{ae}^2}}; \quad \varphi_1 = \tan^{-1}\frac{\omega_1 L_{ae}}{R_e(n)}$$

$$(5.79)$$

with $i = 0$ at $\omega_1 t = \alpha_1$.
  Consequently:

$$A = -B\sin(\alpha_1 - \varphi_1)e^{\frac{R_e(n)\alpha_1}{L_{ae}\omega_1}} < 0 \qquad (5.80)$$

So, $\alpha_1 > \varphi_1$ to secure predominantly positive sinusoidal current for positive voltage polarity.

The angle α_2 within negative polarity of voltage where the current decays to zero is obtained by applying (5.78). The obtained nonlinear equation may be solved numerically for α_2.

It is also evident that the output voltage fundamental V_{1out} is nonlinearly dependent on $(\pi/2-\alpha_1)$ or on $(1-\sin\alpha_1)$.

A robust speed controller is required to overcome this difficulty.

A generic control system for the universal motor is presented in Fig. 5.16.

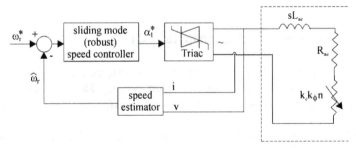

Figure 5.16. Generic control system for universal motors

A speed estimator is required for close loop control. In less demanding applications a well commissioned feedforward open loop control $\alpha_1^*(\omega_r^*)$ with given limited speed ramps may be used to avoid the speed estimator and the speed loop control.

Note: The low cost of universal motor drive at low power levels has secured its presence today in some home appliances and hand-held tools (with natural limited operation life) despite the occurrence of brushless drives. For more details see Reference 6.

5.11. SUMMARY

- Controlled rectifiers, also known as phase delay rectifiers, of various basic configurations, with zero and nonzero source inductances, have been presented in interaction with d.c. brush motors at constant speed.

- Single-phase and three-phase full converters provide for bidirectional power flow for positive output current and positive and negative average output voltages. Consequently, for positive speeds regenerative braking is possible only if the field current direction is changed as negative e.m.f. is required.

- Dual converters with circulating currents, requiring 12 PESs in three phase configurations provide for four-quadrant operation while avoiding the discontinuous current mode at the price of overloading one of the two converters.

- The source inductance produces an overlapping during phase commutation which results in a kind of resistive-like output voltage drop in the rectifier.

- The power factor decreases with a decrease in d.c. output voltage in phase delay rectifiers. Forced commutation or special complex configurations (for high powers) may solve this problem [3-6].

- All controlled rectifiers produce current harmonics and voltage notches on the a.c. source side and special input filters are required to reduce them to acceptable standardized levels.

5.12. PROBLEMS

5.1. A three-phase rectifier with controlled null (Figure 5.17) supplies a load made of a resistance R_s and an inductance L_s. The a.c. supply phase voltage is 120V (rms); $R_s = 10\Omega$; and the transformer ratio $K = w_1/w_2=2$. For $L_s = 0$ and $L_s = \infty$, determine:
  a. The output and transformer secondary voltages and current waveforms for the delay angle $\alpha_1 = \pi / 3$.
  b. The average values of output voltage and current.
  c. The waveform of transformer primary current.

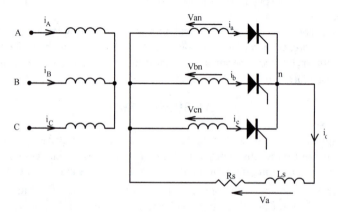

Figure 5.17. Three-phase rectifier with controlled null

5.2. A d.c. series motor is supplied through a single phase semiconverter with a free-wheeling diode (Figure 5.18). Depict the waveforms of the induced voltage e_g and motor voltage V_a for discontinuous and continuous current.

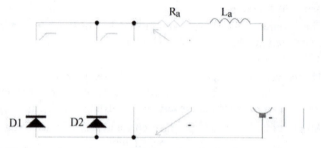

Figure 5.18. D.c. series motor supplied through a single-phase semiconverter

5.3. For problem 5.2 find analytical expressions of armature current for discontinuous and continuous modes for constant speed, with saturation neglected. Find the condition to determine the minimum value of L_a for which the current is still continuous. Determine an expression of average output voltage for the continuous current.

5.4. A single-phase full converter [7] uses power transistors and is controlled through pulse width modulation with a sinusoidal carrier with n pulses per semiperiod (Figure 5.19). Obtain the expressions of average output voltage, current harmonics, power factor (PF), displacement factor (DF) and current harmonics factor (HF).

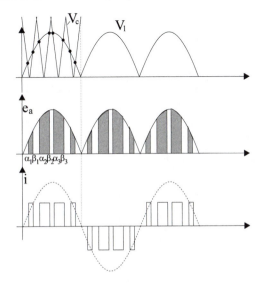

Figure 5.19. Single-phase full converter with PWM control

5.13. SELECTED REFERENCES

1. *N. Mohan, T.M. Underland, W.P. Robbins*, Power electronics, Second edition, Chapter 6, Wiley, 1995.
2. *G.K. Dubey*, Power semiconductor controlled drives, Chapter 3, Prentice Hall, 1989.
3. *T. Ohnuki, O. Miyashita, T. Haneyoski, E. Ohtsuji*, "High power factor PWM rectifiers with an analog PWM prediction controller", IEEE Trans, vol.PE - 11.no.3., 1996, pp.460-465.
4. *M.O. Eissa, S. Leeb, G.C. Verghese, A.M. Stankovic*, "Fast controller for unity power factor PWM rectifier", I.B.I.D., no.1, 1996, pp.1-6.
5. *N. Akagi*, "New trends in active filters for power conditioning", IEEE Trans. vol. IA - 32, no.6.,1996, pp.1312-1327.
6. *I.O. Krah, I. Holtz*, "Total compensation of line-side switching harmonics in converter-fed A.C. locomotives", I.B.I.D. vol.31, no.6., 1995, pp.1264-1273.
7. *M. Malinowski, M. Kazmierkowski*, "Control of three phase PWM rectifiers", Chapter 11

in the book "Control in power electronics", Academic Press, 2002, editors: M. Kazmierkowski, R. Krishnan, F. Blaabjerg.

8.   ***A. di Gerlando, R. Perini,*** "A model of the operation analysis of high speed universal motor with triac regulated mains voltage supply", Proc. of Symposium on Power Electronic and Electric Drives, automation motion, Ravello, Italy, 2002, pp. C407 – C412.

Chapter 6

CHOPPER-CONTROLLED D.C. BRUSH MOTOR DRIVES

6.1. INTRODUCTION

The d.c. chopper is a d.c. to d.c. power electronic converter (PEC) with forced commutation. It is used for armature voltage control in d.c. brush motor drives. D.c. sources to supply d.c. choppers are batteries or diode rectifiers with output filters so typical for urban electric transportation systems or to low power d.c. brush motor drives. Thyristors, bipolar power transistors, MOSFETs or IGBTs are used in d.c. choppers.

The basic configurations are shown in Table 6.1 and they correspond to single, two- or four-quadrant operation.

Table 6.1. Single-phase chopper configurations for the d.c. brush motors

Type	Chopper configuration	ea-Ia characteristics	Function
First-quadrant (step-down) choppers			Va = V0 for S1 on Va = 0 for S1 off and D1 on
Second quadrant, regeneration (step-up) chopper			Va = 0 for S2 on Va = V0 for S2 off and D2 on
Two quadrant chopper			ea = e0 for S1 or D2 on ea = e0 for S2 or D1 on ia>0 for S1 or D1 on ia<0 for S2 or D2 on
Two quadrant chopper			Va = +V0 for S1 & S2 on Va = −V0 for S1 & S2 off and D1 & D2 on
Four quadrant chopper			S4 on & S3 off S1 & S2 operated Va>0 ia - reversible S2 on & S1 off S3 & S4 operated Va<0 ia - reversible

119

The first-quadrant chopper (Figure 6.1) is operated by turning on the PES for the interval t_{on}, when the supply voltage is connected to the load. During the interval t_{off}, when the main switch is off, the load current flows through the freewheeling diode D_1. The output voltage e_a is shown in Figure 6.1.

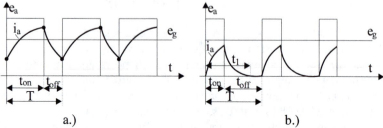

a.)   b.)

Figure 6.1. First-quadrant chopper operation

a.) continuous mode; b.) discontinuous mode

The average voltage V_{av} is

$$V_{av} = e_a \cdot \frac{t_{on}}{T} \le V_0 \qquad (6.1)$$

That is, a step-down chopper.

Constant frequency (constant T) control is preferred in order to improve the input filter operation and reduce the possibility of discontinuous current mode (Figure 6.1b) operation.

The voltage equation for constant speed is

$$V_0 = R_a \cdot i_a + L_a \cdot \frac{di_a}{dt} + e_g; \quad e_g = K_e \cdot \lambda_p \cdot n \quad \text{for } 0 \le t \le t_{on} \qquad (6.2)$$

$$0 = R_a \cdot i_a + L_a \cdot \frac{di_a}{dt} + e_g; \quad t_{on} \le t \le t_1; \quad i_a(t_1) = 0, \qquad (6.3)$$

$$t_1 < T \text{ for discontiuous mode}$$

For continuous current mode $t_1 = T$ and $i_a(T) = i_a(0) \ne 0$ for steady state. For the d.c. brush series motor

$$e_g = K_{ei} \cdot i_a \cdot n + K_{rem} \cdot n \qquad (6.4)$$

In (6.4) K_{rem} refers to the remnant flux while the magnetization curve of the machine is considered linear.

The average output voltage for the discontinuous mode may be determined noting that the motor voltage is then zero

$$V_{av} = V_0 \frac{t_{on}}{T} + e_g \cdot \frac{T - t_1}{T}; \qquad t_1 \leq T \tag{6.5}$$

The output current expressions are obtained from (6.2)-(6.3)

$$i_a = A \cdot e^{-t \cdot \frac{R_a}{L_a}} + \frac{V_0 - e_g}{R_a}; \qquad \text{for } 0 \leq t \leq t_{on} \tag{6.6}$$

$$i_a' = A' \cdot e^{-(t - t_{on}) \cdot \frac{R_a}{L_a}} - \frac{e_g}{R_a}; \text{ for } 0 \leq t \leq t_1 \begin{cases} t_1 = T \text{ for continuous current} \\ t_1 < T \text{ for discontinuous current} \end{cases} \tag{6.7}$$

The continuity condition is

$$i_a(t_{on}) = i_a'(t_{on}) \tag{6.8}$$

The average output current i_{av} is

$$i_{av} = \frac{\displaystyle\int_0^{t_{on}} i_a \, dt + \int_{t_{on}}^{t_1} i_a' \, dt}{T} \tag{6.9}$$

For the second-quadrant chopper (Table 6.1b) the d.c. motor e.m.f. e_g with S_2 on produces a current rise in inductance L_a

$$R_a \cdot i_a + L_a \cdot \frac{di_a}{dt} = -e_g; \qquad \text{for } 0 \leq t \leq t_{on}; \quad i_a(0) = 0 \tag{6.10}$$

When S_2 is turned off, the energy stored in the inductor is sent back to the source as long as $V_0 > V_a$

$$V_0 - e_g = -R_a \cdot i_a' - L_a \cdot \frac{di_a'}{dt}; \quad t_{on} \leq t \leq T \tag{6.11}$$

with the solution

$$i_a = -\frac{e_g}{R_a} + B \cdot e^{-t \cdot \frac{R_a}{L_a}} + i_{a0} \tag{6.12}$$

$$i_a' = +\frac{V_0 - e_g}{R_a} + B' \cdot e^{-(t - t_{on}) \cdot \frac{R_a}{L_a}} \tag{6.13}$$

The boundary conditions are

$$i_a(t_{on}) = i_a'(t_{on}), i_a(0) = i_{a0} \text{ and } i_a'(T) = i_{a0} \tag{6.14}$$

It is thus possible with $e_g < V_0$ to retrieve the energy back from the d.c. brush motor by using the inductor L_a as an energy sink (Figure 6.2).

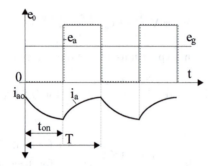

Figure 6.2. Second-quadrant chopper operation

The two-quadrant chopper (Table 6.1c,d) is a combination of one first and one second-quadrant chopper. Finally two-quadrant choppers are combined to obtain a four-quadrant chopper.

As the chopper is an on-off switch, the source current is chopped (Figure 6.3). This makes the peak input power demand high. Also, the supply current (Figure 6.3) has harmonics which produce voltage fluctuations, signal interference, etc.

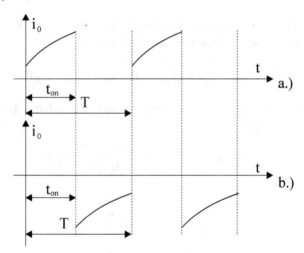

Figure 6.3. Source current waveforms

a.) first-quadrant operation, b.) second-quadrant operation.

An LC input filter (Figure 6.4) will provide a path for the ripple current such that only (approximately) the average current is drawn from the supply.

The nth harmonic current i_n in the supply (Figure 6.4b) is

$$i_n = \frac{X_c / n}{\left(nX_L - X_C / n\right)} I_{sn} = \frac{I_{sn}}{\left(nf_{ch} / f_r\right)^2 - 1} \qquad (6.15)$$

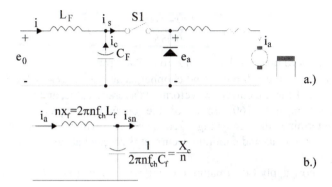

Figure 6.4. First-quadrant chopper with LC input filter

a.) basic circuit, b.) equivalent circuit for nth harmonic

where f_{ch} is the chopping frequency (f_{ch} = 1 / T) and f_r is the resonance frequency of the filter $f_r = 1/2\pi\sqrt{LC}$. To avoid resonance $f_{ch} \geq (2-3)f_r$. Given the variety of configurations in Table 6.1 we will proceed directly with numerical examples in parallel with theoretical development to facilitate quick assimilation of knowledge.

6.2. THE FIRST-QUADRANT (STEP-DOWN) CHOPPER

A d.c. brush motor with permanent magnet excitation with the data R_a = 1Ω, $K_e\lambda_p$ = 0.055 Wb / rpm, is fed through a first-quadrant chopper (Table 6.1a) from a 120 Vd.c. supply at a constant (ideal) armature current of 10A.

Determine:

a. the range of duty cycle α from zero to maximum speed;
b. the range of speed.

Solution:

a. The average output voltage V_a is

$$V_a = V_0 \cdot \alpha = 120\alpha \qquad (6.16)$$

At standstill n = 0 and thus

$$V_a = R_a i_a = 1 \cdot 10 = 10V$$

$$\text{Thus } \alpha_{min} = \frac{(V_a)_{n=0}}{V_0} = \frac{10}{120} = \frac{1}{12} \qquad (6.17)$$

For maximum speed, the voltage is 120V (α = 1). Consequently, α varies from 1/12 to 1.

b. The voltage equation for maximum speed is

$$V_{a_{max}} = R_a I_a + K_e\lambda_p n_{max} \qquad (6.18)$$

$$n_{max} = \frac{120 - 1 \cdot 10}{0.055} = 2000 \, \text{rpm} \qquad (6.19)$$

The speed range is thus from zero to 2000 rpm.

c.   For the d.c. brush motor and chopper as above and $\alpha = 0.3$, calculate the actual armature current waveform, its average value, and voltage average value at n = 1600 rpm, for the chopping frequency f_{ch} = 50Hz. Determine the chopping frequency for which the limit between discontinuous and continuous current is reached at same t_{on} as above.
Solution:
We now apply the armature current expressions (6.6)-(6.7) first for f_{ch} = 50Hz.

The turn-on time interval $t_{on} = \dfrac{1}{f_{ch}} \cdot \alpha = \dfrac{0.3}{50} = 6 \cdot 10^{-3} \, \text{s} = 6 \, \text{ms}$ $\qquad (6.20)$

with $T = \dfrac{1}{f_{ch}} = \dfrac{1}{50} = 0.02 \text{s} = 20 \, \text{ms}$

$$i_a = A \cdot e^{-t\frac{1}{5 \cdot 10^{-3}}} + \frac{120 - 0.055 \cdot 1600}{1} = A \cdot e^{-200t} + 32$$
$$i_a' = A' \cdot e^{-200\left(t - 6 \cdot 10^{-3}\right)} - \frac{88}{1} \qquad (6.21)$$

Assuming discontinuous current mode

$$\left(i_a\right)_{t=0} = 0; \quad A = -32 \qquad (6.22)$$

Also $\left(i_a'\right)_{t_{on}} = \left(i_a\right)_{t_{on}}; \quad A' - 88 = 32\left(1 - e^{-200 \cdot 6 \cdot 10^{-3}}\right) = 22.36$
$$A' = 110.36 \qquad (6.23)$$

The current i_a' becomes zero at $t = t_1$

$$A' e^{-200\left(t_1 - 6 \cdot 10^{-3}\right)} - 88 = 0; \quad 110.36 \cdot e^{-200\left(t_1 - 6 \cdot 10^{-3}\right)} - 88 = 0$$
$$t_1 = 7.132 \cdot 10^{-3} \, \text{s} < T = 20 \, \text{ms} \qquad (6.24)$$

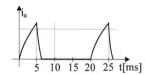

5   10 15 20 25 t[ms]
Figure 6.5. Discontinuous current

Thus indeed the current is discontinuous.
The average current i_{av} is:

$$i_{av} = \frac{1}{T}\left[\int_0^{t_{on}} i_a \, dt + \int_{t_{on}}^{t_1} i_a' \, dt\right] =$$

$$\frac{1}{2 \cdot 10^{-2}}\left[32\int_0^{6 \cdot 10^{-3}}\left(1 - e^{-200t}\right)dt + \int_{6 \cdot 10^{-3}}^{7.132 \cdot 10^{-3}}\left(110.36 \cdot e^{-200\left(t - 6 \cdot 10^{-3}\right)} - 88\right)dt\right] \quad (6.25)$$

$$i_{av} = 7.8452 \text{ A}$$

The chopping frequency for the limit between the discontinuous and continuous current is obtained for

$$t_1 = T_c = 7.132 \cdot 10^{-3}; \quad f_{ch}' = \frac{1}{T_c} = \frac{1}{7.132 \cdot 10^{-3}} = 140.2 \text{ Hz} \quad (6.26)$$

In this case α becomes

$$\alpha_c = \frac{t_{on}}{T_c} = \frac{6 \cdot 10^{-3}}{7.132 \cdot 10^{-3}} = 0.8412 \quad (6.27)$$

As the current is discontinuous the average voltage is from (6.5)

$$V_{av} = V_0 \frac{t_{on}}{T} + e_g \frac{\left(T - t_1\right)}{T} = 120 \cdot 0.3 + 88 \cdot \frac{20 - 7.132}{20} \quad (6.28)$$
$$= 40 + 56.619 = 96.619 \text{ V}$$

6.3. THE SECOND-QUADRANT (STEP-UP) CHOPPER FOR GENERATOR BRAKING

A d.c. brush motor with PM excitation is fed through a second-quadrant chopper for regenerative braking (Table 6.1b).

The motor data are $R_a = 1\Omega$; $L_a = 20$mH; $e_g = 80$V (given speed). The supply voltage V_0 is 120Vd.c. and $t_{on} = 5 \cdot 10^{-3}$ s.

Determine:
a. The waveform of motor current for zero initial current
b. The waveform of source current
c. The maximum average power generated
Solution:
a. The current waveforms are as shown in Figure 6.2 and 6.3 with $i_{a0} = 0$.

$$i_a = -\frac{e_g}{R_a} + B \cdot e^{-t\frac{R_a}{L_a}}; \quad 0 < t < t_{on} \quad (6.29)$$

b. $$i_a' = \frac{V_0 - e_g}{R_a} + B' \cdot e^{-\left(t - t_{on}\right)\frac{R_a}{L_a}}; \quad t_{on} < t < T \quad (6.30)$$

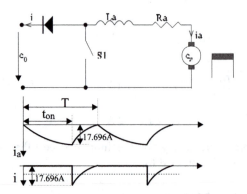

Figure 6.6. The step-up second-quadrant chopper.

The boundary conditions are

$$\left(i_a\right)_{t=0} = 0; \quad \left(i_a\right)_{t=t_{on}} = \left(i_a'\right)_{t=t_{on}}; \quad \left(i_a'\right)_{t=T} = 0 \qquad (6.31)$$

The unknowns are B, B' and T. Consequently,

$$B = \frac{e_g}{R_a} = \frac{80}{1} = 80 \qquad (6.32)$$

$$80 \cdot e^{-5 \cdot 10^{-3} \cdot 50} - 80 = \frac{(120 - 80)}{1} + B'; \quad B' = -57.696 \qquad (6.33)$$

$$40 - 57.696 \cdot e^{-\left(T - t_{on}\right) \cdot 50} = 0; \quad T - t_{on} = 7.326 \cdot 10^{-3} \, s$$
$$\text{thus } T = (7.326 + 5) \cdot 10^{-3} \, s = 12.326 \cdot 10^{-3} \, s \qquad (6.34)$$

Note that the source current occurs during the S_1 turn-off and is negative, proving the regenerative operation.
The average source current i_{av} is

$$i_{av}' = \frac{1}{T} \int_{t_{on}}^{T} i_a' \, dt = \frac{1}{T} \left[\frac{e_0 - e_g}{R_a} \left(T - t_{on}\right) - \frac{L_a}{R_a} B'\left(e^{-\left(T - t_{on}\right) \frac{R_a}{L_a}} - 1 \right) \right] =$$

$$= \frac{1}{12.326 \cdot 10^{-3}} \left[\frac{40}{1} \cdot 7.326 \cdot 10^{-3} + 20 \cdot 10^{-3} \cdot 57.696 \cdot \left(e^{-7.326 \cdot 10^{-3} \cdot 50} - 1 \right) \right]; \ (6.35)$$

$$i_{av}' = -4.938 \, A$$

The average power regenerated P_{av} is

$$P_{av} = -i_{av}' \cdot V_a = 4.938 \cdot 120 = 592.56 \, W \qquad (6.36)$$

6.4. THE TWO-QUADRANT CHOPPER

Consider a two-quadrant chopper (Figure 6.7) supplying a d.c. brush motor with separate excitation. The load current varies between $I_{max} > 0$ and $I_{min} < 0$.

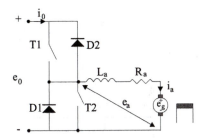

Figure 6.7. Two-quadrant chopper supplying

a separately excited d.c. brush motor

Determine:

a. The voltage and current waveforms for the load current varying between an $I_{max} > 0$ and $I_{min} < 0$ with $I_{max} > |I_{min}|$.

b. Derive the expression of the conducting times t_{d2} and t_{d1} of the diodes D_1 and D_2 for case a.

Solution:

a. Let us first draw the load current which varies from a positive maximum to a negative minimum (Figure 6.8). The conduction intervals for each of the four switches T_1, D_2, D_1, T_2 are

$$T_1 \text{ for } t_{d2} < t < t_c;$$
$$T_2 \text{ for } t_{d1} < t < T; \qquad (6.37)$$
$$D_1 \text{ for } t_c < t < t_{d1};$$
$$D_2 \text{ for } 0 < t < t_{d2}$$

b. The equations for the current are:

$$V_0 - e_g = R_a i_a + L_a \frac{di_a}{dt}; \quad \text{for } 0 < t < t_c \qquad (6.38)$$

$$-e_g = R_a i_a + L_a \frac{di_a}{dt}; \quad \text{for } t_c < t < T \qquad (6.39)$$

with the solutions:

$$i_a = \frac{V_0 - e_g}{R_a} + A \cdot e^{-t \frac{R_a}{L_a}}; \quad 0 < t \le t_c \qquad (6.40)$$

$$i_a' = -\frac{e_g}{R_a} + A' \cdot e^{-(t-t_c)\frac{R_a}{L_a}}; \quad t_c < t \leq T \tag{6.41}$$

with the boundary conditions $i_a(0) = i_{min}$, $i_a'(T) = i_{min}$ and $i_a(t_c) = i_a'(t_c)$.

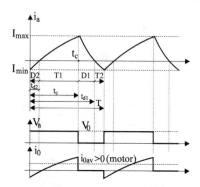

Figure 6.8. Voltage and current waveforms
of two-quadrant chopper-fed d.c. motor

The unknowns A, A' and t_c are obtained from

$$A = I_{min} - \frac{e_0 - e_g}{R_a} \tag{6.42}$$

$$A' = I_{max} + \frac{e_g}{R_a} \tag{6.43}$$

$$T - t_c = -\frac{L_a}{R_a} \cdot \ln\left[\left(A' - \frac{e_0}{R_a}\right) / A\right] \tag{6.44}$$

Consider a d.c. brush motor whose data are $e_0 = 120V$, $R_a = 1\Omega$, $L_a = 5mH$, $e_g = 80V$, $I_{max} = 5A$, $I_{min} = -2A$ and chopping frequency $f_{ch} = 0.5kHz$. For the two-quadrant chopper as above calculate:

a. The $t_c / T = \alpha_{on}$ ratio.
b. The conducting intervals of the 4 switches.
    Solution:
a. The constants A, A', t_c expressions developed above ((6.42)-(6.44)) yield

$$A = I_{min} - \frac{e_0 - e_g}{R_a} = -2 - \frac{120 - 80}{1} = -42 \text{ A} \tag{6.45}$$

$$A' = I_{max} + \frac{e_g}{R_a} = 5 + \frac{80}{1} = 85\,A \tag{6.46}$$

$$T = \frac{1}{f_{ch}} = \frac{1}{500}\,s = 0.002s = 2\,ms \tag{6.47}$$

$$T - t_c = -\frac{L_a}{R_a} \cdot \ln\left[\left(A' - \frac{e_0}{R_a}\right)/A\right] =$$

$$-\frac{5 \cdot 10^{-3}}{1} \cdot \ln\left[\left(85 - \frac{120}{1}\right)/(-42)\right] = 0.9116 \cdot 10^{-3}\,s \tag{6.48}$$

$$t_c = 2 \cdot 10^{-3} - 0.5116 \cdot 10^{-3} = 1.0884 \cdot 10^{-3}\,s$$

b. The conducting interval of D_2, t_{d2}, corresponds to $i_a = 0$

$$\frac{V_0 - e_g}{R_a} + A \cdot e^{-t_{d2}\frac{R_a}{L_a}} = 0$$

$$t_{d2} = -\frac{L_a}{R_a} \cdot \ln\left[\frac{V_0 - e_g}{-A \cdot R_a}\right] = \frac{-5 \cdot 10^{-3}}{1} \ln\left(\frac{120 - 80}{-(-42) \cdot 1}\right) = 0.2439 \cdot 10^{-3}\,s$$
$$\tag{6.49}$$

Thus the main switch T_1 conducts for a time interval

$$t_c - t_{d2} = (1.0884 - 0.2439) \cdot 10^{-3} = 0.84445 \cdot 10^{-3}\,s \tag{6.50}$$

To calculate the conducting time of the diode D_1 we apply the condition $i_a'(t_{d1}) = 0$

$$\frac{-e_g}{R_a} + A' e^{-(t_{d1} - t_c)\frac{R_a}{L_a}} = 0 \tag{6.51}$$

$$t_{d1} - t_c = -\frac{L_a}{R_a} \cdot \ln\left[\frac{e_g}{A' R_a}\right] = \frac{-5 \cdot 10^{-3}}{1} \ln\left(\frac{80}{85 \cdot 1}\right) = 0.303 \cdot 10^{-3}\,s \tag{6.52}$$

Consequently, the diode D_1 conducts for 0.303 ms. Finally, the static switch T_2 conducts for the time interval

$$T - t_{d1} = (2 - 1.0884 - 0.303) \cdot 10^{-3} = 0.6084 \cdot 10^{-3}\,s \tag{6.53}$$

Note: As seen above, the two-quadrant operation of the chopper resides in the variation of t_c / T as the main switches command signals last t_c and, respectively, $T - t_c$ intervals though they conduct less time than that, allowing the diodes D_2 and D_1 to conduct. The two-quadrant chopper has the advantage of natural (continuous) transition from motor to generator action.

6.5. THE FOUR-QUADRANT CHOPPER

A d.c. brush motor with separate excitation is fed through a four-quadrant chopper (Table 6.1e). Show the waveforms of voltage and current in the third and fourth quadrants.

Solution:

The basic circuit of a four-quadrant chopper is shown in Figure 6.9.

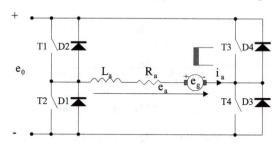

Figure 6.9. D.c. brush motor fed through a four-quadrant chopper

If T_4 is on all the time, T_1-D_1 and T2-D2 provide first- and (respectively) second-quadrant operations as shown in previous paragraphs. With T_2 on all the time and T_3-D_3 and, respectively, T_4-D_4 the third- and fourth-quadrant operations is obtained (Figure 6.10). So, in fact, we have 2 two-quadrant choppers acting in turns.

However, only 2 out of 4 main switches are turned on and off with the frequency f_{ch} while the third main switch is kept on all the time and the fourth one is off all the time.

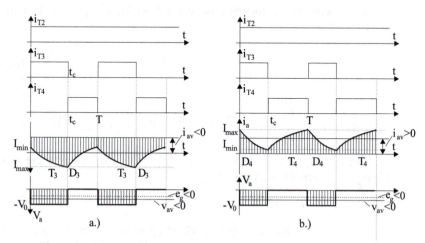

Figure 6.10. Four-quadrant chopper supplying a d.c. brush motor

a.) Third quadrant: iav<0, Vav<0; b.) Fourth quadrant: iav>0, Vav<0.

Four-quadrant operation is required for fast response reversible variable speed drives.

As expected, discontinuous current mode is also possible but it should be avoided by increasing the switching frequency f_{ch} or adding an inductance in series with the motor.

Let us assume that:

A d.c. brush motor, fed through a four-quadrant chopper, works as a motor in the third quadrant (reverse motion). The main data are $V_0 = 120V$, $R_a = 0.5\Omega$, $L_a = 2.5mH$, rated current $I_{an} = 20A$; rated speed $n_n = 3000$ rpm; separate excitation.

a. Calculate the rated e.m.f., e_g, and rated electromagnetic torque, T_e.
b. For $n = -1200$ rpm and rated average current ($i_{av} = -I_{an}$) determine the average voltage V_{av}, $t_c / T = \alpha_{on}$, and maximum and minimum values of motor current I_{max} and I_{min} for 1kHz switching frequency.

Solution:

a. The motor voltage equation for steady state is:

$$V_{av} = R_a i_a + e_g \tag{6.54}$$

for rated values $V_{av} = V_0 = 120V$, $i_a = i_{an} = 20A$, thus

$$e_{gn} = K_e \lambda_p n_n = V_{av} - R_a i_a = 120 - 20 \cdot 0.5 = 110 \text{ V} \tag{6.55}$$

$$K_e \lambda_p = \frac{e_{gn}}{n_n} = \frac{110}{50} = 2.2 \text{ Wb} \tag{6.56}$$

b. The motor equation in the third quadrant is

$$V_{av} = R_a i_a + e_g = 0.5 \cdot (-20) + 2.2 \cdot (-20) = -54 \text{ V}, \tag{6.57}$$

the conducting time t_c for T_3 (Figure 6.10a) is

$$\frac{t_c}{T} = \frac{V_{av}}{-V_0} = \frac{-54}{-120} = 0.45 \tag{6.58}$$

$$t_c = T \cdot 0.45 = \frac{1}{f_{ch}} \cdot 0.45 = \frac{1}{10^3} \cdot 0.45 = 0.45 \cdot 10^{-3} \text{s} \tag{6.59}$$

Form ((6.40)-(6.41)) the motor current variation (Figure 6.10a) is described by

$$i_a = \frac{V_0' - e_g}{R_a} + A \cdot e^{-t \frac{R_a}{L_a}}; \quad 0 < t \le t_c \tag{6.60}$$

$$i_a' = -\frac{e_g}{R_a} + A' \cdot e^{-(t-t_c)\frac{R_a}{L_a}}; \quad t_c < t \le T \tag{6.61}$$

The current continuity condition ($i_a(t_c) = i_a'(t_c)$) provides

$$t_c = -\frac{L_a}{R_a} \cdot \ln\left[\left(A' - \frac{V_0'}{R_a}\right) / A\right] \tag{6.62}$$

The second condition is obtained from the average current expression

$$i_{av} = \frac{1}{T}\left[\int_0^{t_c} i_a dt + \int_{t_c}^T i_a' dt\right] =$$

$$\frac{1}{T}\left\{\frac{V_0' - e_g}{R_a} t_c - \frac{e_g}{R_a}(T - t_c) + \frac{L_a}{R_a}\left[\left(1 - e^{-t_c\frac{R_a}{L_a}}\right)A + A'\left(1 - e^{-(T-t_c)\frac{R_a}{L_a}}\right)\right]\right\}$$

$$\tag{6.63}$$

From (6.62) and (6.63) we obtain:

$$\left(A' - \frac{V_0'}{R_a}\right) / A = e^{-t_c\frac{R_a}{L_a}};$$

$$V_0' = -V_0; \quad e_g = K_e\lambda_p n = 2.2 \cdot (-20) = -44 \text{ V} \tag{6.64}$$

$$\left(A' + \frac{(-120)}{0.5}\right) / A = e^{-0.45 \cdot 10^{-3}\frac{0.5}{2.5 \cdot 10^{-3}}} = 0.914$$

$$-20 = 10^3 \{\frac{-120 - (-44)}{0.5} 0.45 \cdot 10^{-3} - \frac{(-44)}{0.5}0.55 \cdot 10^{-3}$$

$$+ \frac{2.5 \cdot 10^{-3}}{0.5}\left[\left(1 - e^{-0.45 \cdot 10^{-3}\frac{0.5}{2.5 \cdot 10^{-3}}}\right)A + A'\left(1 - e^{-0.55 \cdot 10^{-3}\frac{0.5}{2.5 \cdot 10^{-3}}}\right)\right]\} \tag{6.65}$$

$$-20 = -20 + 0.43A + 0.5205A' \tag{6.66}$$

$$0.43A + 0.5205A' = 0 \tag{6.67}$$

$$A' + 240 = 0.914A \tag{6.68}$$

$$A = 137.62; \ A' = -113.92 \tag{6.69}$$

Now we may calculate $I_{min} = i_a(0)$

$$I_{min} = A + \frac{V_0' - e_g}{R_a} = 137.92 + \frac{-120 - (-44)}{0.5} = -15.08 \text{ A} \qquad (6.70)$$

Also $I_{max} = i_a'(t_c)$

$$I_{max} = A' - \frac{e_g}{R_a} = -113.92 + \frac{-(-44)}{0.5} = -25.92 \text{ A} \qquad (6.71)$$

6.6. THE INPUT FILTER

A first-quadrant chopper with an L-C input filter supplies a d.c. brush motor with PM excitation under constant current start-up.

a. Demonstrate that maximum rms ripple current in the chopper current i_{ch} occurs at a duty cycle $\alpha = 0.5$.

b. For $f_{ch} = 400$ Hz, $I_a = 100$ A, the rms fundamental (a.c.) current allowed in the supply is 10% of d.c. supply current. Capacitors of 1mF which can take 5 A rms ripple current are available. Determine L_f and C_f of the filter for $f_{ch} > 2f_r$.

c. For case b. calculate the d.c., first and third harmonics of the supply current.

Solution:

a. The chopper configuration (Figure 6.11) provides square current pulses for i_s of width α.

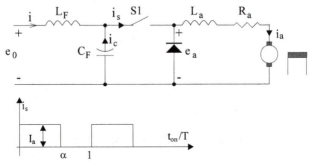

Figure 6.11. First-quadrant chopper with LfCf filter

a.) basic circuit, b.) chopper input current

Thus, the d.c., I_{sdc}, rms I_{srms} and ripple $I_{sripple}$ components of chopper currents are:

$$I_{sdc} = I_a \cdot \alpha \qquad (6.72)$$

$$I_{srms} = \sqrt{\int_0^\alpha I_a{}^2 d\alpha} = I_a \sqrt{\alpha} \tag{6.73}$$

$$I_{sripple} = \sqrt{\left(I_a \sqrt{\alpha}\right)^2 - \left(I_a \alpha\right)^2} = I_a \sqrt{\left(\alpha - \alpha^2\right)} \tag{6.74}$$

The maximum ripple current is obtained for $\dfrac{dI_{sripple}}{d\alpha} = 0$ which leads to $\alpha = 1/2$. The filter design will be performed for this worst case.

b.   To design the filter we need first the chopper current harmonics content which, for $\alpha = 0.5$, is

$$i_s = I_{sdc} + \frac{4}{\pi} \cdot \frac{I_a}{2n}(\sin \omega t + \sin 3\omega t + \sin 5\omega t) \tag{6.75}$$

with

$$I_{sdc} = I_a \cdot \alpha = 100 \cdot \frac{1}{2} = 50 \text{ A} \tag{6.76}$$

$$\left(i_{s_1}\right)_{rms} = \frac{4 \cdot 100}{\pi \cdot 2 \cdot 1} \cdot \frac{1}{\sqrt{2}} = 45.1733 \text{ A} \tag{6.77}$$

$$\left(i_{s_3}\right)_{rms} = \frac{4 \cdot 100}{\pi \cdot 2 \cdot 3} \cdot \frac{1}{\sqrt{2}} = 15.05 \text{ A} \tag{6.78}$$

$$\left(i_{s_5}\right)_{rms} = \frac{4 \cdot 100}{\pi \cdot 2 \cdot 5} \cdot \frac{1}{\sqrt{2}} = 9.03 \text{ A} \tag{6.79}$$

The input (source) a.c. current is not to surpass 10% of d.c. input current, that is $I_1 = I_{sdc} \cdot \dfrac{10}{100} = 50 \cdot \dfrac{10}{100} = 5 \text{ A}$.

According to Equation (6.15) for n = 1 we obtain

$$I_1 = \frac{x_C}{x_L - x_C} \cdot I_{s_1}; \quad 5 = \frac{x_C}{x_L - x_C} \cdot 45 \quad \text{or} \quad x_L = 10x_C \tag{6.80}$$

Also, the fundamental capacitor current I_{C1} is

$$I_{C1} = \frac{x_L}{x_L - x_C} \cdot I_{s_1} = \frac{10x_C}{10x_C - x_C} \cdot 45 = 50 \text{ A} \tag{6.81}$$

As each 1mF capacitor can take 5A, 10 such capacitors in parallel are needed and thus $C_f = 10$mF. On the other hand, the reactance x_L is

$$x_L = 10 x_C = 10 \cdot \frac{1}{2 \cdot \pi \cdot 400 \cdot 10 \cdot 10^{-3}} = 0.398 \ \Omega \qquad (6.82)$$

$$L_f = \frac{x_L}{2 \cdot \pi \cdot 400} = \frac{0.398}{2 \cdot \pi \cdot 400} = 0.15847 \cdot 10^{-3} \, H \qquad (6.83)$$

The filter resonance frequency f_r is

$$f_r = \frac{1}{2 \cdot \pi \sqrt{L_f C_f}} = \frac{1}{2 \cdot \pi \sqrt{0.15847 \cdot 10^{-3} \cdot 10^{-2}}} = 127 \ Hz \qquad (6.84)$$

The ratio between the chopper switching frequency f_{ch} and the filter resonance frequency f_r is

$$f_{ch} / f_r = 400/127 = 3.15$$

c. The same a.c. current components (6.15) (after filtering) are

$$I_1 = \frac{I_{s_1}}{\left(f_{ch} / f_r \right)^2 - 1} \approx \frac{45}{3.15^2 - 1} = 5 \ A \qquad (6.85)$$

$$I_3 = \frac{I_{s_1}}{\left(3 f_{ch} / f_r \right)^2 - 1} \approx \frac{15}{(3 \cdot 3.15)^2 - 1} = 0.17 \ A \qquad (6.86)$$

$$I_5 = \frac{I_{s_1}}{\left(5 f_{ch} / f_r \right)^2 - 1} \approx \frac{9}{(5 \cdot 3.15)^2 - 1} = 0.036 \ A \qquad (6.87)$$

As noted the $L_f C_f$ filter produces a drastic reduction of source-current harmonics.

6.7. DIGITAL SIMULATION THROUGH MATLAB/SIMULINK

Simulation results of a d.c. motor drive with a four-quadrant chopper are presented. The motor model was integrated in a block (DC brush motor in Figure 6.12). Changing of motor parameters is done by clicking on their block. A dialogue box appears and you can change them by modifying their default values. This motor model block includes the possibility of adding an extra inductance in the motor circuit (L_{add}).

The drive system consists of PI speed controller (Ki = 1, Ti = 0.1), PI torque controller (Ki = 50, Ti = 0.0005) and motor blocks. The study examines the system behavior during starting, load perturbation (at 0.2s) and speed reversal with no load (at 0.5 s) and another load perturbation (at 0.6 s).

The integration step (10 μs) can be modified from the Simulink's *Simulation / Parameters*. The chopper frequency is 20 kHz and this block input is the t_c / T_e ratio.

To find out the structure of each block presented above, unmask it (*Options/Unmask*). Each masked block contains a brief description of that block (inputs / outputs / parameters).

The block diagram of the electric drive system is presented in Figure 6.12.

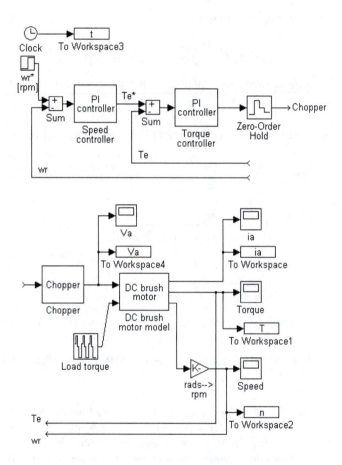

Figure 6.12. D.c. brush motor drive block diagram

The motor used for this simulation has the following parameters: V_{dc} = 120V, I_n = 20A, n_n = 3000rpm, R_s = 0.5Ω, L_a = 0.0025H, J = 0.001kgm^2, $K_e\lambda_p$ = 2.2Wb.

The Figures (6.13-6.16) represent the speed (Figure 6.13), torque (Figure 6.14) responses and current (Figure 6.15) and voltage (Figure 6.16) waveforms, for the *starting process and load torque (8Nm) applied at 0.2s, and reversal with no load at 0.5s and load torque applied at 0.6s.*

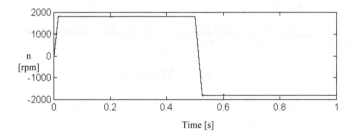

Figure 6.13. Speed transient response

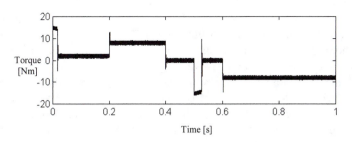

Figure 6.14. Torque response

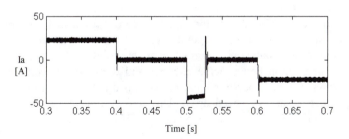

Figure 6.15. Current waveform (ia).

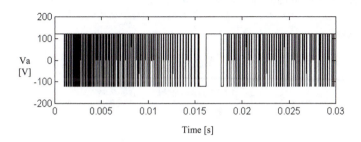

Figure 6.16. Voltage waveform (Va).

Fast response with rather low current ripple is obtained due to the rather high (20 kHz) switching frequency for a 5 ms electrical time constant d.c. brush PM motor four-quadrant drive.

6.8. SUMMARY

- In this chapter, single, two- or four-quadrant chopper basic configurations in interaction with the d.c. brush motor (for constant speed) have been introduced.
- In all configurations the PESs have been considered ideal - with instantaneous hard commutation (switching).
- Resonant d.c.-d.c. converters have not been treated. The interested reader is invited to study the literature [1].
- Choppers, as rectifiers, face the same problem of continuous-discontinuous current mode switching.
- Under discontinuous current mode, for low voltage low inductance motors and low switching frequency (thyristors), the dependence of output voltage on the on/off time ratios of PESs renders the control sluggish. Increasing the switching frequency or including a series inductance solves the problem.
- The input current has harmonics and thus an input filter is required to reduce them to acceptable limits.
- As the current in the motor pulsates, care must also be exercised in assessing the armature copper losses and the eventual derating of the motor for a certain application when chopper-fed.
- The interaction between the chopper and the motor at constant speed discussed in this chapter should be continued with closed-loop control to produce a variable speed drive. This will be done in the next chapter, though the Matlab simulation paragraph 6.7 anticipates it.

6.9. PROBLEMS

6.1. A step-down (buck) first-quadrant chopper is shown in Figure 6.17.

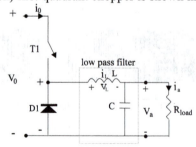

Figure 6.17. Step down (BUCK) chopper

a. Explain the principle and draw the inductance voltage and current waveforms for continuous current mode.

b. Calculate the output voltage and current for a resistive load as a function of duty ratio α, considering ideal switching and zero losses.

6.2. For problem 6.1 find α for the limit between continuous and discontinuous current mode and the average output voltage for discontinuous current mode. Plot the average output voltage versus average output current both for continuous and discontinuous modes.

6.3. A step-up (boost) first-quadrant chopper is shown in Figure 6.18. Considering ideal elements determine:

a. The voltage and current waveforms for continuous current mode.

b. The boundary between continuous and discontinuous current mode.

c. The output voltage ripple for continuous current operation.

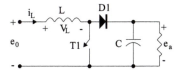

Figure 6.18. Step-up (boost) chopper

6.4. A four-quadrant chopper supplying a d.c. brush motor with PM excitation is controlled through PWM with bipolar voltage switching, Figure 6.19, where (T_{A+}, T_{B-}) and (T_{A-}, T_{B+}) are controlled simultaneously. Determine:

a. The voltage and current waveforms for positive control voltage $V_{control}$.

b. The voltage and current waveforms for negative control V_{con}.

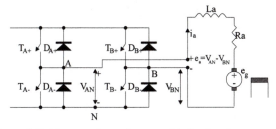

Figure 6.19. Four-quadrant chopper with d.c. brush motor

6.5. A d.c. series brush motor has the data: L_{at} = 20 mH, R_{at} = 1 Ω, rated voltage V_n = 120 V, rated current I_{an} = 10 A, rated speed n_n = 1800 rpm. The remnant flux induced voltage at rated speed is e_{grem} = 5V. When fed from a first-quadrant step-down chopper, at 120 Vdc with α = 0.3, f_{ch} = 50 Hz at rated speed, calculate the armature waveform, average current and average voltage.

6.6. Solve problem 6.5. with Matlab/Simulink or PSpice simulation programs.

6.10. SELECTED REFERENCES

1. **N. Mohan, T.M. Undeland, W.P. Robbins**, Power electronics, Second edition, Chapter 7, Wiley, 1995.

Chapter 7

CLOSED-LOOP MOTION CONTROL IN ELECTRIC DRIVES

7.1. INTRODUCTION

By motion control we mean torque, speed or position control. Motion control systems are characterized by precision, response quickness and immunity to parameter detuning, torque and inertia perturbations and energy conversion rates. Motion control through electric motors and power electronic converters (PECs) may be approached by the theory and practice of linear and nonlinear, continuous or discrete control systems. Control systems is a field of science in itself and in what follows only a few solutions of practical interest for motion control will be introduced with some application examples, from simple to complex.

To simplify the treatment and to include the PM d.c. brush motor control, the latter will be considered for various motion control systems. PM d.c. brush motors are characterized by a low electrical time constant, $\tau_e = L/R$, of a few milliseconds or less. The armature (torque) current is fully decoupled from the PM field because of the orthogonality of the armature and PM fields, both at standstill and for any rotor speed.

As shown in later chapters, vector control of a.c. motors also decouples flux and torque control if the orientation to the flux linkage is secured and flux amplitude is kept constant. Consequently, vector controlled a.c. motors are similar to d.c. brush motors and thus the application of various motion control systems to the d.c. motor holds notable generality while also eliminating the necessity of a separate chapter on closed-loop control of brushless motors.

To start with, the cascaded motion linear control is presented through numerical examples.

7.2. THE CASCADED MOTION CONTROL

A typical motion control system comprises three loops: one for torque, one for speed and one for position (Figure 7.1.) Ideally the system has independent position and speed sensors though, in general, only one is used — the speed sensor for speed control and the position one for position control — with the speed in the latter case estimated from the position information.

The PM d.c. brush motor equations are

$$V = Ri + L\frac{di}{dt} + \lambda_{PM}\omega_r \qquad (7.1)$$

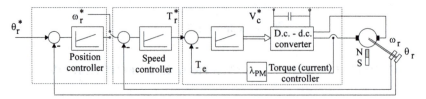

Figure 7.1. Typical cascaded motion control

$$J\frac{d\omega_r}{dt} = T_e - T_L \qquad (7.2)$$

$$\frac{d\theta_r}{dt} = \omega_r \qquad (7.3)$$

$$T_e = \lambda_{PM}I \qquad (7.4)$$

7.2.1. The torque loop

The torque control for constant flux linkage (λ_{PM}) means (with core loss neglected) armature current control for a PM d.c. brush motor. Fast current control also provides for fast current protection. The design of the torque loop implies knowing the load torque T_L. Alternatively, T_L = constant and, after the design is completed, the influence of load torque perturbations on loop stability is investigated and adequate corrections added, if required.

For constant (or zero) load torque the PM d.c. brush motor current / voltage transfer function, from (7.1)-(7.4), becomes (see eqn. 4.43):

$$H_V(s) = \frac{i(s)}{V(s)} = \frac{s\tau_{em}}{\left(s^2\tau_{em}\tau_e + s\tau_{em} + 1\right)R} \qquad (7.5)$$

where
$$\tau_{em} = \frac{JR}{\lambda_{PM}^2} \qquad (7.6)$$

is the electromechanical time constant of the motor.

The superaudible frequency chopper may be modeled through its gain K_c as its delay may be neglected. The torque constant $K_T = T_e / I = \lambda_{PM}$ and the current sensor amplification is K_I.

The typical torque controller is PI type with the gain K_{SI} and the time constant τ_{si}. The block diagram is shown in Figure 7.2. The theory of linear systems offers numerous design approaches to PI controllers [1] both for continuous and discrete implementation.

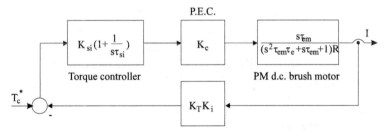

Figure 7.2. PI torque loop for a PM d.c. brush motor

In what follows we are using the critical frequency ω_c and phase margin φ_c constraints for the open-loop transfer function $A(s)$ of the system in Figure 7.2.

$$A(s) = \frac{K_{si}(1+s\tau_{si})}{s\tau_{si}R} \frac{K_C K_T K_i s\tau_{em}}{s^2 \tau_{em}\tau_e + s\tau_{em} + 1} \qquad (7.7)$$

The critical frequency ω_c should be high — up to 1...2 kHz — to provide fast torque (current) control.

Let us take as a numerical example a PM d.c. brush motor with the data: $V_n = 110$ V, $P_n = 2$ kW, $n_n = 1800$ rpm, $R = 1\ \Omega$, $L = 20$ mH, $K_T = 1.1$ Nm/A, $\tau_{em} = 0.1$ sec., $K_C = 25$ V/V, $K_i = 0.5$ V/A, critical frequency $f_c = 500$ Hz, and the phase margin $\varphi_c = 47°$.

The phase margin φ_c of $A(s)$ from (7.7), for the critical frequency $\omega_c = 2\pi f_c$, is

$$\varphi_c = 180° + \text{Arg}[A(j\omega_c)] =$$
$$= 180° + \tan^{-1}(\omega_c \tau_{si}) - \tan^{-1}\left(\frac{\omega_c \tau_{em}}{1 - \omega_c^2 \tau_{em}\tau_e}\right) \qquad (7.8)$$

Consequently

$$\tan^{-1}(\omega_c \tau_{si}) = 180° + 47° + \tan^{-1}\frac{2\pi 500 \cdot 0.1}{1 - (2\pi 500)^2 \cdot 0.1 \cdot 0.02} = 46° \qquad (7.9)$$

And thus,

$$\tau_{si} = \frac{\tan 46°}{2\pi 500} = 0.3075\ \text{ms} \qquad (7.10)$$

The gain of the torque controller K_{si} may be calculated from the known condition:

$$|A(j\omega_c)| = 1 \qquad (7.11)$$

Finally,

$$K_{si} = \frac{0.3075 \cdot 10^{-3} \cdot 1}{25 \cdot 1.1 \cdot 0.5 \cdot 0.1} \cdot \sqrt{\frac{\left(10^3 \pi 0.1\right)^2 + \left(1 - 10^6 \pi^2 0.1 \cdot 0.01\right)^2}{1 + 0.965^2}} = 2.205 \quad (7.12)$$

7.2.2. The speed loop

In numerous applications, speed control is required. The torque loop is still there to limit the current, quicken the response and reduce the gain in the current loop. The block diagram of a system for speed control, with the torque (current) loop included, is shown in Figure 7.3.

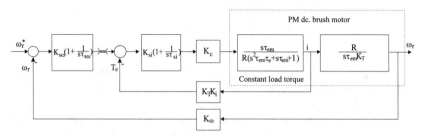

Figure 7.3. Speed control with torque (current) inner loop

Assuming that the torque loop has been already designed as above, only the speed-loop design will be discussed here. The block diagram in Figure 7.3 may be restructured as in Figure 7.4, where K_ω is the gain of the speed sensor.

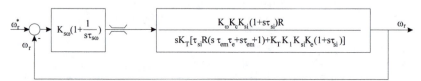

Figure 7.4. Simplified speed-loop block diagram

For the speed-loop design we may proceed as above for the torque loop, but with a critical frequency $f_{c\omega} = 100 Hz$ and a phase margin $\varphi_{c\omega} = 60°$ with $K_\omega = 0.057$ V sec/rad. The final results are $\tau_{s\omega} = 2.87$ ms and $K_{s\omega} = 2.897 \cdot 10^3$. The amplification $K_{s\omega}$ is rather high but the phase margin is ample and produces a reasonably low overshoot and well-damped oscillatory response.

In a similar manner we may proceed to design and add the position loop for completion of the cascaded controller. It is known that such cascaded linear controllers do not excel in terms of robustness to parameter, inertia or load changes. Additional measures are required, and they will be treated later on in this chapter.

For now, however, we are treating the digital position control through a single digital filter rather than a cascaded controller, to broaden the spectrum of motion control practical approaches.

7.2.3. Digital position control

The basic configuration of a digital position control system is shown in Figure 7.5.

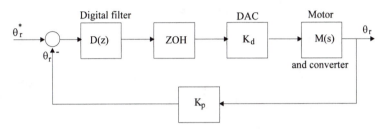

Figure 7.5. Basic digital position control system

Figure 7.5 features the motor-converter transfer function, a digital-analog converter (DAC), a sample-and-hold (ZOH) and digital filter D(z) which represents, in fact, the position controller. Only position feedback is included as obtained from an encoder with N pulses per revolution which, through the division of the two 90° phase-shifted pulse trains, produces 4N pulses per revolution and a bit for motion direction.

Suppose only an 8-bit DAC with $V_d = \pm 10V$ is used. Consequently its amplification K_d is

$$K_d = \frac{2V_d}{2^n} = \frac{20}{2^8} = \frac{20}{256} \, V/pulse \qquad (7.13)$$

The amplification of the encoder K_p is

$$K_p = \frac{4N}{2\pi} \, pulse / rad \qquad (7.14)$$

The position control is performed in a DSP with the discretization period of T. The position error $\varepsilon(KT)$ is

$$\varepsilon(KT) = \theta_r^*(KT) - \theta_r(KT) \qquad (7.15)$$

The position controller filters this error to produce the command Y(KT) which is periodically applied through the DAC

$$V_c = K_d \cdot Y(KT) \qquad (7.16)$$

The command voltage V_c is kept constant during the discretization period T, the effect being known as ZOH. The digital filter may be expressed by finite difference equations of the type [2]

$$Y(KT) = 2\varepsilon(KT) - \varepsilon((K-1)T) \tag{7.17}$$

z-transform may be used to model (7.17). The z-transform shifting property is

$$Y(K-m) \rightarrow z^{-m}f(z) \quad \text{with} \quad f(K) \rightarrow f(z) \tag{7.18}$$

Consequently, (7.17) becomes

$$Y(z) = 2\varepsilon(z) - z^{-1}\varepsilon(z) \tag{7.19}$$

The D(z) is thus

$$D(z) = \frac{Y(z)}{\varepsilon(z)} = \frac{2z-1}{z} \tag{7.20}$$

Let us now prepare for a numerical example by specifying the PM d.c. brush motor-converter transfer function with the electrical time constant τ_e neglected as it is very low for an axial airgap configuration motor, for example,

$$M(s) = \frac{\theta_s(s)}{V_c(s)} = \frac{K_c}{s(1 + s\tau_{em})K_T} \tag{7.21}$$

The data for this new numerical example are $K_T = 0.1$Nm/A, $R = 1\Omega$, $L = 0$H, $J = 10^{-3}$ kgm^2, $K_c = 5$ V/V with the critical frequency $\omega_c = 125$ rad/sec and the phase margin $\varphi_{cp} = 45°$.

The problem may be solved adopting a continual filter (controller) and finally "translating" it into a discrete form.

The motor-converter transfer function M(s) is

$$M(s) = \frac{50}{s(1 + s \cdot 0.1)}; \quad \tau_{em} = \frac{RJ}{K_T^2} = \frac{1 \cdot 10^{-3}}{0.1^2} = 0.1\,\text{s} \tag{7.22}$$

The DAC amplification is $K_d = 10/128$ (8 bit). The encoder produces N = 500 pulses/revolution, which provides in fact 4N = 2000 pulses/revolution, and has the amplification K_p

$$K_p = \frac{4N}{2\pi} = \frac{2000}{2\pi} = 318 \text{ pulses / rad} \tag{7.23}$$

The discretization (sampling) time is $T = 10^{-3}$ sec. Thus the ZOH transfer function is

$$ZOH(s) \approx e^{-\frac{sT}{2}} = e^{-s \cdot 5 \cdot 10^{-4}} \tag{7.24}$$

The various transfer functions in Figure 7.5 are united in a unique transfer function H(s)

$$H(s) = K_d \cdot K_p \cdot ZOH(s) \cdot M(s) \tag{7.25}$$

or

$$H(s) = \frac{1242.8 \cdot e^{-s \cdot 5 \cdot 10^{-4}}}{s(1 + s \cdot 0.1)} \tag{7.26}$$

The phase angle of H(s) for ω_c is

$$\varphi_H = \frac{\omega_c T}{2} \cdot \frac{180°}{\pi} - 90° - \tan^{-1}(0.1 \cdot 125) = -179° \tag{7.27}$$

To produce a phase margin φ_c of 45° the digital filter G(z) has to produce a phase anticipation φ_D of 44°. A tentative G(s) lead-lag filter is

$$G(s) = K\frac{s + \omega_1}{s + \omega_2} \tag{7.28}$$

The maximum anticipation is obtained for $\omega_c = (\omega_1\omega_2)^{0.5}$. The filter phase angle φ_D is

$$\varphi_D = \tan^{-1}\left(\frac{\omega_c}{\omega_1}\right) - \tan^{-1}\left(\frac{\omega_c}{\omega_2}\right) \tag{7.29}$$

with $\omega_c = 125$ rad/sec and $\varphi_D = 44°$, $\omega_1 = 53.3$ rad/sec and $\omega_2 = 293.15$rad/sec.

Finally, the amplification K of G(s) is obtained from the condition that the amplitude of the open-loop transfer function of the system |H(s) G(s)| be equal to unity for critical frequency. With the above data, K = 2.96.

To discretize the G(s) transfer function, the correspondence between z and s is considered

$$s = \frac{2}{T}\frac{z-1}{z+1} = 2000\frac{z-1}{z+1} \tag{7.30}$$

Consequently G(z) is

$$G(z) = 2.65\frac{z - 0.947}{z - 0.744} = \frac{Y(z)}{\varepsilon(z)} \tag{7.31}$$

or

$$Y(z) - 0.744 \cdot z^{-1}Y(z) = 2.65\varepsilon(z) - 2.51 \cdot z^{-1}\varepsilon(z) \tag{7.32}$$

Finally the output of the digital filter Y(k) is

$$Y(k) = 0.744 \cdot Y(k-1) + 2.65\varepsilon(k) - 2.51 \cdot \varepsilon(k-1) \qquad (7.33)$$

7.2.4 Positioning precision

It is known that the positioning precision is influenced by the friction torque. Let us consider a friction torque Tf = 0.05Nm.

The speed is small around the target position and thus the motion induced voltage of the motor may be neglected

$$V = R \cdot I = R \frac{T_f}{K_T} = \frac{1 \cdot 0.05}{0.1} = 0.5 \text{ V} \qquad (7.34)$$

The command voltage of the static power converter V_c is

$$V_c = \frac{V}{K_c} = \frac{0.5}{5} = 0.05 \text{ V} \qquad (7.35)$$

The DAC input N_c is

$$N_c = \frac{V_c}{K_d} = \frac{0.05}{\dfrac{10}{128}} = 0.64 \text{ pulses} \qquad (7.36)$$

The gain of the digital filter K_o is obtained for Z = 1

$$K_0 = 2.65 \left(\frac{1-0.94}{1-0.7441} \right) = 0.548 \qquad (7.37)$$

Consequently, the input of the digital filter N_e is

$$N_e = \frac{N_c}{K_0} = \frac{0.64}{0.548} = 1.33 \text{ pulses} \qquad (7.38)$$

The closest integer is N_{ei} = 2 pulses and thus the position error $\Delta\theta_r$ caused by the above friction torque is

$$\Delta\theta_r = \frac{N_e \cdot 360°}{4N} = \frac{2 \cdot 360}{2000} = 0.36° \qquad (7.39)$$

The positioning precision may be improved through diverse methods such as increasing the encoder number of pulses per revolution N or by a feedforward signal proportional to the torque perturbation. Optical encoders with up to 20,000 pulses/revolution are available today.

Industrial position controllers in digital implementation are, in general, of P type but contain a speed feedforward signal and an inner PI speed-loop besides the torque loop (Figure 7.6) [4].

It should be noticed that both ω_r^* and $\hat{\omega}_r$ — estimated reference speed and actual speed — are calculated through the position time derivative and thus contain noise and errors.

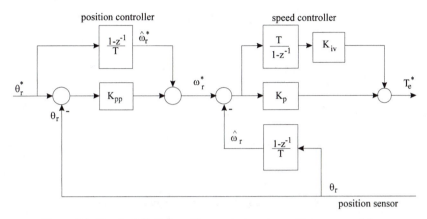

Figure 7.6. Standard digital position control system with inner speed-loop

In general such a controller responds well to slow dynamics commands, well below the frequency of the position controller (0.5 to 10 Hz). For this reason, in order to follow correctly the targeted position during transients, in the presence of torque and inertia changes, state-space control is used where the speed and acceleration are estimated adequately from the measured position.

7.3. STATE-SPACE MOTION CONTROL

State-space motion control which makes use of a few variables has the advantage that the scales are directly related to the capacity to reject torque perturbations (to mechanical rigidity). In such systems there is a separation of position tracking commands from those related to perturbation rejection. To yield zero position tracking errors a positive torque reaction is used. That does not affect the stability to perturbations but poses problems to torque dynamics and estimation precision (Figure 7.7).

It should be noted that the friction torque coefficient b_p (Figure 7.7) as well as the inertia $\hat{J}$ have to be known with good precision. The estimated $(\hat{\omega}_r)$ and reference (ω_r^*) speed and the reference acceleration $(\dot{\omega}_r^*)$ are all given as time functions

$$\hat{\omega}_r = \frac{\theta_r(k) - \theta_r(k-1)}{T} \qquad (7.40)$$

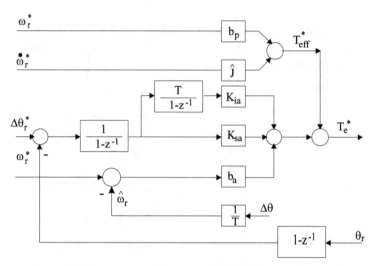

Figure 7.7. State-space position controller with zero tracking error

The speed estimation with zero delay may be done through an observer (Figure 7.8) that more complicated than the simple reference equation in (7.40).

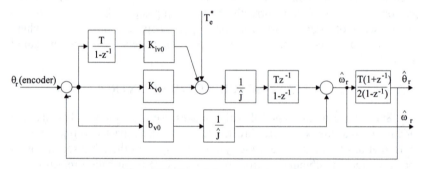

Figure 7.8. Zero delay speed observer

However, even such an observer is sensitive to torque dynamics and to torque estimation precision. The immunity to torque perturbations in position control may be further increased, especially at low speeds, through an acceleration feedback. The required acceleration observer is similar to the speed observer in Figure 7.8 but is also provided with a speed-loop whose reference is the output of the speed observer itself. Consequently, the position control with observers and loops for speed and acceleration thus obtained (Figure 7.9) shows an increased dynamics rigidity. Even in this case problems remain related to the needed knowledge of instantaneous torque and inertia $\hat{J}$ (J_a is only an active inertia). In what follows the estimation of torque perturbations is approached.

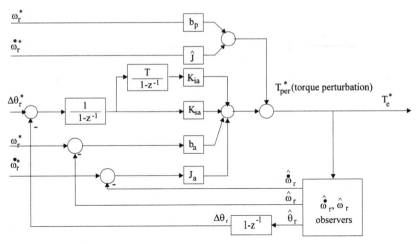

Figure 7.9. Position controller with observers and loops for speed and acceleration
and reference torque feedforward signal

7.4. TORQUE PERTURBATION OBSERVERS

The response of the cascaded motion control systems, designed for given (constant) torque and rated motor parameters, proved to be strongly influenced by torque, inertia and parameter variations. Increasing the gain would make the system more rigid but it could excite low frequency mechanical resonant modes which may lead to instabilities. The reference torque feedforward signal produces similar effects. Motion control (in robotics, etc.) requires, on the other hand, even more increased dexterity and thus a variable dynamic rigidity. A robot that rotates a cam shaft requires a force loop while another that plants electronic components implies refined motion control. In these latter cases there is no urgent need for control rigidity but the robustness of response is also lost if the gains would be reduced in the standard controller. This is how a variable rigidity controller becomes necessary. To reach this goal use is made of total perturbation observers (of torque and parameters) or partial ones (of torque) [5].

To simplify the presentation a single degree of freedom system is considered. Its motion equation is

$$J \frac{d\omega_r}{dt} = T_e - T_L; \quad T_L = T_{in} + T_{L0} + T_{fr} \tag{7.41}$$

where T_{in} is the inertia torque; T_{Lo} is the load torque; and T_{fr} is the friction torque.

The electromagnetic torque of the motor T_e (of the PM d.c. brush or of vector controlled a.c. motors) has the simple expression

$$T_e = K_T \cdot I^* \tag{7.42}$$

where I^* is the reference torque current and, consequently, the torque (current) controller is considered to be hyper-rapid in response. The parameters in (7.41-7.42) are J and K_T, and they deviate from their rated values J_n and K_{Tn}

$$J = J_n + \Delta J \tag{7.43}$$

$$K_T = K_{Tn} + \Delta K_T \tag{7.44}$$

In terms of torque, the variations of ΔJ and ΔK_T are $\Delta Js\omega_r$ and $\Delta K_T I^*$, respectively. Consequently the torque perturbation T_{per} is

$$T_{per} = T_L + \Delta Js\omega_r - \Delta K_T \cdot I^* \tag{7.45}$$

Making use of (7.43)-(7.45) in (7.41) leads to

$$T_{per} = K_{Tn} \cdot I^* - J_n s\omega_r \tag{7.46}$$

Based on (7.46) the torque perturbation may be rather simply calculated. To avoid the time derivative in (7.46), a change is operated and a low pass filter added (Figure 7.10). If the critical frequency, a, of the low pass filter is sufficiently high, the estimated torque perturbation T_{per} is very close to the actual one. The torque perturbation T_{per} may also be used as a positive reaction (Figure 7.11) to replace the same quantity but recalculated with the reference values.

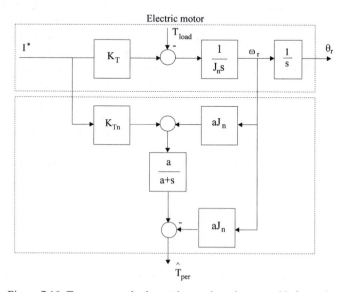

Figure 7.10. Torque perturbation estimator based on speed information

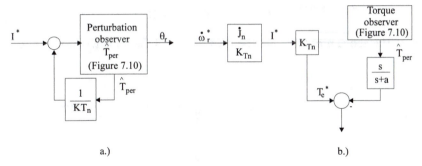

a.)   b.)

Figure 7.11. Torque perturbation reaction

a.) positive reaction; b.) negative reaction at the reference T_e^*.

7.5. PATH TRACKING

Making use of torque perturbation (T_{per}) leads to increased control rigidity and thus to the possibility of close trajectory (path) tracking. For path tracking the reference position in two (three) previous time steps is known and thus reference acceleration and speed signals may be calculated and used in the control system.

Stabilizing the position response is accomplished by at least two poles with negative real parts which could be made dependent on the gains of the position and speed-loops K_1 and K_2. This solution plus the inverse system may lead to a robust path tracking control system (Figure 7.12) — for multivariable systems.

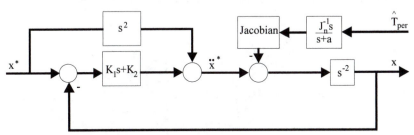

Figure 7.12. Position tracking multivariable system (J - Jacobian matrix)

The relationship between the actual and reference position (Figure 7.12) is

$$x = x^* - \text{Jacobian} \frac{s \hat{T}_{per}}{(s+a)J_n(s^2 + K_1 s + K_2)} \tag{7.47}$$

7.6. FORCE CONTROL

The above procedure may be used to control the force without its direct measurement (Figure 7.13).

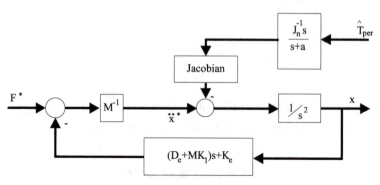

Figure 7.13. Multivariable force control

The rigidity K_e and the viscosity coefficient D_e model the medium where the force acts. The speed feedback amplification K_1 provides for stability. It may well be said that the external loop in Figure 7.13 adjusts the motion rigidity by generating the reference acceleration $\overset{\bullet\bullet}{x}{}^*$ while the interior loop provides for force tracking, yielding a robust response.

Applying such robust motion control systems also implies, as shown above, speed calculation from the measured position [5]. Results obtained at low speed show remarkable performance (Figure 7.14) [5].

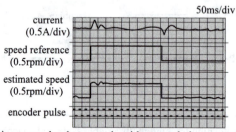

Figure 7.14. Motion control at low speeds with a speed observer with average speed
input (4N = 5000 pulses / rev, with a pole z = 0.3) [5]

We should also mention that so far inertia was considered known. In reality even this parameter has to be estimated and this is why the block diagrams so far contain the estimated inertia $\hat{J}$ [4-6]. To estimate $\hat{J}$ we may use the motion equation in a few moments in time

$$J \overset{\bullet}{\omega}_r(k) = K_{Tn} \cdot I(k) + T_{per}(k) \tag{7.48}$$

$$J\overset{\bullet}{\omega}_r(k-1) = K_{Tn} \cdot I(k-1) + T_{per}(k-1) \tag{7.49}$$

Making use of the acceleration and perturbation torque estimators we may calculate — through averaging methods (such as the least-square recursive method) — in real time, the inertia. Consequently, a fully robust motion controller performance is obtained (Figure 7.15) [5].

Though the series of procedures for robust control presented under the umbrella of state-space control seems to exhaust the subject, other less complex but high performing nonlinear control methods have been proposed. Among them we will treat in some detail the variable structure (sliding-mode) systems, fuzzy control systems and neural-network control systems.

$\hat{J}$:0.00588[kgm^2/div], ω:50.0[rpm/div]
$\dot{T}_{dis}$:2.0[Nm/div], time:250[ms/div]

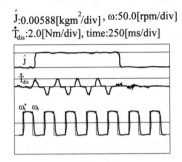

Figure 7.15. Inertia identification

(based on the time interval between encoder pulses T_2 = 100μsec,

with the speed estimation time T_1 = 5ms; observer poles: z = 0.6, M = 50) [5]

7.7. SLIDING-MODE MOTION CONTROL

Sliding-mode motion controllers force the system to move on a given surface in the state-space (plane) [7]. The simplest "surface" in the state plane is a straight line. The equation of the sliding mode functional s_s is

$$s_s = c\varepsilon - \frac{d\varepsilon}{dt}; \quad \varepsilon = x^* - x \tag{7.50}$$

where ε is the error of the controlled variable x. For $s_s = 0$ a straight line is obtained.

According to Figure 7.16, in the initial moment the system starts from point A with an error ε_i = OA and its derivative is zero. At A, $s_s > 0$. By forcing the command variable (voltage or current) in the positive direction the motor has to be capable of jumping up to the straight line $s_s = 0$.

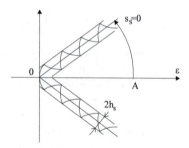

Figure 7.16. Sliding-mode functional

Once this is done, the positive and negative commands ($\pm V$) alternate depending on the sign of s_s (at a constant switching frequency or with a hysteresis band h_s) to run the system to origin where a new steady-state point is reached

$$V_c = +V_0 \quad \text{for} \quad s_s > h_s$$
$$V_c = 0 \quad \text{for} \quad |s_s| < h_s \qquad (7.51)$$
$$V_c = -V_0 \quad \text{for} \quad s_s < -h_s$$

Such a simple forcing process done through the switching of the command variable between three discrete values leads, however, to oscillations around the target position (chattering) unless the switching frequency is high enough. The system's behavior seems to depend only on the constant c and thus only the limits within which the motor parameters and load torque vary are known in order to allow for sliding along the straight line $s_s = 0$. However the sliding-modes existence conditions have to be fulfilled also [7]

$$s_s \cdot \dot{s}_s < 0 \qquad (7.52)$$

The simplifications introduced by the sliding-mode robust control seem notable but operation without chattering requires some additions such as a perturbation observer [8].

The block diagram of a motion (position) control system with sliding-modes for a robot, is shown in Figure 7.17.

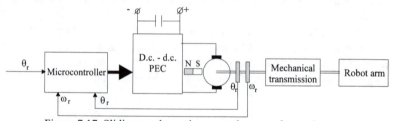

Figure 7.17. Sliding-mode motion control system for a robot arm

The mathematical model of the PM d.c. brush motor for position control, accounting for the mechanical transmission ratio K_{mT}, the friction torque coefficient D and the total load torque T_{per}, has a second order only if the motor inductance is considered zero

$$\ddot{\theta}_r = \frac{-(K_E K_T + DR)\dot{\theta}_r}{RJ} + \frac{K_E K_{mT} V_a}{RJ} - \frac{K_{mT}}{J} T_{per} \qquad (7.53)$$

D - friction coefficient;
J - total inertia referred to the motor shaft;

K_E - PM motion induced voltage coefficient ($E = K_E \dot{\theta}_r$);
K_T - motor torque coefficient ($T_e = K_T I$; $K_T = \lambda_{PM}$);
V_a - motor armature voltage.
Two new variables x_1 and x_2 are introduced

$$x_1 = \overset{*}{\theta}_r - \theta_r \text{ and } x_2 = \frac{dx_1}{dt} \qquad (7.54)$$

with

$$V = aV_a; \quad a = \frac{K_E K_{mT}}{JR}$$
$$b = \frac{DR + K_E K_T}{JR}; \quad f = \frac{K_{mT} T_{per}}{J} \qquad (7.55)$$

Equation (7.53) becomes

$$\dot{x}_1 = x_2$$
$$\dot{x}_2 = bx_2 - V + f \qquad (7.56)$$

The sliding-mode command voltage is chosen in the form [9]

$$V_c = \gamma x_1 + K_f \, \text{sgn}(s_s) \qquad (7.59)$$

with

$$\gamma = \begin{cases} \alpha \text{ if } s_s x_1 > 0 \\ \beta \text{ if } s_s x_1 < 0 \end{cases} \qquad (7.60)$$

The state functional is given by (7.50) while the sliding-mode existence condition is given by (7.52). The existence condition (7.52) leads to the inequalities

$$\alpha > c(b - c) \qquad (7.61)$$

$$\beta < c(b - c) \qquad (7.62)$$

$$K_f > (f)_{max} \qquad (7.63)$$

These conditions secure safe jumping of the system to the straight line of the sliding-mode functional. We call K_f a dither signal, and it should be greater than a quantity proportional to the largest torque perturbation. Unfortunately once the straight line is reached, the term in K_f in the command voltage (7.59) causes undesirable vibrations.

Figures (7.18-7.19) show results obtained with such a control system implemented on a DSP (NEC: μPD77230, 32 bit) with a sampling time of 200μs, 20kHz switching frequency, for a SCARA robot with two axes [8].

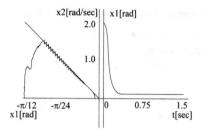

Figure 7.18. Sliding-mode control (7.59) with low K_f

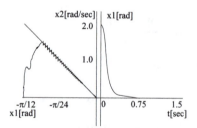

Figure 7.19. Sliding-mode control with large K_f

The perturbation, which in the first case (K_f small, Figure 7.18) produces a nonzero steady-state position error, is caused by pulling down the robot arm with 300g through a spring (about 20% loading). The undesirable mechanical vibrations are notable. Elimination of such vibrations is accomplished through the contribution of a perturbation observer with a reduction of K_f. If stepwise perturbations are assumed, the equations are:

$$\dot{T}_{per} = 0$$
$$\dot{x}_2 = \frac{K_{mT} T_{per}}{J} - b\hat{x}_2 - V \qquad (7.64)$$

It may be proved that the system of (7.64) is observable and, to calculate (observe) T_{per} and x_2, a complete observer may be used

$$\dot{T}_{per} = K_1\left(x_2 - \hat{x}_2\right) \tag{7.65}$$

$$\dot{x}_2 = \frac{K_{mT}\hat{T}_{per}}{J} - b\hat{x}_2 - V + K_2\left(x_2 - \hat{x}_2\right) \tag{7.66}$$

The two amplifications K_1 and K_2 may be calculated by the method of pole allocation. Complex poles with negative real part are assumed. The positive perturbation reaction V_0 becomes

$$V_0 = \frac{K_{mT}T_{per}}{J} \tag{7.67}$$

Finally, the complete command voltage Vcc is

$$V_{cc} = V_c + V_0 = \gamma x_1 + K_f \, sgn(s_s) + V_0 \tag{7.68}$$

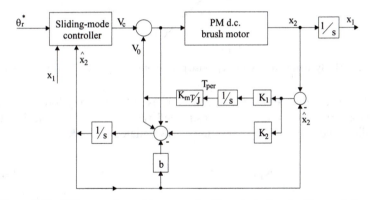

Figure 7.20. Sliding-mode position control with perturbation feedforward signal

The stability of such a nonlinear system may be studied through Lyapunov's theorem. A Lyapunov function, $V(s_s, \Delta x_2, \Delta T_{per})$, is first defined as

$$V\left(s_s, \Delta x_2, \Delta T_{per}\right) = \frac{1}{2}s_s^2 + \frac{1}{2}\left[\frac{1}{K_1}\left(\Delta T_{per}\right)^2 + \frac{J}{K_{mT}}\left(\Delta x_2\right)^2\right] \tag{7.69}$$

The sufficient condition for stability is satisfied if $\dot{V} < 0$

$$\dot{V} = (c-b)\left|\hat{s}_s\right|^2 - \left[c(c-b)+\gamma\right]\hat{s}_s \, \Delta \hat{x}_1 -$$
$$- \left[K_f - K_2\Delta\hat{x}_2 \, sgn\left(\hat{s}_s\right)\right]\left|\hat{s}_s\right| - \left(b+K_2\right)\frac{J}{K_{mT}}\left(\Delta \hat{x}_2\right)^2 < 0 \tag{7.70}$$

If each term in (7.70) is negative, then $\dot{V} < 0$

$$c \leq b; \qquad \alpha > c(b-c)$$

$$\beta < c(b-c); \qquad K_f > K_2 \left| \Delta \hat{x}_2 \right|_{max} \qquad (7.71)$$

$$K_1 > 0; \qquad K_2 > -b$$

In case of a high step-torque perturbation ($K_{mT} T_{per} / J = 0.15$, $K_f = 0.01$), K_f becomes too small to fulfill (7.71) and thus the controller regains stability only after Δx_2 increases notably (Figure 7.21) [8].

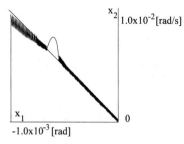

Figure 7.21. System response to step-wise perturbation ($K_{mT} T_{per} / J = 0.15$, $K_f = 0.01$)

Because the perturbation observer is of the step-form a ramp perturbation causes errors in the estimated speed x_2 and thus the last two stability conditions in (7.71) have to be reformulated as [8]

$$K_f > \frac{K_2}{K_1} T_{per}^{ramp}; \qquad T_{per}^{ramp}(s) = \frac{T_{per}^{ramp}}{s} \qquad (7.72)$$

The stability analysis for this case leads to new conditions related to K_f

$$cT < 1 \qquad (7.73)$$

$$TK_f < \left| c\hat{x}_1(k+1) + \hat{x}_2(k+1) \right| \qquad (7.74)$$

The same control system may be extended to position tracking when the reference position, speed and acceleration, θ_r^*, $\dot{\theta}_r^*$, $\ddot{\theta}_r^*$ are known time functions. In this case we invert the system (7.54) with (7.55)-(7.56)

$$V = \frac{1}{a}\left(b\dot{\theta}_r + \ddot{\theta}_r^* + hx_2 + V_{cc} \right) \qquad (7.75)$$

Comparing (7.53) with (7.75) yields:

$$\dot{x}_2 = -hx_2 - V_{cc} + \frac{K_{mT} T_{per}}{J} \tag{7.76}$$

Eqn. (7.75) represents the new control law which embeds (7.68) and adds three new terms for good path tracking. Figure 7.22 presents a triangle drawn by the robot SCARA (as a plotter) in the presence of a 20% torque perturbation (obtained through 200g hanged on the robot arm through a mechanical spring).

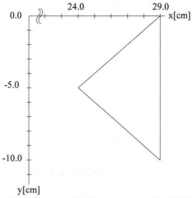

Figure 7.22. Triangular trajectory in the presence of a 20% torque perturbation

The maximum speed is 0.5m/sec [8], a = 2550, b = –2550, c = 40, h = 100, the observer poles are $p_{1,2} = Lb(-\cos\gamma + j\sin\gamma)$ with L = 30, $\gamma = 30°$.

Practical experience has shown that if in the trajectory planning high accelerations ($\ddot{\theta}_r$) are allowed, vibrations occur in the system's response. A quadratic function E, to minimize along the trajectory, may be chosen [9]

$$E = \int_0^t \left\{ \left[f(x,y) \right]^2 + A_1 \left(\ddot{\theta}_{r1}^* \right)^2 + A_2 \left(\ddot{\theta}_{r2}^* \right)^2 \right\} J \cdot dt \tag{7.77}$$

$$A_1 > 0; \quad A_2 > 0$$

with $\qquad\qquad x = L_1 \cos\theta_{r1}^* + L_2 \cos\left(\theta_{r1}^* + \theta_{r2}^*\right) \tag{7.78}$

$$y = L_1 \sin\theta_{r1}^* + L_2 \sin\left(\theta_{r1}^* + \theta_{r2}^*\right) \tag{7.79}$$

where f(x,y) is the required trajectory in plane.

The scope of the optimization process is to determine θ_{r1}^*, $\dot{\theta}_{r1}^*$, $\ddot{\theta}_{r1}^*$ and θ_{r2}^*, $\dot{\theta}_{r2}^*$, $\ddot{\theta}_{r2}^*$ along the two axes as time functions such that the penalty function is minimum. Through variational methods combined with spline trajectory approximations, the above problem may be solved iteratively. Figure 7.23 shows the results obtained with SCARA robot in drawing a circle without and with acceleration reduction [8].

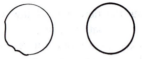

Figure 7.23. The circle drawn by SCARA robot without and with the reduction of

reference acceleration (total time 1.6 sec.)

Far from exhausting the subject, the sliding-mode motion control with torque perturbation compensation, as applied to robotics, merely suggests the complexity of motion robust control even with the motor electrical time constant (constants) considered zero.

For systems with vague mathematical models, new motion control systems such as fuzzy systems and neural network systems have been recently proposed. Their first implementation has been done on PM d.c. brush motors [10] and on a.c. motors [11, 12].

7.8. MOTION CONTROL BY FUZZY SYSTEMS

Fuzzy systems stem from a logic which treats vaguely known plants by membership functions (MF) with values between 0 and 1.

In fuzzy sets, based on fuzzy logic, an object (variable) has a membership degree to a given set with values between 0 and 1.

A fuzzy variable has linguistic values, for example, LOW, MEDIUM, HIGH which may be defined through bell shape (Gauss) membership functions with gradual variation (Figure 7.24).

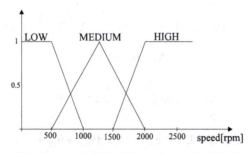

Figure 7.24. Fuzzy sets defined through membership functions

The membership functions are, in general, either symmetric or nonsymmetric, triangular or trapezoidal.

At 800rpm (Figure 7.24), for example, the variable belongs with 50% (MF = 0.5) to LOW and with 50% (MF = 0.5) to MEDIUM. All possible values of a variable constitute the universe of discourse.

The properties of Boolean theory remain valid for fuzzy sets. Reunion (corresponding to OR) is

$$\mu_{AUB}(x) = \max\left[\mu_A(x), \mu_B(x)\right] \tag{7.80}$$

Intersection (corresponding to AND) writes

$$\mu_{A \cap B}(x) = \min\left[\mu_A(x), \mu_B(x)\right] \tag{7.81}$$

The negation (corresponding to NOT) is

$$\mu_A(x) + \mu_{\Gamma A}(x) = 1 \tag{7.82}$$

Triangular membership functions OR, AND and NOT are illustrated in Figure 7.25.

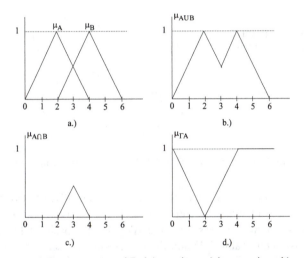

Figure 7.25. a.) Fuzzy sets A and B; b.) reunion; c.) intersection; d.) negation

While conventional control systems are based on the mathematical model of the plant, fuzzy control is based on the intuition and experience of the human operator. And thus, besides self-tuning adaptive systems, reference model adaptive systems, variable structure systems, for plants with vaguely known model, fuzzy control is clearly opportune and adequate.

In essence, implicitly, fuzzy motion control is self-adaptive and thus its robustness becomes apparent.

The command law in fuzzy systems is of the form

$$\text{IF } x = A \text{ AND } y = B \text{ THEN } z = C \tag{7.83}$$

where x, y, z are fuzzy variables where the universe of discourse is A, B, C. For example, for the speed control of a PM d.c. brush motor, the fuzzy variables are the speed error $E = \omega_r^* - \omega_r$ and its derivative $CE = \dot{\omega}_r^* - \dot{\omega}_r$. Let us remember that the same variables have been used to express the sliding-mode functional. The rule number one may be of the form: if E is zero (ZE) and the error derivative CE is negative small (NS), then the increment dV_1 of the command voltage is negative small (NS).

The linguistic variables ZE, NS and ΔV are defined through membership symmetric functions (Figure 7.26).

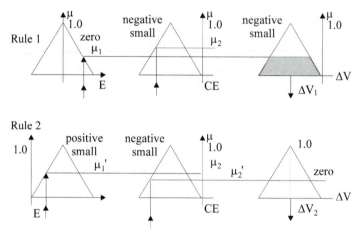

Figure 7.26. The MAX-MIN (SUP-MIN) Fuzzy rule composition

The command law ΔV_1 may be obtained graphically, but two rules may emerge, the second with the output ΔV_2. The effective control will be an average between two outputs ΔV_1 and ΔV_2. First a basis of fuzzy rules is usually constituted (Table 7.1) and the corresponding membership functions are described (Figure 7.27).

Table 7.1.

CE \ E	NB	NM	NS	Z	PS	PM	PB
NB	NVB	NVB	NVB	NB	NM	NS	Z
NM	NVB	NVB	NB	NM	NS	Z	PS
NS	NVB	NB	NM	NS	Z	PS	PM
Z	NB	NM	NS	Z	PS	PM	PB
PS	NM	NS	Z	PS	PM	PB	PVB
PM	NS	Z	PS	PM	PB	PVB	PVB
PB	Z	PS	PM	PB	PVB	PVB	PVB

Z - zero, PS - positive small, PM - positive medium, PB - positive big, PVB - positive very big
NS - negative small, NM - negative medium, NB - negative big, NVB - negative very big

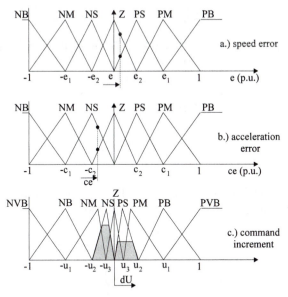

Figure 7.27. The fuzzy membership functions for speed control

The composition operation refers to the calculus of the command law. Among the composition methods we mention the MAX-MIN (SUP MIN) and MAX-DOT. The MAX-MIN method already illustrated in Figure 7.26 is written in the form

$$\mu_u(u) = SUP_x \left[\min \mu_x(x) \cdot \mu_R(x,u) \right] \qquad (7.84)$$

Consequently, the membership function for each rule is given by the minimum value (MIN) and the combined output is given by the supreme maximum of all rules. The general structure of a single input fuzzy motion controller is given in Figure 7.28.

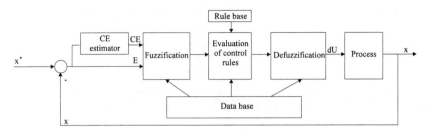

Figure 7.28. Typical single input fuzzy control system

The command variable (voltage or current in our example) is determined based on the two variables e and ce (in relative values) calculated from the absolute values E and CE by division through the corresponding amplifications.

Fuzzification may be done by taking definite values of the variables to calculate the membership function in these cases (Figures 7.26-7.27). Defuzzification may be done in quite a few ways one of which is the centroid method by which the command variable U_0 is obtained through the gravity center of the fuzzy function U (Figure 7.26).

$$U_0 = \frac{\int U \mu_u(u) \cdot du}{\int \mu_u(u) \, du} \qquad (7.85)$$

and the height method, where the centroid of each function first and, then, a weighted medium value of the heights is calculated

$$U_0 = \frac{\sum_{i=1}^{n} U_i \mu_u(u_i)}{\sum_{i=1}^{n} \mu_u(u_i)} \qquad (7.86)$$

Though the fuzzy control looks simple, mathematical implication increases the implementation time. Recently neural-fuzzy sets that allow the rapid selection of the membership functions, rules and command signals have been introduced.

Besides the rule-based method (for fuzzy systems) a relational method has been proposed in [12]. According to the relational method, regions look for linear combinations of input functions with G and H values.

The rule number one may be defined: IF W is MEDIUM and H is MEDIUM, then

$$I_s = A_{01} + A_{11}W + A_{21}H \qquad (7.87)$$

The coefficients A_{ij} may be determined through linear regression and then adjusted through observation and simulation. Finally the linear equation of the outputs (7.87) is defuzzified, that is a weighted average of the components I_{s1} and I_{s2} by the membership functions μ_1 and μ_2 is calculated

$$I_s = \frac{I_{s1}\mu_1 + I_{s2}\mu_2}{\mu_1 + \mu_2} \qquad (7.88)$$

To control the speed of a PM d.c. brush motor (with the rule based method, Table 7.1) the command signal is

$$U(k) = U(k-1) + GU \cdot du \qquad (7.89)$$

where GU is the voltage amplification and du the command voltage increment expressed in relative values and calculated during defuzzification through the height method. For a PM d.c. brush motor of 1.84kW, 1800rpm, $J = 0.0465$ kgm^2, R = 0.6 Ω, L = 8 mH, $K_E = 0.55$ V sec/rad, $T_s = k_L \omega_r^2$, $K_T = 2.78 \times 10^{-4}$ Nm sec^2/rad, the system's response to a step perturbation (Figure

7.29) and to a step-torque increase in inertia (Figure 7.30) demonstrates a remarkable robustness.

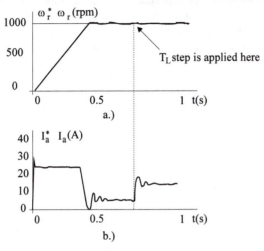

Figure 7.29. A fuzzy controller; the response to torque perturbation

after step speed response: a.) speed, b.) current [10].

If for high speeds the performance is good, at low speed problems could occur and may be solved through a new set of fuzzy rules and commands.

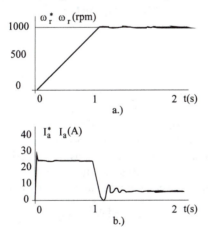

Figure 7.30. Fuzzy speed controller; the response to inertia variation:

a.) speed, b.) current [10]

7.9. MOTION CONTROL THROUGH NEURAL NETWORKS

Neural networks (NN) represent systems of interconnections between artificial neurons which emulate the neuron system of the human brain. NN constitutes a more general form of artificial intelligence than the expert systems or fuzzy sets [11]. The model of an artificial neuron looks as in Figure 7.31.

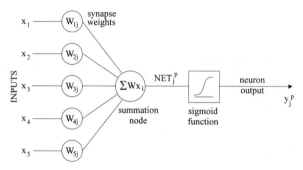

Figure 7.31. The structure of an artificial neuron

Each input continuous signal is passed through weighting synapses — positive (exciters) or negative (inhibiters) — towards the summation mode, then, through a nonlinear transfer function, to output. The transfer function may be step-wise, limit form, sign form (sgn), linear with a limit or nonlinear, of the sigmoid form, for example

$$y = \frac{1}{1+e^{-\alpha_s x}} \qquad (7.90)$$

For large α_s y gets close to a step function and varies between zero (for x = $-\infty$) to 1 (for x = $+\infty$). Though it is still a mystery how the neurons are interconnected in the human brain, more than sixty models of artificial neural networks have been proposed to be applied in science and technology. They may be classified in neural networks with positive and negative reaction. The first ones are dominant so far. Such a positive reaction model with three layers — input, hidden (screened) and output — is shown on Figure 7.32.

The relationships between the input and output variables are of the form

$$[V]_b = [W]_{ba}[X]_a \qquad (7.91)$$

$$[Y]_c = [W]_{cb}[V]_b \qquad (7.92)$$

The twenty-five weights $W_{i,j}$ (on Figure7.32) have to be calculated through an iterative educational process by comparing a large number of desired types of output signals with the actual ones.

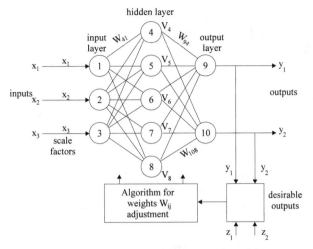

Figure 7.32. Positive reaction neural model

The back propagation educational algorithm, which has received rather wide acceptance lately, starts with a few initial values and, based on a penalty function (such as the minimum of the sum of squared errors), continues through the descent gradient method and changes the weights $W_{i,j}$.

For the case of a single layer (Figure7.31) the network equations, based on (7.91) - (7.92), become

$$NET_j^p = \sum_1^N W_{ij}X_i \qquad (7.93)$$

$$Y_i^p = f_i\left(NET_j^p\right) \qquad (7.94)$$

For the pth input signal, the squared error E_p for all the neurons of input layer is

$$E_p = \frac{1}{2}\left(d^p - y^p\right)^2 = \frac{1}{2}\sum_{j=1}^s\left(d_j^p - y_j^p\right)^2 \qquad (7.95)$$

For all the input signals the resultant error E is

$$E = \sum_{p=1}^P E_p = \frac{1}{2}\sum_{p=1}^P\sum_{j=1}^s\left(d_j^p - y_j^p\right)^2 \qquad (7.96)$$

The weights $W_{i,j}$ will be reduced until E reaches a minimum

$$W_{ij}(t+1) = W_{ij}(t) + \eta\left(\frac{\partial E_p}{\partial W_{ij}(t)}\right) + \alpha\left[W_{ij}(t) - W_{ij}(t-1)\right] \qquad (7.97)$$

The iterative procedure propagates the error backwards and provides for a global minimum by adequately choosing the educational rate η and the coefficient α

$$\eta(t+1) = U \cdot \eta(t); \quad U > 0 \tag{7.98}$$

Special education algorithms with back propagation are now available.

7.10. NEURO-FUZZY NETWORKS

The neuro-fuzzy networks do the calculations, required in the fuzzy systems, by the rule-based or relational methods through which the rules and membership functions are identified.

For the fuzzy speed control (Figure7.27) a neuro-fuzzy network is given in Figure 7.33.

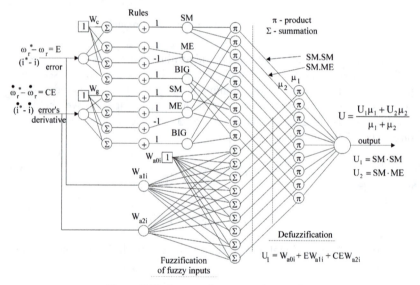

Figure 7.33. A neuro-fuzzy speed controller

In essence, the fuzzy control, with the relational method used for command signal calculation, is done through the neuro-fuzzy controller. The quantities W_c and W_g create the space between the membership functions and, respectively, their ramp. Nine rules occur. The linear functions (for the relational method) are given on the lower part of Figure 7.33 while defuzzification is shown on the right side of the same figure.

Besides positive reaction, negative reaction neural networks [14] have been proposed.

7.10.1. Applications of neural networks

It is also possible to use NN (neural networks) to selectively eliminate harmonics for pulse width modulation in static power converters in the sense that the NN may be educated through the voltage chopping angles. Education of NN may be done "on line" or "off line" with applications in diagnosis and monitoring and in intelligent motion control. Also flux linkage, torque, or active power may be estimated through NN, based on measured voltages and currents [15]. Other applications include a.c. current controllers for a.c. motors [16] or d.c. brush motor control [17].

The NN may be implemented through software in microcontrollers or on special hardware. A few dedicated IC chips for the scope are already on the market such as INTEL 8017 NxETANN, Microdevice MD 1220NBS and Neural Semiconductor NUSU32.

7.11. SUMMARY

- This chapter presented a synthesis of classical — cascaded — and of nonlinear motion closed-loop control systems for electrical drives.
- The cascaded-torque, speed controllers are presented in detail with the PECs simulated by simple gains.
- The digital position controller performed through a single digital filter is illustrated by an example where the PEC is simulated by a gain and a zero order hold.
- The nonlinear motion control systems — a science in itself — is introduced here only through a few methods such as: state-space control with torque and speed observers, sliding mode control, fuzzy control, and neural networks control with recent results from literature, all related to the PM d.c. brush motor drives. Our introduction to this subject is only a starting point for the interested reader who should pursue the selected literature at the end of this chapter.
- Though applied only to d.c. brush motors the issues in this chapter are far more general and may be applied to vector-controlled a.c. motors.

7.12. PROBLEMS

7.1. Calculate the response of the current controller in Section 7.2.1 for a step increase of reference current of 10A.

7.2. A d.c. brush motor with the data $R_a = 0.25 \ \Omega$, rated current 20 A, rated (source) voltage $e_0 = 110$ V, rated speed $n_n = 1800$ rpm is fed through a power transistor chopper in the continuous (constant) current mode. The inertia J = 0.1 kgm2. The mechanical time constant $\tau_m = 2\pi J/B = 10$ sec. The electrical time constant $\tau_e = 40$ ms and thus the electrical phenomena are much faster than the mechanical ones. The drive has a proportional (P) current controller with a current sensor having $K_I = 0.5$ V/A ratio and a PI speed controller. Design the speed controller — find

K_{sn} and τ_{sn} — for a damping ratio of $1 / \sqrt{2}$ and a natural frequency ω_n = 10rad/s. The speed sensor gain is $K_n = 0.1$ Vs/rot.

7.3. For the speed controller d.c. brush motor system in problem 7.2 find the speed-load torque function and show that the steady-state error in speed to a torque perturbation is zero.

7.4. A d.c. brush series motor has the data: total (armature + field) resistance $R_{at} = 1\ \Omega$, total inductance $L_{at} = 40$ mH, rated voltage $V_n = 120$ V, rated speed $n_n = 1800$ rpm, rated current 10 A. The remnant flux induced voltage at rated speed is $e_{rem} = 5$ V and the magnetizing curve is otherwise linear and $J = 10^{-2}$ kgm^2.

    a.   Determine the open-loop transfer functions $i_a(p) / e_a(p)$, $n(p) / i_a(p)$ for zero load torque variation, after linearization around the rated point (e_a - terminal voltage).

    b.   Determine the open-loop transfer functions $i_a(p) / T_L(p)$, $n(p) / i_a(p)$ for zero input voltage variation, after linearization around the rated point.

7.13. SELECTED REFERENCES

1.   **B. Kisacanin, G.C. Agarwal,** Linear Control Systems, Kluwer Academic Press, 2001

2.   **P. Katz,** Digital Control Systems Using Microprocessors, Prentice Hall, 1982

3.   **H.F. Vanlandingham,** Introduction to Digital Control Systems, Macmillan, 1985

4.   **R. Lorenz, T.A. Lipo, D.W. Novotny,** Motion Control with Induction Motors, IEEE Proc., vol.82, no.8, 1994, pp.1215-1240

5.   **K. Ohnishi, N. Matsui, Y. Hori,** Estimation, Identification and Sensorless Control in Motion Control Systems, IBID., pp.1253-1265

6.   **I. Awaya** et al, New Motion Control With Inertia Identification Using Disturbance Observer Record of IEEE - IECON 1992, vol.1, pp.77-80

7.   **V.I. Utkin,** Variable Structure System With Sliding Mode, IEEE Trans., vol.AC - 27, no.2, 1977, pp.212-222

8.   **A Kawamura, H. Itoh, K. Sakamoto,** Chattering Reduction of Disturbance Observer Based Sliding Mode Control, IEEE Trans., vol.IA - 30, no.2, 1994, pp.456-461

9.   **K. Sakamoto, A. Kawamura,** Trajectory Planning Using Optimum Solution of Variational Problem, Power Conversion Conference, Yokohama 1993, pp.666-671

10.   **G.D. Sousa, B.K. Bose,** A Fuzzy Set Theory Based Control of a Phase - Controlled Converter D.C. Machine Drive, IEEE Trans., vol.IA - 30, no.1, 1994, pp.34-44

11.   **B.K. Bose,** Expert System, Fuzzy Logic And Neural Network Applications in Power Electronics and Motion Control, IEEE Proc., vol.82, no.8, 1994, pp.1303-1323

12.   **T. Takagi, M. Sugero,** Fuzzy Identification Systems and Its Applications to Modeling and Control, IEEE Trans., vol.SMC - 15, no1, 1985, pp.116-132

13.   **L.A. Zadeh,** Fuzzy Sets, Informat. Contr., vol.8, 1965, pp.338-343

14.   **California Scientific Software,** Introduction to Neural Networks, Grass Valley CA, 1991

15.   **M.G. Simoes, B.K. Bose,** Feedback Signal Estimation by Neural Network, Record of IEEE - IAS - 1994 Annual Meeting;

16.   **M.R. Buhland, R.D. Lorenz,** Design and Implementation of Neural Networks for Digital Current Regulation of Inverter Drive, Record of IEEE-IAS - 1991 Annual Meeting, Part 1, pp.415-423

17.   **S. Weerasooriya, M.A. El-Sharkawi,** Identification and Control of a D.C. Motor Using Back - Propagation Neural Networks, IEEE Trans., vol.EC - 6, no.6, 1991, pp.663-669.

Chapter 8

INDUCTION MOTORS FOR DRIVES

The induction motor is considered to be the workhorse of industry. It is an a.c. motor, either three-phase or (for low powers) single phase. Industrial (conventional) induction motors are supplied from constant voltage and frequency industrial power grids for rather constant speed operation. For variable speed drives induction motors are fed from PECs at variable voltage amplitude and frequency.

Like the d.c. motor, an induction motor consists of a stator (the fixed part) and a rotor (the moving part) mounted on mechanical bearings and separated from the stator by an airgap.

8.1. THE STATOR AND ITS TRAVELING FIELD

The stator consists essentially of a magnetic core made up of punchings (laminations) — 0.1 mm to 0.5mm thick — carrying slot-embedded coils. These coils are interconnected in a certain fashion to constitute the so-called a.c. armature (primary) winding (Figure 8.1).

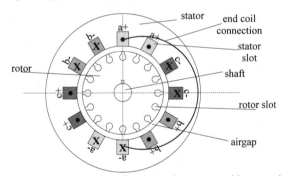

Figure 8.1. Cross section of an induction motor with two poles

The three-phase windings may be placed in slots in a single layer (Figure 8.1) or in two layers. All the coils are, in general, identical and the span is close or equal to what is called the pole pitch τ or the half-period of m.m.f. of that coil or phase. The number of poles per periphery is denoted as 2p.

Under each pole there are three zones, one for each phase. Each phase zone per pole contains q slots (q = 2–8, integer in general). The consecutive phase zones of the same polarity, a_+a_+, b_+b_+ in Figure 8.1. are spatially shifted by a geometrical angle of $2\pi/3p$ which corresponds to 2/3 of pole pitch of the winding.

As one pole pitch corresponds to a semiperiod (180° electrical degrees), the electrical angle α_e is related to the mechanical angle α_g by

$$\alpha_e = p \cdot \alpha_g \tag{8.1}$$

The m.m.f. of each phase coil has a stepwise waveform (Figure 8.2) which is assimilated by a sinusoidal distribution with the semiperiod τ (pole pitch). There are also harmonics which, in general, produce parasitic torques augmented by the rotor and stator slot openings.

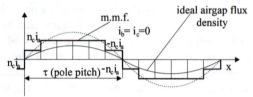

Figure 8.2. The m.m.f. and airgap flux density of phase a.

The phase current is sinusoidal and thus the phase m.m.f. fundamental $F_{a1}(x, t)$ may be written as:

$$F_{a1}(x,t) = F_{a1m} \cdot \sin\frac{\pi}{\tau}x \cdot \sin\omega_1 t \tag{8.2}$$

$$i_a(t) = I\sqrt{2} \sin\omega_1 t \tag{8.3}$$

As the other two phases b, c are space shifted by an electrical angle of $2\pi/3$ and their currents i_b, i_c (in time) by $2\pi/3$, the m.m.f. of phase b and c are

$$F_{b1}(x,t) = F_{a1m} \cdot \sin\left(\frac{\pi}{\tau}x - \frac{2\pi}{3}\right) \cdot \sin\left(\omega_1 t - \frac{2\pi}{3}\right) \tag{8.4}$$

$$F_{c1}(x,t) = F_{a1m} \cdot \sin\left(\frac{\pi}{\tau}x + \frac{2\pi}{3}\right) \cdot \sin\left(\omega_1 t + \frac{2\pi}{3}\right) \tag{8.5}$$

The resultant stator m.m.f. fundamental $F_{s1}(x,t)$ is the sum of F_{a1}, F_{b1} and F_{c1}

$$F_{s1}(x,t) = \frac{3}{2}F_{a1m} \cdot \cos\left(\frac{\pi}{\tau}x - \omega_1 t\right) \tag{8.6}$$

This is evidently a wave traveling along the rotor periphery with the linear speed U_s obtained from

$$\frac{\pi}{\tau}x - \omega_1 t = \text{const.} \tag{8.7}$$

Its increment should be zero

$$\frac{\pi}{\tau}dx - \omega_1 dt = 0$$

Finally

$$U_s = \frac{dx}{dt} = \tau \cdot \frac{\omega_1}{\pi} = 2\tau f_1 \qquad (8.8)$$

where f_1 - the primary frequency.

As the airgap is uniform, neglecting the slot openings, the airgap flux density (for zero rotor currents) $B_{g1}(x,t)$ is

$$B_{g1}(x,t) \approx \mu_0 \cdot \frac{F_{s1}(x,t)}{g_e} \qquad (8.9)$$

where g_e is an equivalent airgap accounting globally for the slot openings and stator and rotor core magnetic saturation.

Consequently, the three-phase stator winding produces — for zero rotor current — a traveling field in the airgap, with the linear speed $U_s = 2\tau f_1$. As the peripheral speed is related to the angular speed n_1 and the stator bore D_i by

$$U_s = \pi \cdot D_i \cdot n_1 = 2p \cdot \tau \cdot n_1 \qquad (8.10)$$

the angular speed of the traveling field n_1 is

$$n_1 = f_1 / p \qquad (8.11)$$

n_1 is also called the synchronous speed since for this speed of the rotor no voltages are induced in the rotor windings.

8.2. THE CAGE AND WOUND ROTORS ARE EQUIVALENT

The rotor consists of a laminated core with uniform slotting accommodating either aluminum (copper) bars short-circuited by end-rings (the squirrel cage), Figure 8.3.a., or a three-phase winding (as in the stator) connected to some copper rings and fixed brushes, the wound rotor (Figure 8.3.b).

It has been demonstrated that a symmetric cage (with round bars) may be modeled by an equivalent three-phase winding, that is, a wound rotor.

A PEC or variable resistor may be connected to the wound rotor brushes.

8.3. SLOT SHAPING DEPENDS ON APPLICATION AND POWER LEVEL

Stator slots are either semiclosed (Figure 8.4a) for low and medium power (hundreds of kW) or open (Figure 8.4b) above hundreds of kW when preformed coils are introduced in slots. Rotor slots for wound rotors are in general semiclosed if the stator ones are open to allow for a rather small airgap (below 2mm) even for high powers (MW and above). Cage rotor slot shapes depend on the power speed level and starting torque requirements in constant frequency fed (industrial) applications.

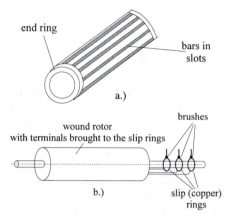

Figure 8.3. Induction motor rotors: a.) cage-type rotor; b.) wound rotor

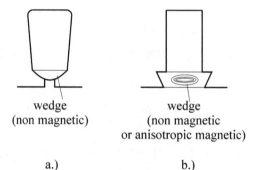

Figure 8.4. Stator or wound rotor slots
a.) for low power (semiclosed slots), b.) for high power (open slots)

Round semiclosed slots (Figure 8.5a.) do not exhibit notable skin effect at start and may be used for constant frequency fed low power low starting torque motors or for variable speed when skin effect is to be avoided.

Skin effect is the concentration of current in the rotor bar towards the upper part of the rotor bar at high rotor frequency (beginning with constant frequency f_1, at standstill when the rotor frequency $f_2 = f_1$). The consequence is an apparent increase in rotor resistance and a less important slot leakage inductance reduction.

Double cages are used in the medium power range to reduce the starting current and increase the starting torque (Figure 8.5e, f). Skin effect (deep bar) or double cages imply higher rotor resistance and losses at rated speed and thus are to be avoided in variable speed drives. In variable speed (frequency) drives, slots as in Figure 8.5g are proposed to reduce the rotor surface losses.

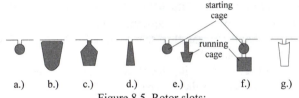

Figure 8.5. Rotor slots:
a.) semiclosed - round for low power and variable frequency motors,
b.) closed - for low noise or high speed motors,
c.) semiclosed with moderate skin effect and starting torque - constant frequency,
d.) semiclosed with high skin effect for high starting torque - constant frequency,
e,f.) double cage - for superhigh starting torque - constant frequency,
g.) for high speed inverter-fed motors.

We entered this discussion as many existing motors are now provided with PECs and thus care must be exercised about performance. Moreover, for constant frequency (speed) operation, the efficient motor category has been introduced — though at lower starting torque and higher starting current. These efficient motors have a short pay-back period due to power loss reduction and are also more adequate in PEC fed variable speed drives.

8.4. THE INDUCTANCE MATRIX

An electric machine is a system of electric and magnetic circuits that are coupled magnetically and electrically. It may be viewed as an assembly of resistances, self-inductances and mutual inductances. We now briefly discuss these inductances.

As a symmetrical rotor cage is equivalent to a three-phase winding we will consider the wound rotor induction motor case (Figure 8.6).

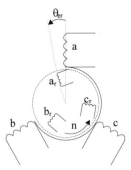

Figure 8.6. Three-phase induction motor with equivalent wound rotor

There are six circuits (phases) and each has a self-inductance and mutual inductances. Let us notice that the self-inductance of phases a, b, c, a_r, b_r, c_r do not depend on rotor position (slot openings are neglected). They have a main part L_{ms}, L_{mr} which corresponds to the flux paths that cross the airgap and embrace all windings on the stator and rotor, and leakage parts L_{ls} and L_{lr}

which correspond to the flux paths in the slots, coil end connections, mostly in air, which do not embrace rotor and stator windings.

Let us notice that the airgap flux density of each phase m.m.f is basically sinusoidal along the rotor periphery, and any coupling inductance between phases varies cosinusoidally with the electrical angle of the two windings. Also, mutual inductances on the stator L_{ab}, L_{bc}, L_{ac} and, respectively, on the rotor $L_{a_r b_r}$, $L_{a_r c_r}$, $L_{b_r c_r}$, though referring mainly to flux paths through the airgap, do not depend on rotor position

$$L_{ab} = L_{bc} = L_{ca} = L_{ms} \cos \frac{2\pi}{3} = -\frac{L_{ms}}{2} \qquad (8.12)$$

$$L_{a_r b_r} = L_{b_r c_r} = L_{c_r a_r} = L_{mr} \cos \frac{2\pi}{3} = -\frac{L_{mr}}{2} \qquad (8.13)$$

Finally, the stator-rotor coupling inductance L_{sr} depends on rotor position through $\cos\theta_{er}$ as the airgap flux density in the airgap produced by any stator phase is sinusoidal

$$L_{sr}(\theta_{er}) = L_{ms} \cdot \frac{W_{re}}{W_{se}} \cos \theta_{er} \qquad (8.14)$$

W_{re}, W_{se} are the equivalent turns ratio in the rotor and stator phases

$$\frac{W_{re}}{W_{se}} = \sqrt{\frac{L'_{mr}}{L_{ms}}} = K_{rs} \qquad (8.15)$$

as self-inductances depend on the number of turns squared.

Consequently, $L_{sr}(\theta_{er})$ is

$$L_{sr}(\theta_{er}) = L_{srm} \cos \theta_{er} \qquad (8.16)$$

$$L_{srm} = \sqrt{L_{ms} \cdot L'_{mr}} \qquad (8.17)$$

The inductances may be now assembled into the so-called matrix inductance $\left[L_{a,b,c,a_r^r,b_r^r,c_r^r}(\theta_{er}) \right]$

$$
\left[L_{a,b,c,a_r^r,b_r^r,c_r^r} \left(\theta_{er} \right) \right] =
\begin{array}{c c}
 & \begin{array}{cccccc} a & b & c & a_r & b_r & c_r \end{array} \\
\begin{array}{c} a \\ b \\ c \\ a_r \\ b_r \\ c_r \end{array} &
\left[
\begin{array}{cccccc}
L_{aa} & L_{ab} & L_{ac} & L_{a,a} & L_{b,a} & L_{c,a} \\
L_{ab} & L_{bb} & L_{bc} & L_{a,b} & L_{b,b} & L_{c,b} \\
L_{ac} & L_{bc} & L_{cc} & L_{a,c} & L_{b,c} & L_{c,c} \\
L_{a,a} & L_{a,b} & L_{a,c} & L_{a,a_r} & L_{a,b_r} & L_{a,c_r} \\
L_{b,a} & L_{b,b} & L_{b,c} & L_{a,b_r} & L_{b,b_r} & L_{b,c_r} \\
L_{c,a} & L_{c,b} & L_{c,c} & L_{a,c_r} & L_{b,c_r} & L_{c,c_r}
\end{array}
\right]
\end{array}
\tag{8.18}
$$

where

$$
L_{aa} = L_{bb} = L_{cc} = L_{ls} + L_{ms}; \qquad L_{ab} = -L_{ms}/2 ; \quad L_{ac} = -L_{ms}/2 ;
$$

$$
L_{a,a} = L_{b,b} = L_{c,c} = L_{srm} \cos\theta_{er} ; \qquad L_{a,a_r} = L_{b,b_r} = L_{c,c_r} = L^r_{lr} + L^r_{mr} ;
$$

$$
L_{c,a} = L_{a,b} = L_{b,c} = L_{srm} \cos(\theta_{er} - 2\pi/3);
$$

$$
L_{c,b} = L_{a,c} = L_{b,a} = L_{srm} \cos(\theta_{er} + 2\pi/3);
$$

$$
L_{a,b_r} = -L^r_{mr}/2 ; \qquad L_{b,c_r} = -L^r_{mr}/2 .
$$

8.5. REDUCING THE ROTOR TO STATOR

It is useful to replace the actual rotor winding by an equivalent one with the same performance (and losses) but having the same number of turns per phase as in the stator.

Evidently, for this case, the maximum mutual inductance becomes equal to $L_{ms} = L_{mr}$

$$
L_{srm} = L_{ms} \tag{8.19}
$$

Denoting the currents and voltages in the rotor with i^r_{ar}, i^r_{br}, i^r_{cr}, V^r_{ar}, V^r_{br}, V^r_{cr} and the ones reduced to the stator as i_{ar}, i_{br}, i_{cr}, V_{ar}, V_{br}, V_{cr}

$$
\frac{i_{ar}}{i^r_{ar}} = \frac{i_{br}}{i^r_{br}} = \frac{i_{cr}}{i^r_{cr}} = K_{rs} \tag{8.20}
$$

To conserve the rotor input power

$$
\frac{V_{ar}}{V^r_{ar}} = \frac{V_{br}}{V^r_{br}} = \frac{V_{cr}}{V^r_{cr}} = \frac{1}{K_{rs}} \tag{8.21}
$$

For equal winding losses and leakage magnetic energy

$$
\frac{r_r}{r^r_r} = \frac{L_{lr}}{L^r_{lr}} = \frac{1}{K^2_{rs}} \tag{8.22}
$$

So the new inductance matrix $\left[L_{a,b,c,a_r,b_r,c_r} \left(\theta_{er} \right) \right]$ is similar to that in (8.18), but with $L^r_{mr} \to L_{ms}, L_{srm} \to L_{ms}$.

Note: For a cage-rotor the expression of the rotor to stator reduction factor K_{rs} of (8.15) is more complicated but, as we do not have access to the cage current, dealing directly with the reduced parameters is common practice. For cage motors, K_{rs} is thus required only for motor design purposes.

We may now pursue the mathematical model in phase coordinates (variables).

8.6. THE PHASE COORDINATE MODEL GOES TO 8th ORDER

In matrix form, with rotor reduced to the stator, the voltage current equations — in stator coordinates for the stator and in rotor coordinates for the rotor — are

$$[V] = [r] \cdot [i] + \frac{d}{dt}[\lambda] \tag{8.23}$$

$$[\lambda] = \left[L_{a,b,c,a_r,b_r,c_r}(\theta_{er}) \right] \cdot [i] \tag{8.24}$$

$$[r] = \text{Diag}[r_s, r_s, r_s, r_r, r_r, r_r] \tag{8.25}$$

$$[V] = [V_a, V_b, V_c, V_{ar}, V_{br}, V_{cr}]^T \tag{8.26}$$

$$[i] = [i_a, i_b, i_c, i_{ar}, i_{br}, i_{cr}]^T \tag{8.27}$$

Using (8.24) in (8.23) — with θ_{er} variable in time in any case — we obtain

$$[V] = [r] \cdot [i] + [L] \frac{d[i]}{dt} + \frac{d[L]}{d\theta_{er}} \cdot [i] \cdot \frac{d\theta_{er}}{dt} \tag{8.28}$$

where
$$\frac{d\theta_{er}}{dt} = \omega_r = p\Omega_r \tag{8.29}$$

Ω_r is the mechanical angular speed ($\Omega_r = 2\pi n$).

In the absence of magnetic saturation multiplying (8.28) by $[i]^T$, yields

$$[i]^T \cdot [V] = [i]^T \cdot [r] \cdot [i] + \frac{d}{dt} \frac{1}{2} \left[[L] \cdot [i] \cdot [i]^T \right] + \frac{1}{2} [i]^T \cdot \frac{d[L]}{d\theta_{er}} \cdot [i] \cdot \omega_r \quad (8.30)$$

The first right term represents the winding losses, the second, the stored energy variation, and the third, the electromagnetic power P_e

$$P_e = T_e \cdot \Omega_r = \frac{1}{2}[i]^T \cdot \frac{d[L]}{d\theta_{er}}[i] \cdot \omega_r \tag{8.31}$$

Finally, the electromagnetic torque T_e is

$$T_e = \frac{1}{2} p \cdot [i]^T \cdot \frac{d[L]}{d\theta_{er}} [i] \qquad (8.32)$$

The motion equations are

$$\frac{J}{p} \frac{d\omega_r}{dt} = T_e - T_{load}; \quad \frac{d\theta_{er}}{dt} = \omega_r \qquad (8.33)$$

An 8th order nonlinear model with time variable coefficients (inductances) has been obtained, while still neglecting the core losses. Only numerical methods provide a solution to it. This complex model is to be used directly only in special cases, when the computation effort is justified. Not so for electric drives, in general.

Complex (space-phasor) variables are introduced to obtain a model with position (time) independent coefficients.

8.7. THE SPACE-PHASOR MODEL

Let us introduce first the following notations [1-4]

$$a = e^{j\frac{2\pi}{3}}; \quad \cos\frac{2\pi}{3} = Re[a]; \quad \cos\frac{4\pi}{3} = Re[a^2];$$

$$\cos\left(\theta_{er} + \frac{2\pi}{3}\right) = Re[a \cdot e^{j\theta_{er}}]; \quad \cos\left(\theta_{er} + \frac{4\pi}{3}\right) = Re[a^2 \cdot e^{j\theta_{er}}]; \quad (8.34)$$

Using (8.34) in the flux expression (8.24), the phase a and a_r flux linkages λ_a and λ_{ar} write

$$\lambda_a = L_{ls} \cdot i_a + L_{ms} \cdot Re[i_a + a \cdot i_b + a^2 \cdot i_c] + L_{ms} \cdot Re[(i_{ar} + a \cdot i_{br} + a^2 \cdot i_{cr}) e^{j\theta_{er}}]$$

$$(8.35)$$

$$\lambda_{ar} = L_{lr} \cdot i_{ar} + L_{ms} \cdot Re[i_{ar} + a \cdot i_{br} + a^2 \cdot i_{cr}] + L_{ms} \cdot Re[(i_a + a \cdot i_b + a^2 \cdot i_c) e^{-j\theta_{er}}]$$

$$(8.36)$$

We now introduce the following complex variables as space-phasors

$$\overline{i_s}^s = \frac{2}{3} \cdot (i_a + a \cdot i_b + a^2 \cdot i_c) \qquad (8.37)$$

$$\overline{i_r}^r = \frac{2}{3} \cdot (i_{ar} + a \cdot i_{br} + a^2 \cdot i_{cr}) \qquad (8.38)$$

where

$$\text{Re}\left[\bar{i}_s^s\right] = i_a - \frac{1}{3} \cdot \left(i_a + i_b + i_c\right) = i_a - i_0 \tag{8.39}$$

$$\text{Re}\left[\bar{i}_r^r\right] = i_{ar} - \frac{1}{3} \cdot \left(i_{ar} + i_{br} + i_{cr}\right) = i_{ar} - i_0 \tag{8.40}$$

In symmetric transient and steady-state regimes and symmetric windings

$$i_a + i_b + i_c = 0; \quad i_{ar} + i_{br} + i_{cr} = 0 \tag{8.41}$$

With definitions (8.37)-(8.38), Equations (8.35)-(8.36) become

$$\lambda_a = L_{ls} \cdot \text{Re}\left(\bar{i}_s^s\right) + L_m \cdot \text{Re}\left(\bar{i}_s^s + \bar{i}_r^s \cdot e^{j\theta_{er}}\right); \quad L_m = \frac{3}{2}L_{ms} \tag{8.42}$$

$$\lambda_{ar}^r = L_{lr} \cdot \text{Re}\left(\bar{i}_r^r\right) + L_m \cdot \text{Re}\left(\bar{i}_r^r + \bar{i}_s^s \cdot e^{-j\theta_{er}}\right) \tag{8.43}$$

If to (8.42)-(8.43) we add similar equations for phases b, b$_r$ and c, c$_r$, (8.23) becomes

$$\overline{V}_s^s = r_s \cdot \bar{i}_s^s + \frac{d\overline{\lambda}_s^s}{dt} = r_s \cdot \bar{i}_s^s + L_s \cdot \frac{d\bar{i}_s^s}{dt} + L_m \frac{d\left(\bar{i}_r^r e^{j\theta_{er}}\right)}{dt} \tag{8.44}$$

$$\overline{V}_r^r = r_r \cdot \bar{i}_r^r + \frac{d\overline{\lambda}_r^r}{dt} = r_r \cdot \bar{i}_r^r + L_r \cdot \frac{d\bar{i}_r^r}{dt} + L_m \frac{d\left(\bar{i}_s^s e^{-j\theta_{er}}\right)}{dt} \tag{8.45}$$

with $$L_s = L_{ls} + L_m; \quad L_r = L_{lr} + L_m \tag{8.46}$$

and $$\overline{V}_s^s = \frac{2}{3} \cdot \left(V_a + a \cdot V_b + a^2 \cdot V_c\right); \quad \overline{V}_r^r = \frac{2}{3} \cdot \left(V_{ar} + a \cdot V_{br} + a^2 \cdot V_{cr}\right) \tag{8.47}$$

The complex variables $\overline{V}_s^s, \overline{V}_r^r, \bar{i}_s^s, \bar{i}_r^r, \overline{\lambda}_s^s, \overline{\lambda}_r^r$, are still represented in their respective coordinates (stator for stator, rotor for rotor). We may now use a rotation of the complex variables by the angle θ_b in the stator and by $\theta_b - \theta_{er}$ in the rotor to obtain a unique coordinate system at some speed ω_b

$$\omega_b = \frac{d\theta_b}{dt} \tag{8.48}$$

$$\overline{\lambda}_s^s = \overline{\lambda}_s^b \cdot e^{j\theta_b}; \quad \bar{i}_s^s = \bar{i}_s^b \cdot e^{j\theta_b}; \quad \overline{V}_s^s = \overline{V}_s^b \cdot e^{j\theta_b} \tag{8.49}$$

$$\overline{\lambda}_r^r = \overline{\lambda}_r^b \cdot e^{j(\theta_b - \theta_{er})}; \quad \bar{i}_r^r = \bar{i}_r^b \cdot e^{j(\theta_b - \theta_{er})}; \quad \overline{V}_r^r = \overline{V}_r^b \cdot e^{j(\theta_b - \theta_{er})} \tag{8.50}$$

With the new variables, (8.44)-(8.45) become

$$\overline{V}_s = r_s \cdot \overline{i}_s + \frac{d\overline{\lambda}_s}{dt} + j \cdot \omega_b \cdot \overline{\lambda}_s$$

$$\overline{V}_r = r_r \cdot \overline{i}_r + \frac{d\overline{\lambda}_r}{dt} + j \cdot (\omega_b - \omega_r) \cdot \overline{\lambda}_r \qquad (8.51)$$

with

$$\overline{\lambda}_s = L_s \cdot \overline{i}_s + L_m \cdot \overline{i}_r \qquad (8.52)$$

$$\overline{\lambda}_r = L_r \cdot \overline{i}_r + L_m \cdot \overline{i}_s \qquad (8.53)$$

Notice that for clarity we dropped the superscript b.
The torque should be calculated from (8.42) with the above notations

$$T_e = \frac{3}{2} \cdot p \cdot \mathrm{Re}\!\left(j \cdot \overline{\psi}_s \cdot \overline{i}_s^*\right) = -\frac{3}{2} \cdot p \cdot \mathrm{Re}\!\left(j \cdot \overline{\psi}_r \cdot \overline{i}_r^*\right) \qquad (8.54)$$

Equations (8.51)-(8.54) together with the equations of motion (8.33) constitute the complex variable or space-phasor model of the induction machine with single rotor cage and with core loss neglected.

We may now decompose in plane the space-phasors along two orthogonal axes d and q moving at speed ω_b [5]

$$\overline{V}_s = V_d + j \cdot V_q; \quad \overline{i}_s = i_d + j \cdot i_q; \quad \overline{\lambda}_s = \lambda_d + j \cdot \lambda_q$$

$$\overline{V}_r = V_{dr} + j \cdot V_{qr}; \quad \overline{i}_r = i_{dr} + j \cdot i_{qr}; \quad \overline{\lambda}_r = \lambda_{dr} + j \cdot \lambda_{qr} \qquad (8.55)$$

With (8.55) the two voltage equations (8.51) become

$$V_d = r_s \cdot i_d + \frac{d\lambda_d}{dt} - \omega_b \cdot \lambda_q$$

$$V_q = r_s \cdot i_q + \frac{d\lambda_q}{dt} + \omega_b \cdot \lambda_d$$

$$V_{dr} = r_r \cdot i_{dr} + \frac{d\lambda_{dr}}{dt} - (\omega_b - \omega_r) \cdot \lambda_{qr} \qquad (8.56)$$

$$V_{qr} = r_r \cdot i_{qr} + \frac{d\lambda_{qr}}{dt} + (\omega_b - \omega_r) \cdot \lambda_{dr}$$

$$T_e = \frac{3}{2} p \left(\lambda_d i_q - \lambda_q i_d\right) = \frac{3}{2} p L_m \left(i_q i_{dr} - i_d i_{qr}\right)$$

Also from (8.49)-(8.50) and (8.47)

$$\begin{bmatrix} V_d \\ V_q \\ V_0 \end{bmatrix} = [P(\theta_b)] \cdot \begin{bmatrix} V_a \\ V_b \\ V_c \end{bmatrix} \qquad (8.57)$$

$[P(\theta_b)]$ is the Park transformation

$$[P(\theta_b)] = \frac{2}{3} \cdot \begin{bmatrix} \cos(-\theta_b) & \cos\left(-\theta_b + \frac{2\pi}{3}\right) & \cos\left(-\theta_b - \frac{2\pi}{3}\right) \\ \sin(-\theta_b) & \sin\left(-\theta_b + \frac{2\pi}{3}\right) & \sin\left(-\theta_b - \frac{2\pi}{3}\right) \\ \frac{1}{2} & \frac{1}{2} & \frac{1}{2} \end{bmatrix} \quad (8.58)$$

The inverse of Park transformation is

$$[P(\theta_b)]^{-1} = [P(\theta_b)]^{T} \quad (8.59)$$

A similar transformation is valid for rotor quantities with θ_b–θ_{er} replacing θ_b in (8.58).

It may be proved that the homopolar components V_0, i_0, V_{0r}, i_{0r} have separate equations and do not interfere in the energy conversion process in the motor.

$$\overline{V}_0 = r_s \cdot \overline{i}_0 + \frac{d\overline{\lambda}_{0s}}{dt}; \quad \overline{\lambda}_0 \approx L_{ls} \cdot \overline{i}_0$$

$$\overline{V}_{0r} = r_r \cdot \overline{i}_{0r} + \frac{d\overline{\lambda}_{0r}}{dt}; \quad \overline{\lambda}_{0r} \approx L_{lr} \cdot \overline{i}_{0r} \quad (8.60)$$

Equations (8.56)-(8.60) represent the dq0 model of the induction machine that operates with real (not complex) variables. The complex variable (space-phasor) and d-q model are equivalent as they are based on identical assumptions (symmetric sinusoidally distributed windings and constant airgap). The speed of the reference system ω_b is arbitrary as the airgap is uniform.

Up to now we rushed through equations to quickly obtain the complex variable (d-q) model of the induction machine; we now insist on some graphical representations to facilitate the assimilation of this new knowledge.

Example 8.1. The space-phasor of sinusoidal symmetric currents
Consider three symmetrical sinusoidal currents and show how their complex space-phasor $\overline{i}_s^{\,s}$ varies in time through 6 instants. Give a graphical description of this process in time.
Solution:
The three-phase currents may be written as

$$i_{a,b,c} = I\sqrt{2} \cdot \cos\left(\omega_1 t - (i-1) \cdot \frac{2\pi}{3}\right); i = 1,2,3 \quad (8.61)$$

The space-phasor in stator coordinates $\overline{i}_s^{\,s}$ is (8.37)

$$\bar{i}_s^s = \frac{2}{3}I\sqrt{2}\left[\cos\omega_1 t + e^{j\frac{2\pi}{3}}\cos\left(\omega_1 t - \frac{2\pi}{3}\right) + e^{j\frac{4\pi}{3}}\cos\left(\omega_1 t - \frac{4\pi}{3}\right)\right] \quad (8.62)$$

with $\qquad e^{j\frac{2\pi}{3}} = \cos\frac{2\pi}{3} + j\cdot\sin\frac{2\pi}{3}; \; e^{j\frac{4\pi}{3}} = \cos\frac{4\pi}{3} + j\cdot\sin\frac{4\pi}{3} \qquad (8.63)$

(8.62) becomes $\qquad \bar{i}_s^s = I\sqrt{2}[\cos\omega_1 t + j\cdot\sin\omega_1 t] = i_d + j\cdot i_q \qquad (8.64)$

The position of the space-phasor for $\omega_1 t = 0, \dfrac{\pi}{3}, \dfrac{2\pi}{3}, \pi, \dfrac{4\pi}{3}, \dfrac{5\pi}{3}$ is shown in Figure 8.7.

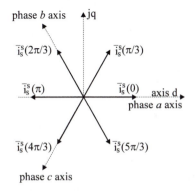

Figure 8.7. The space-phasor of sinusoidal three-phase currents

It should be noticed that the time "produces" the instantaneous values of currents while the definition of the space-phasor shows that each phase current instantaneous value is placed along the axis of the corresponding phase. So, in fact, the space-phasor travels in the d-q stator plane with the electrical angular speed equal to ω_1.

The same concept is applied to flux linkages and thus their space-phasor is related to the traveling field in a.c. machines.

Also note that

$$\text{Re}\left[\bar{i}_s^s\right] = i_a; \; \text{if } i_a + i_b + i_c = 0 \qquad (8.65)$$

8.8. THE SPACE-PHASOR DIAGRAM FOR ELECTRICAL TRANSIENTS

The space-phasor model equations (8.51)-(8.53) may be represented on a space-phasor diagram in the d-q plane, with axes d and q rotating at speed $\omega_b = \omega_1$ (Figure 8.8). Let us consider $\overline{V}_r^b = 0$.

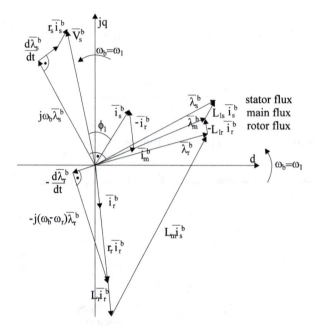

Figure 8.8. The space-phasor diagram of induction motor valid for transients (for steady d/dt = 0) in synchronous coordinates (cage rotor: $\overline{V}_r^b = 0$).

As ω_1 is the frequency of the actual stator voltages at steady-state, for this latter case d/dt = 0. In general, during steady-state

$$d/dt = j(\omega_1 - \omega_b). \tag{8.66}$$

Also for steady-state the rotor flux equation (8.52) becomes

$$\overline{V}_r^b = r_r \cdot \overline{i}_r^b + j(\omega_1 - \omega_b) \cdot \overline{\lambda}_r^b + j(\omega_b - \omega_r) \cdot \overline{\lambda}_r^b =$$
$$= r_r \cdot \overline{i}_r^b + j(\omega_1 - \omega_r) \cdot \overline{\lambda}_r^b \tag{8.67}$$

Only for $\overline{V}_r^b = 0$ (short-circuited, cage, rotor) the rotor current and flux space-phasors are *orthogonal to each other*.

For this case the torque expression (8.54) becomes

$$T_e = \frac{3}{2} p \cdot \lambda_r^b \cdot i_r^b; \quad \left|\overline{\lambda}_r^b\right| = ct. \tag{8.68}$$

8.9. ELECTRICAL TRANSIENTS WITH FLUX LINKAGES AS VARIABLES

Equations (8.51)-(8.53) at constant speed allow for the elimination of rotor and stator currents $\overline{i}_s^b$ and $\overline{i}_r^b$.

With

$$\bar{i}_s{}^b = \sigma^{-1}\left(\frac{\bar{\lambda}_s{}^b}{L_s} - \frac{\bar{\lambda}_r{}^b \cdot L_m}{L_s L_r}\right) \qquad (8.69)$$

$$\bar{i}_r{}^b = \sigma^{-1}\left(\frac{\bar{\lambda}_r{}^b}{L_r} - \frac{\bar{\lambda}_s{}^b \cdot L_m}{L_s L_r}\right) \qquad (8.70)$$

$$\sigma = 1 - \frac{L_m{}^2}{L_s L_r} \qquad (8.71)$$

(8.51)-(8.53) become

$$\tau_s{}'\frac{d\bar{\lambda}_s{}^b}{dt} + \left(1 + j \cdot \omega_b \cdot \tau_s{}'\right) \cdot \bar{\lambda}_s{}^b = \tau_s{}'\overline{V}_s{}^b + K_r \bar{\lambda}_r{}^b \qquad (8.72)$$

$$\tau_r{}'\frac{d\bar{\lambda}_r{}^b}{dt} + \left(1 + j \cdot \left(\omega_b - \omega_r\right) \cdot \tau_r{}'\right) \cdot \bar{\lambda}_r{}^b = \tau_r{}'\overline{V}_r{}^b + K_s \bar{\lambda}_s{}^b \qquad (8.73)$$

with

$$K_s = \frac{L_m}{L_s}; \qquad K_r = \frac{L_m}{L_r}$$
$$\tau_s{}' = \tau_s \cdot \sigma; \qquad \tau_r{}' = \tau_r \cdot \sigma \qquad (8.74)$$
$$\tau_s = \frac{L_s}{r_s}; \qquad \tau_r = \frac{L_r}{r_r}$$

While τ_s, τ_r are the stator and rotor time constants, $\tau_s{}'$ and $\tau_r{}'$ are transient time constants of the rotor and stator.

The structural diagram corresponding to (8.72)-(8.73) is shown on Figure 8.9.

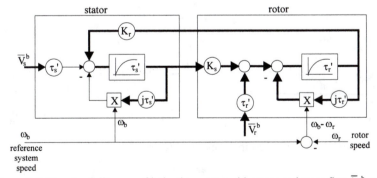

Figure 8.9. Structural diagram of induction motor with stator and rotor flux $\bar{\lambda}_s{}^b$ and $\bar{\lambda}_r{}^b$ as variables; random speed (ω_b) coordinates.

8.10. COMPLEX EIGENVALUES FOR ELECTRICAL TRANSIENTS

Equations (8.72)-(8.73) imply a second-order complex variable system with only two complex eigenvalues corresponding to the determinant

$$\begin{vmatrix} \tau_s's+1+j\omega_b\tau_s' & -K_r \\ -K_s & \tau_r's+1+j(\omega_b-\omega_r)\tau_r' \end{vmatrix} = 0 \qquad (8.75)$$

The complex eigenvalues $\underline{s}_{1,2}$ from (8.75) depend on the reference system speed ω_b and on rotor speed ω_r, but their real part is, in general, negative suggesting attenuated periodic response [4,6].

Example 8.2. For an induction motor with the data $r_s = 0.5\Omega$, $r_r = 0.60\Omega$, $L_s = L_r = 0.08H$, $L_m = 0.075H$, $p = 2$ pole pairs, calculate the complex eigenvalues for constant speed, in synchronous coordinates ($\omega_1 = 2\pi60$) for n = 0 and n = 1800 rpm.

Solution:

The value of $\omega_b = \omega_1 = 2\pi60$ rad/s. σ, τ_s', τ_r', K_s, K_r are now calculated from (8.71)-(8.74)

$$\sigma = 1-\frac{L_m^2}{L_sL_r} = 1-\frac{0.075^2}{0.08^2} = 0.1211 \qquad (8.76)$$

$$\tau_s' = \sigma\tau_s = \sigma\frac{L_s}{r_s} = 0.1211\cdot\frac{0.08}{0.5} = 0.01937\,s \qquad (8.77)$$

$$\tau_r' = \sigma\tau_r = \sigma\frac{L_r}{r_r} = 0.1211\cdot\frac{0.08}{0.6} = 0.01614\,s \qquad (8.78)$$

$$K_s = \frac{L_m}{L_s} = \frac{0.075}{0.08} = 0.9375 \qquad (8.79)$$

$$K_r = \frac{L_m}{L_r} = \frac{0.075}{0.08} = 0.9375 \qquad (8.80)$$

Also
$$\omega_r = 2\pi pn = 4\pi n \qquad (8.81)$$

Now we rewrite (8.75) in a canonical form

$$\underline{s}^2 \tau_s' \tau_r' + \underline{s}[(\tau_s' + \tau_r') + j\tau_s' \tau_r'(2\omega_b - \omega_r)]$$
$$+ (1 + j \cdot \omega_b \cdot \tau_s')(1 + j \cdot (\omega_b - \omega_r) \cdot \tau_r') - K_s K_r =$$
$$\underline{s}^2 \cdot 0.01937 \cdot 0.01614 +$$
$$\underline{s}[(0.01937 + 0.01614) + j \cdot 0.01937 \cdot 0.01614 \cdot (2 \cdot 376.8 - 4\pi n)]$$
$$+ (1 + j \cdot 376.8 \cdot 0.01937) \cdot (1 + j \cdot (376.8 - 4\pi n) \cdot 0.01614)$$
$$- 0.9375^2 = 0$$

(8.82)

It is evident that the roots s depend on speed n. The solving of the second order equation is straightforward.

8.11 ELECTRICAL TRANSIENTS FOR CONSTANT ROTOR FLUX

By constant rotor flux we mean, in general, constant amplitude at $(\omega_1 - \omega_b)$ frequency. For synchronous coordinates it would mean zero frequency. So constant rotor flux means that in (8.72)-(8.73)

$$\frac{d\overline{\lambda}_r^b}{dt} = j \cdot (\omega_1 - \omega_b) \cdot \overline{\lambda}_r^b$$

(8.83)

Using (8.83) in (8.73)

$$\overline{\lambda}_r^b = \frac{\tau_r' \overline{V}_r^b + K_s \overline{\lambda}_s^b}{1 + jS\omega_1 \tau_r'}$$

(8.84)

It is more convenient to use synchronous coordinates for equations (8.72):

$$\tau_s' \frac{d\overline{\lambda}_s^b}{dt} + (1 + j \cdot \omega_1 \cdot \tau_s') \cdot \overline{\lambda}_s^b = \tau_s' \overline{V}_s^b + K_r \overline{\lambda}_r^b$$

(8.85)

Equations (8.84)-(8.85) lead to a simplification in the structural diagram of Figure 8.9 as the derivative term in the rotor disappeares (Figure 8.10).

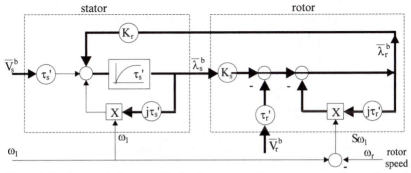

Figure 8.10. Structural diagram of induction motor with constant rotor flux and speed in synchronous coordinates $(\omega_b = \omega_1)$

The system's order is reduced and thus the stator flux presents only one complex eigenvalue as may be inferred from (8.84)-(8.85).

8.12. STEADY-STATE: IT IS D.C. IN SYNCHRONOUS COORDINATES

Steady-state means, in general, that the three-phase voltages are symmetric and sinusoidal:

$$V_{a,b,c} = V\sqrt{2} \cdot \cos\left(\omega_1 t - (i-1) \cdot \frac{2\pi}{3} \right) ; i = 1, 2, 3 \qquad (8.86)$$

The voltage space-phasor in random coordinates is

$$\overline{V}_s^{\ b} = \frac{2}{3}\left[V_a(t) + V_b(t) \cdot e^{j\frac{2\pi}{3}} + V_c(t) \cdot e^{-j\frac{2\pi}{3}} \right] \cdot e^{-j\theta_b} \qquad (8.87)$$

with (8.86)

$$\overline{V}_s^{\ b} = V\sqrt{2}\left[\cos(\omega_1 t - \theta_b) + j \cdot \sin(\omega_1 t - \theta_b) \right] \qquad (8.88)$$

For steady-state

$$\theta_b = \omega_b t + \theta_0 \qquad (8.89)$$

Consequently

$$\begin{aligned} \overline{V}_s^{\ b} &= V\sqrt{2}\left[\cos[(\omega_1 - \omega_b)t - \theta_0] + j \cdot \sin[(\omega_1 - \omega_b)t - \theta_0] \right] = \\ &= V\sqrt{2} \cdot e^{j[(\omega_1 - \omega_b)t - \theta_0]} \end{aligned} \qquad (8.90)$$

It is obvious that, for steady-state, the currents in the model must have the voltage frequency, which is: $(\omega_1 - \omega_b)$. Once again for steady-state we use (8.51), with d/dt = $j \cdot (\omega_1 - \omega_b)$ as inferred in (8.66).

Consequently from (8.51)

$$\begin{aligned} \overline{V}_{so}^{\ b} &= r_s \cdot \overline{i}_{so}^{\ b} + j \cdot \omega_1 \cdot \overline{\lambda}_{so}^{\ b} \\ \overline{V}_{ro}^{\ b} &= r_r \cdot \overline{i}_{ro}^{\ b} + j \cdot S\omega_1 \cdot \overline{\lambda}_{ro}^{\ b}; \quad S = 1 - \omega_r / \omega_1 \end{aligned} \qquad (8.91)$$

S is known as slip.

So the form of steady-state equations is independent of the reference speed ω_b. What counts is the primary (stator) frequency ω_1 and the actual rotor currents frequency $S\omega_1$.

Notice that, for synchronous coordinates, steady-state means d/dt = j(ω_b– ω_1) = 0, that is *d.c. quantities*.

In the flux equations (8.52)-(8.53) we may separate the main (airgap) flux linkage $\overline{\lambda}_m$

$$\overline{\lambda}_s^{\ b} = L_{ls} \cdot \overline{i}_s^{\ b} + \overline{\lambda}_m; \quad \overline{\lambda}_m = L_m \cdot \left(\overline{i}_s^{\ b} + \overline{i}_r^{\ b} \right) = L_m \cdot \overline{i}_m^{\ b} \qquad (8.92)$$

$$\overline{\lambda}_r^{\ b} = L_{lr} \cdot \overline{i}_r^{\ b} + \overline{\lambda}_m \qquad (8.93)$$

Equations (8.91)-(8.93) lead to the standard equivalent circuit of Figure 8.11.

Magnetic saturation may be considered through $\lambda_m(i_m)$ functions approximating measurements or field distribution calculations. [7]

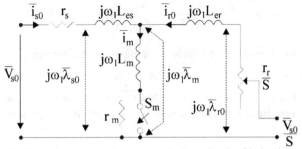

Figure 8.11. Space-phasor steady-state equivalent circuit of induction machine

Note (on core loss): The core loss occurs mainly in the stator as $\omega_1 \gg S\omega_1$; that is, the frequency of magnetic fields is higher in the stator since the rated slip is $S_n = 0.08$-0.005 (decreasing with power). An equivalent resistance r_m — determined from experiments — may be included in the equivalent circuit, Figure 8.11, to account for the core losses during steady-state. For investigating the role of core loss during transients, additional stator windings in the d-q model have to be included (see [7]).

The torque expression (8.54) with (8.91) and $V_{r0}^{\ b} = 0$, yields

$$T_e = -\frac{3}{2} p \cdot Re\left(j\overline{\lambda}_r^{\ b} \cdot \overline{i}_r^{\ b*} \right) = \frac{3}{2} p \cdot \lambda_{r0} \cdot i_{r0} \qquad (8.94)$$

From (8.92)

$$\overline{i}_{r0} = -j \cdot S\omega_1 \cdot \frac{\overline{\lambda}_{r0}}{r_r} \qquad (8.95)$$

Consequently the electromagnetic torque T_e is

$$T_e = \frac{3}{2} p \cdot \frac{\lambda_{r0}^{\ 2}}{r_r} \cdot S\omega_1 \qquad (8.96)$$

Using the equivalent circuit of Figure (8.11) and (8.95) we may obtain the conventional torque expression

$$T_e = \frac{3p}{\omega_1} \frac{V^2 \cdot r_r / S}{\left(r_s + c_1 r_r / S\right)^2 + \omega_1^2 \left(L_{ls} + c_1 L_{lr}\right)^2} \tag{8.97}$$

with
$$c_1 \approx 1 + \frac{L_{ls}}{L_m} \tag{8.98}$$

8.13. NO-LOAD IDEAL SPEED MAY GO UNDER OR OVER CONVENTIONAL VALUE ω_1

The no-load ideal speed (slip S_0) corresponds to zero torque, that is, (8.91), zero rotor current

$$\overline{V_{r0}}^b = j \cdot S_0 \omega_1 \cdot \overline{\lambda_{r0}}^b \tag{8.99}$$

Only for short-circuited rotor windings (or passive impedance at rotor terminals) ($\overline{V_r}^b = 0$), the ideal no-load slip $S_0 = 0$ and $\omega_{r0} = \omega_1$.

When the induction machine is doubly fed ($\overline{V_r}^b \neq 0$), the ideal no-load slip is different from zero and the no-load ideal speed is, in general

$$\omega_{r0} = \omega_1 \left(1 - S_0\right) \tag{8.100}$$

The value and phase shift between $\overline{V_{r0}}^b$ and $\overline{\lambda_{r0}}^b$ could be arranged through a PEC supplying the wound rotor. So S_0 could be either positive or negative. Thus only for $V_r^b = 0$ from (8.96) the torque is positive (motoring), for S>0 and negative (generating) for S<0.

The doubly fed induction motor could operate either as a motor or as a generator below and above $\omega_{r0} = \omega_1$ ($S_0 = 0$), provided that the PEC can produce bidirectional power flow between the wound rotor and the power grid. These doubly fed induction motor drives will be dealt with separately in Chapter 14 dedicated to high power industrial drives.

Example 8.3. For steady-state, calculate the stator voltage, stator flux, current, power factor, torque of an induction motor at 10% ideal no-load (synchronous) speed ω_1 and S = 0.02.
The motor data are $r_s = 0.5\Omega$, $r_r = 0.6\Omega$, $L_s = L_r = 0.08H$, $L_m = 0.075H$, $\lambda_{r0} = 0.8Wb$, $\omega_1 = 2\pi60$ rad/s, p = 2 pole pairs, $V_r^b = 0$ (cage rotor).
Solution:
First we have to calculate the initial conditions, which are implicitly steady-state. Let us use synchronous coordinates

$$\omega_{r0} = \omega_1 \left(1 - S\right) \tag{8.101}$$

So:
$$(\omega_1)_{t=0} = \frac{\omega_{r0}}{1-(S)_{t=0}} = \frac{0.1 \cdot 2 \cdot \pi \cdot 60}{1-0.02} = 38.45 \text{ rad/s} \qquad (8.102)$$

The torque (T_e) is (8.96)

$$(T_e)_{t=0} = \frac{3}{2}p \cdot (\lambda_r^b)^2 \cdot \frac{S\omega_1}{r_r} = \frac{3}{2} \cdot 2 \cdot (0.8)^2 \cdot 0.02 \cdot \frac{2\pi60}{0.6} = 24.115 \text{ Nm} \quad (8.103)$$

From (8.72)-(8.73) with d/dt = 0 and V_r^b = 0, we may now calculate the stator flux $\overline{\lambda}_s^b$

$$\overline{\lambda}_s^b = \overline{\lambda}_r^b \cdot \frac{1+j \cdot S \cdot \omega_1 \cdot \tau_r'}{K_s} \qquad (8.104)$$

Note that the motor parameters are as in example 8.2 and thus τ_r' = 0.01614s, τ_s' = 0.01937s, $K_s = K_r$ = 0.9375, σ = 0.1211.

$$\overline{\lambda}_s^b = \overline{\lambda}_r^b \cdot \frac{1+j \cdot 0.02 \cdot 2\pi60 \cdot 0.01614}{0.9375} = 0.8533 + j0.1038 \quad (8.105)$$

Let us consider the d axis along the rotor flux and thus $\overline{\lambda}_r^b = \lambda_r^b$.

Now from (8.72) with $\dfrac{d\overline{\lambda}_s^b}{dt} = 0$ we may find the stator voltage $\overline{V}_s^b$

$$\overline{V}_s^b = \frac{(1+j\omega_1\tau_s')\overline{\lambda}_s^b - K_r \cdot \overline{\lambda}_r^b}{\tau_s'} =$$
$$= \frac{(1+j \cdot 2\pi60 \cdot 0.01937) \cdot 0.8 \cdot (1.066 + j0.1297) - 0.9375 \cdot 0.8}{0.01937} \quad (8.106)$$
$$= -33.74 + j326.69; \quad V_s^b = 328.4\text{V}$$

The motor phase voltage (r.m.s. value)

$$V = V_s^b / \sqrt{2} = 328.4 / 1.41 = 232.91 \text{ V} \qquad (8.107)$$

The stator current is obtained from (8.69)

$$\overline{i}_s^b = \frac{[\overline{\lambda}_s^b - \overline{\lambda}_r^b \cdot K_r]}{\sigma \cdot L_s} =$$
$$\frac{[0.8 \cdot (1.066 + j0.1297) - 0.8 \cdot 0.9357]}{0.1211 \cdot 0.08} = 10.76 + j10.71 \qquad (8.108)$$

The rotor space-phasor $\overline{i}_r^b$ is (8.95)

$$\overline{i}_r^b = -jS\omega_1 \frac{\lambda_r^b}{r_r} = -j \cdot 0.02 \cdot 2\pi60 \cdot \frac{0.8}{0.6} = -j10.05 \text{ A} \qquad (8.109)$$

The amplitude of the stator current $i_s^b = 15.181\,A$. The results are illustrated by the space-phasor diagram in Figure 8.12.

The power factor angle φ_1 is

$$\varphi_1 = \arg \overline{V}_s^b - \arg \overline{i}_s^b = \frac{\pi}{2} + \tan^{-1}\frac{33.74}{326.69} - \tan^{-1}\frac{10.71}{10.76} = \quad (8.110)$$

$$= 90° + 5.9° - 45.156° = 50.74°$$

Finally, $\cos\varphi_1 = 0.633$.

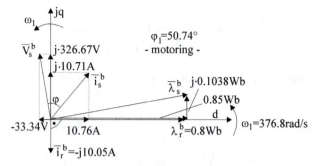

Figure 8.12. Steady-state space-phasor diagram in synchronous coordinates ($\omega_b = \omega_1$): d.c. quantities

Example 8.4. Loss breakdown
A high efficiency induction motor with cage rotor has the data rated power $P_n = 5kW$, rated line voltage (rms) $V_L = 220V$ (star connection), rated frequency $f_1 = 60Hz$, number of pole pairs $p = 2$, core loss (P_{iron}) = mechanical loss $P_{mec} = 1.5\%$ of P_n, additional losses $p_{add} = 1\%P_n$, stator per rotor winding losses $p_{cor}/p_{cos} = 2/3$, rated efficiency $\eta_n = 0.9$ and power factor $\cos\varphi_n = 0.88$.

Calculate all loss components, phase current (rms), then rated slip, speed, electromagnetic torque, shaft torque and stator current (as space-phasors).
Solution:
The loss breakdown diagram of the induction motor is shown in Figure 8.13.

The input power P_{in} is

$$P_{in} = \frac{P_{out}}{\eta_n} = \frac{5000}{0.9} = 5555.55\ W \quad (8.111)$$

The phase current I_n (rms) is

$$I_n = \frac{P_{in}}{\sqrt{3}\cdot V_L \cdot \cos\varphi_n} = \frac{5555.55}{\sqrt{3}\cdot 220\cdot 0.88} = 16.58\ A \quad (8.112)$$

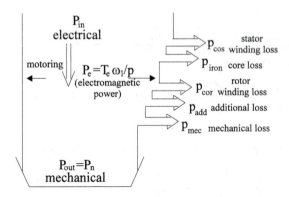

Figure 8.13. Induction motor energy conversion

The total losses $\sum p$ are

$$\sum p = P_{in} - P_{out} = 5555.55 - 5000 = 555.55 \text{ W} \qquad (8.113)$$

Consequently

$$p_{iron} = p_{mec} = 0.015 \cdot 5000 = 75 \text{W} \qquad (8.114)$$

$$p_{add} = 0.01 \cdot 5000 \text{ W} \qquad (8.115)$$

So

$$p_{cos} + p_{cor} = \sum p - p_{iron} - p_{mec} - p_{add} = 555.55 - 75 - 75 - 50 \approx 355 \text{W} \qquad (8.116)$$

$$p_{cos} + \frac{2}{3}p_{cos} = 355 \text{W} \qquad (8.117)$$

$$p_{cos} = 213.3 \text{ W}; \quad p_{cor} = 142 \text{ W} \qquad (8.118)$$

The electromagnetic power P_e is the active power that crosses the airgap

$$P_e = P_{in} - P_{cos} - p_{iron} = T_e \cdot \frac{\omega_1}{p} \qquad (8.119)$$

$$P_e = 5555.55 - 213 - 75 = 5267 \text{ W} \qquad (8.120)$$

$$T_e = \frac{5267}{2\pi 60} \cdot 2 = 27.956 \text{ Nm} \qquad (8.121)$$

The rotor winding loss p_{cor} is

$$p_{cor} = S_n \cdot P_e \qquad (8.122)$$

and thus the rated slip is

$$S_n = \frac{142}{5267} = 0.02696 \qquad (8.123)$$

The rated speed n_n is

$$n_n = \frac{f_1}{p}(1 - S_n) = \frac{60}{2}(1 - 0.02696) = 29.1912\text{rps} = 1751.472\text{rpm}$$

$$(8.124)$$

The shaft torque T_n is calculated directly from mechanical power P_n

$$T_n = \frac{P_n}{2\pi n_n} = \frac{5000}{2\pi \cdot 29.1912} = 27.274\text{Nm} < T_e \qquad (8.125)$$

Finally the amplitude of the stator current, $\bar{i}_s$ in synchronous coordinates (d.c. quantities) is

$$i_s = i_n \sqrt{2} = 16.58 \cdot 1.41 = 23.3778 \text{ A} \qquad (8.126)$$

Note: To calculate all motor parameters, resistances and inductances, more data are required. This is beyond our scope in this example, however.

8.14. MOTORING, GENERATING, A.C. BRAKING

The equivalent circuit for steady-state (Figure 8.11) shows that the active power in the rotor (the electromagnetic power P_e), is

$$P_e = \frac{3}{2} \cdot I_{r0}^2 \cdot \frac{r_r}{S} = T_e \cdot \frac{\omega_1}{p}; \quad S = 1 - \frac{\omega_r}{\omega_1} \qquad (8.127)$$

This expression is valid for the cage rotor ($V_r^b = 0$).

Motoring mode is defined as the situation when the torque has the same sign as the speed

$$P_e > 0, T_e > 0; \quad \omega_r > 0 \Rightarrow 0 < S < 1 \qquad (8.128)$$

For *generating* the torque is negative (S<0) in (8.127) but the speed is positive

$$P_e < 0, T_e < 0; \quad \omega_r > 0 \Rightarrow S < 0 \qquad (8.129)$$

The generator produces braking ($T_e < 0$, $\omega_r > 0$) but the energy transfer direction in the motor is reversed. The energy is pumped back into the power source through the stator.

Braking is obtained when again ($T_e > 0$, $\omega_r < 0$) (or $T_e < 0$, $\omega_r > 0$) but the electromagnetic power is still positive

$$P_e > 0; T_e > 0; \omega_r < 0 \Rightarrow S > 1$$
$$P_e > 0; T_e < 0; \omega_r > 0 \Rightarrow S > 1 \tag{8.130}$$

We may synthesize the results on operation modes as in Table 8.1.

Table 8.1. Operation modes (cage rotor)

S	$-\infty$ - - - - - - - 0 + + + + + 1 + + + + + + $+\infty$
ω_r	$+\infty$ + + + + + ω_1 + + + + + 0 - - - - - - - - - $-\infty$
T_e	0 - - - - - - - 0 + + + + + T_e(start) + + + + 0
P_e	- - - - - - - - - 0 + + + + + + + + + + + + +
mode	Generating	Motoring	a.c. braking

Using the torque expression (8.97) we may find the maximum torque for the critical slip:

$$S_k = \frac{\pm c_1 r_r}{\sqrt{r_s^2 + \omega_1^2 (L_{ls} + c_1 L_{lr})^2}} ; c_1 = 1 + \frac{L_{sl}}{L_m} \tag{8.131}$$

$$T_{ek} = \frac{3p}{\omega_1} \cdot \frac{V^2}{2c_1 \left(r_s \pm \sqrt{r_s^2 + \omega_1^2 (L_{ls} + c_1 L_{lr})^2} \right)} \tag{8.132}$$

The torque slip (speed) curve is shown in Figure 8.14.

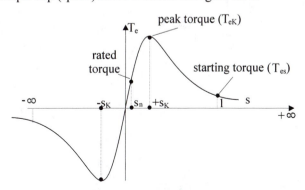

Figure 8.14. Torque/speed curve of induction motors for constant voltage and frequency

Using the definition of critical slip S_k and torque T_{ek} in the torque expression (8.97) we obtain, after some approximations, the Kloss formula

$$\frac{T_e}{T_{ek}} \approx \frac{2}{\frac{S}{S_k} + \frac{S_k}{S}} \tag{8.133}$$

8.15. D.C. BRAKING: ZERO BRAKING TORQUE AT ZERO SPEED

For moderate braking requirements d.c. braking is commonly used in modern electrical drives. To calculate the d.c. braking torque we redraw the space-phasor equivalent circuit in Figure 8.11 but with a d.c. stator current source space-phasor in Figure 8.15.

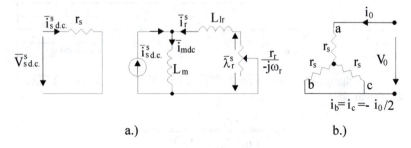

a.)   b.)

Figure 8.15. Equivalent circuit for d.c. braking — stator coordinates — in space-phasors; steady-state.

The electromagnetic torque is still computed from (8.94)

$$T_e = \frac{3}{2}p\lambda_{rdc}i_{rdc} = -\frac{3}{2}p\frac{i_{rdc}^2 \cdot r_r}{\omega_r} \tag{8.134}$$

with

$$\bar{i}_{rdc}^s = \bar{i}_{sdc}^s \cdot \frac{L_m}{L_{lr} - j\frac{r_r}{\omega_r}} \tag{8.135}$$

Finally

$$T_e = -\frac{3}{2}p\frac{L_m^2 \cdot i_{sdc}^2 \cdot \omega_r \cdot r_r}{L_{lr}^2\omega_r^2 + r_r^2} \tag{8.136}$$

The peak torque is obtained for ω_{rk}

$$\omega_{rk} = \frac{r_r}{L_{lr}} \approx \frac{1}{\tau_r'} \tag{8.137}$$

and its value is

$$T_{ek} = -\frac{3}{2}p\frac{L_m^2 \cdot i_{sdc}^2}{2L_{lr}} \tag{8.138}$$

The braking torque may be modified through the d.c. current level in the stator. The PEC can produce the phase connection in Figure 8.15b where the current complex variable in stator coordinates is

$$\bar{i}_{sdc}{}^s = \frac{2}{3}\left(i_a + i_b e^{j\frac{2\pi}{3}} + i_c e^{-j\frac{2\pi}{3}} \right) = \frac{2}{3}\left(i_0 - \frac{i_0}{2} 2\cos\frac{2\pi}{3} \right) = i_0 \quad (8.139)$$

The torque speed curve for d.c. braking is shown in Figure 8.16. Notice also that the rotor kinetic energy is dumped into the rotor resistor and that for zero speed the braking torque is zero. Also, above ω_{rk} (which is fairly small in high efficiency motors), the torque is again rather small.

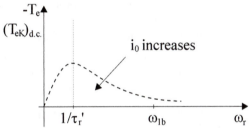

Figure 8.16. D.c. braking torque of induction motors

8.16. SPEED CONTROL METHODS

Variable speed is required in many applications. It has to be performed at high energy conversion rates.

The no-load ideal speed ω_{r0} is ((8.100) with (8.99))

$$\omega_{r0} = \omega_1\left[1 + \operatorname{Im}ag\left(\frac{\overline{V}_{r0}{}^b}{\omega_1 \overline{\lambda}_{r0}{}^b} \right) \right]; \quad \omega_1 = \omega_r + \omega_2 \quad (8.140)$$

Evidently for the cage rotor $\overline{V}_r{}^b = 0$ and thus $\omega_{r0} = \omega_1 = 2\pi f_1$ or

$$n_0 = \frac{\omega_{r0}}{2\pi p} = \frac{f_1}{p} \quad (8.141)$$

ω_2 is the frequency of the rotor current (or of $\overline{V}_{r0}$ through rotor side PEC). There are three essential methods to vary speed by changing the no-load ideal speed (as the rated slip is small) as suggested by (8.140):

- stator frequency f_1 variation;
- pole number (2p) changing;
- *wound rotor supply* (or rotor frequency f_2 variation).

While pole number 2p changing involves either a separate stator winding or a special winding with a switch to change from $2p_1$ to $2p_2$ poles (Dahlander winding) the other two methods require variable frequency either in the stator or in the rotor, obtainable with PECs in the stator and, respectively, in the rotor.

For limited speed variation, the power rating of rotor-side PEC is smaller than that of the PEC in the stator. In high power applications the rotor PEC solution with a wound rotor induction motor is the one preferred for limited speed range ($\pm 30\%$) control.

Stator frequency control is far more frequently used, especially for wide speed control range. However, the level of flux depends on the current in the machine, especially on the magnetization current

$$\overline{I}_m = \overline{I}_s{}^b + \overline{I}_r{}^b \tag{8.142}$$

Consequently, it is crucial to control I_m properly to avoid excessive magnetic saturation, while varying frequency f_1.

So we have to adopt either voltage V_s and frequency f_1 coordinated control or current $\overline{I}_s$ (or $\overline{I}_m$) and frequency f_1 control. In general V_1/f_1 coordinated change is applied. There is an infinity of possibilities to relate $\overline{V}_s(V_1)$ or $\overline{I}_s(\overline{I}_m)$ to frequency f_1 to obtain the desired performance. However, only three main methods reached the markets:

- V_1/f_1 - scalar control;
- constant (controlled) rotor flux (λ_r) vector control;
- constant (controlled) stator flux (λ_s) - vector control.

Here only the torque/speed curves obtainable with the above methods are given. Notice that in all these cases the PEC voltage supplying the induction motor is voltage limited with the maximum voltage reached at base speed ω_b.

8.17. V_1 / f_1 TORQUE/SPEED CURVES

V_1/f_1 control means that:

$$V_1 = V_0 + K_f \cdot f_1 \tag{8.143}$$

We may judge the torque/speed curves obtained in this case through the critical slip S_k and torque T_{ek} of (8.131)-(8.132). The critical slip increases notably with f_1 (ω_1) reduction while the critical torque is only slightly decreased with f_1 decrease at frequencies above 5Hz. Below this value the peak torque decreases dramatically if (8.143) is applied (Figure 8.17). For a safe start $V = V_0 + K \cdot f_1$ is applied to compensate for the stator resistance drop $r_s i_s$

$$V_0 = c_0 r_s i_{sn} \tag{8.144}$$

V_0 is called the voltage boost and amounts to a few percent of rated voltage V_n, higher for low power motors.

Above rated (base) speed the voltage remains constant and invariably the critical torque decreases and, eventually, constant power is preserved up to a maximum frequency f_{1max}. Above base speed, if we neglect the voltage drop $\overline{r_s i_s}$ in (8.91), the stator equation (for steady-state) is

$$\overline{V}_{s0} \approx j\omega_1 \overline{\lambda}_{s0}{}^b ; \qquad \omega_1 > \omega_b \qquad (8.145)$$

So, for constant voltage V_{s0}, the stator flux λ_{s0} and, consequently, the main flux λ_m decrease with speed (frequency) increasing above ω_b. This zone — ω_b to ω_{1max} — is called the flux weakening zone. In many applications constant power is required for a ratio ω_{1max}/ω_b of 2-4. The entire motor design (sizing) depends on this requirement in terms of both electromagnetic and thermal loading.

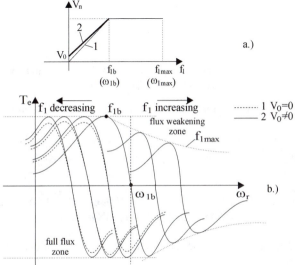

Figure 8.17. Torque/speed curves for $V_1 = V_0 + K_f \cdot f_1$ (V_1/f_1 control):
a.) V_1/f_1 dependences, b.) T_e/ω_r curves;

V_1/f_1 drives are standard with low dynamics applications and moderate speed control range ($\omega_{1b}/\omega_{1min} = 10 - 15$ or so) such as pumps or fans where the load torque is solely dependent on speed and thus an optimal V_1/f_1 relationship may be calculated off line — for maximum efficiency or power factor — and implemented in the drive hardware.

8.18. ONLY FOR CONSTANT ROTOR FLUX TORQUE/SPEED CURVES ARE LINEAR

The torque expression for constant rotor flux $\lambda_r{}^b$ is (8.96)

$$T_e = \frac{3}{2} p \frac{\lambda_{r0}{}^2 (\omega_1 - \omega_r)}{r_r} \qquad (8.146)$$

and it represents a straight line (Figure 8.18). This is ideal for speed (or torque) control purposes and led to (now widely accepted in industry), vector control. It is, however, to be noted that above base frequency ω_{1b}, when the full voltage capability of PEC is reached, the value of rotor flux magnitude may not be maintained any more as the difference between the stator flux (8.145) and rotor flux amplitudes (8.104) is less than 15% in general (lower values for higher powers). So, above ω_{1b}, the torque/speed curves degenerate into the shapes obtained for V_1/f_1 control.

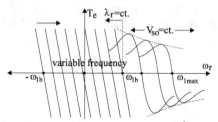

Figure 8.18. Torque/speed curves for constant rotor flux λ_r^b = ct. up to base frequency ω_{1b}; constant voltage and variable frequency above ω_{1b}.

Four-quadrant operation presented in Figure 8.18 is obtainable as with V_1/f_1 control.

Negative frequency ω_1 means negative (reverse) m.m.f. wave speed, to be obtained by changing the stator phase voltage sequence (from a b c to a c b).

8.19. CONSTANT STATOR FLUX TORQUE/SPEED CURVES HAVE TWO BREAKDOWN POINTS

Let us note from (8.194) that the stator-rotor flux relationship for steady-state (cage rotor) is

$$\bar{\lambda}_{s0} = \bar{\lambda}_{r0} \frac{\left(1 + jS\omega_1\tau_r'\right)}{K_s} \tag{8.147}$$

Consequently the torque T_e from (8.146) becomes

$$T_e = \frac{3}{2}p\frac{K_s^2}{r_r}\frac{S\omega_1\lambda_{s0}^2}{1+\left(S\omega_1\tau_r'\right)^2} \tag{8.148}$$

This expression has extreme (critical) values for

$$\left(S\omega_1\right)_k = \pm\frac{1}{\tau_r'} \tag{8.149}$$

$$T_{ek} = \frac{3}{2}p\frac{K_s^2\lambda_s^2}{2r_r\tau_r'} \tag{8.150}$$

So the peak torque is independent of frequency as long as λ_s amplitude may be realized, that is, below base (rated) frequency.

Above base frequency, according to (8.145), the approximate peak torque would be

$$\left(T_{ek}\right)_{\omega_1 > \omega_{1b}} = \frac{3}{2} p \frac{K_s^2}{2r_r\tau_r'} \cdot \frac{V_{s0}^2}{\omega_1^2} \tag{8.151}$$

The torque/speed curves are shown in Figure 8.19 for four-quadrant operation.

The peak torque is safely provided even for zero speed. Still the departure from linearity in the torque/speed curve has to be dealt with when investigating electric drive transients and stability.

Note: In principle PECs can also provide constant main (airgap) flux variable frequency control. So far this operation mode has not reached wide markets in variable speed drives.

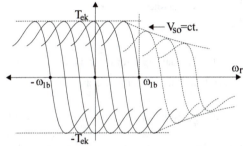

Figure 8.19. Torque/speed curves for constant stator flux amplitude λ_s up to ω_{1b} and constant voltage above ω_{1b}.

8.20. THE SPLIT-PHASE INDUCTION MOTOR

The split-phase induction motor is provided with two orthogonal stator windings: the main winding and the auxiliary winding.

This induction motor is, in general, fed from a single phase a.c. supply of constant or variable voltage and frequency.

So, even for variable speed, it requires, essentially, a single phase PWM inverter.

Connecting a capacitor in series with the auxiliary winding causes the current I_a to lead the current in the main winding I_m and thus the two orthogonal placed windings produce an elliptical magnetic field in the airgap which has a traveling component. This explains the safe selfstarting from the capacitor auxiliary to the main winding direction of rotation.

Switching the capacitor from auxiliary to main winding leads to the reversal of traveling field component in the airgap and thus the direction of motion may be reversed. For this latter case the two windings should be identical (same number of turns and slots). For unidirectional motion, a higher than unity turns ratio a $=W_a/W_m$ is typical with $I_a < I_m$, as in many cases the auxiliary winding is turned off after starting the motor.

It is also common to use two capacitors: one (larger) for starting, and a smaller one for running.

It goes without saying that the power factor of the capacitor split-phase IM is very good, due to the presence of the capacitor.

Typical connections for dual capacitors IM are shown in Fig. 8.20.

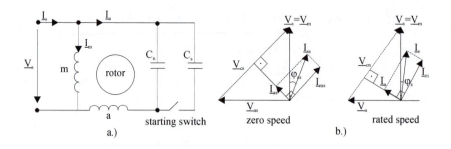

Figure 8.20. The dual capacitor IM: a.) Equivalent scheme,

b.) Phasor diagrams for zero and rated load.

The capacitor voltage is rectangular to auxiliary winding current. For symmetry conditions – 90° electrical degrees between the I_a and I_m and equal Ampere turns $W_m I_m = W_a I_a$ – a pure traveling field is produced; once at zero speed with the capacitor C_s and once at rated speed (load) with capacitor $C_n < C_s$ ($C_s / C_n \cong 4 - 6$).

At any other speed (slip S) the magnetic field in the airgap will have an additional (undesirable) inverse component which produces a braking torque and additional losses.

As done for the three phase motor, the m.m.f. of stator windings fundamental components are:

$$F_m(\theta_{es}, t) = F_{1m} \cos(\omega_1 t) \cos\theta_{es}$$
$$F_a(\theta_{es}, t) = -F_{1a} \cos(\omega_1 t + \gamma_i) \sin\theta_{es}$$

(8.152)

The ratio of the two m.m.f. amplitudes is:

$$\frac{F_{1a}}{F_{1m}} = \frac{W_m k_{Wm1}}{W_a k_{Wa1}} \frac{I_m}{I_a} = \frac{1}{a} \frac{I_m}{I_a}$$

(8.153)

The total m.m.f. $F(\theta_{es}, t)$ may be decomposed in forward and backward waves:

$$F(\theta_{es}, t) = F_m(\theta_{es}, t) + F_a(\theta_{es}, t) =$$

$$F_{1m} \cos(\omega_1 t) \cos \theta_{es} + F_{1a} \cos(\omega_1 t + \gamma_i) \sin \theta_{es} = \qquad (8.154)$$

$$\frac{1}{2} F_m C_f \sin(\theta_{es} - \omega_1 t - \beta_f) + \frac{1}{2} F_m C_b \sin(\theta_{es} + \omega_1 t - \beta_b)$$

With:

$$C_f = \sqrt{\left(1 + \frac{F_{1a}}{F_{1m}} \sin \gamma_i\right)^2 + \left(\frac{F_{1a}}{F_{1m}}\right)^2 \cos^2 \gamma_i} \qquad (8.155)$$

$$C_b = \sqrt{\left(1 - \frac{F_{1a}}{F_{1m}} \sin \gamma_i\right)^2 + \left(\frac{F_{1a}}{F_{1m}}\right)^2 \cos^2 \gamma_i} \qquad (8.156)$$

$$\sin \beta_f = \frac{1 + \dfrac{F_{1a}}{F_{1m}} \sin \gamma_i}{C_f}; \quad \sin \beta_b = \frac{1 - \dfrac{F_{1a}}{F_{1m}} \sin \gamma_i}{C_b} \qquad (8.157)$$

It is evident that for $F_{1a}/F_{1m} \cong 1$ and the phase shift $\gamma_i = 90^0$, $C_b = 0$ and thus the backward field (and torque) is zero. The slip is $S_+ = (\omega_1 - \omega_r) / \omega_1$ for the direct m.m.f. component and $S_- = 2\text{-}S = (-\omega_1 - \omega_r) / (-\omega_1)$ for the backward (inverse) m.m.f. component.

For the steady state the symmetrical $(+ / -)$ component model is straightforward, while for transients the dq model in stator coordinates is very practical.

The $+ / -$ model

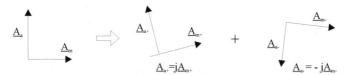

Figure 8.21. The $+ / -$ model decomposition;

The superposition principle is used:

$$\underline{A}_m = \underline{A}_{m+} + \underline{A}_{m-}; \quad \underline{A}_a = \underline{A}_{a+} + \underline{A}_{a-} \qquad (8.158)$$

And

$$\underline{A}_{m+} = \frac{1}{2}(\underline{A}_m - j\underline{A}_a); \underline{A}_{m-} = \underline{A}^*_{m+} \qquad (8.159)$$

Now the machine behaves like two separate fictitious machines for the two components:

$$\underline{V}_{m+} = \underline{Z}_{m+}\underline{I}_{m+}; \underline{V}_{m-} = \underline{Z}_{m-}\underline{I}_{m-}$$
$$\underline{V}_{a+} = \underline{Z}_{a+}\underline{I}_{a+}; \underline{V}_{a-} = \underline{Z}_{a-}\underline{I}_{a-}$$
(8.160)

$$\underline{V}_m = \underline{V}_{m+} + \underline{V}_{m-}; \ \underline{V}_a = \underline{V}_{a+} + \underline{V}_{a-}$$
(8.161)

The $+ / -$ impedances $\underline{Z}_{m\pm}$ represent the total forward/backward impedances of the machine on a per phase basis, when the rotor cage is reduced to the main (m) and respectively, auxiliary (a) winding (Figure 8.22)

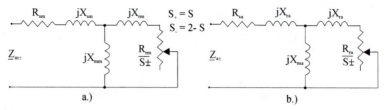

Figure 8.22. Equivalent symmetrical $(+ / -)$ impedance: a.) reduced to main winding; b.) reduced to auxiliary winding;

The relationship of $\underline{V}_m$ and $\underline{V}_a$ voltages to the source voltage $\underline{V}_s$ are:

$$\underline{V}_s = \underline{V}_m; \underline{V}_a + (\underline{I}_{a+} + \underline{I}_{a-})\underline{Z}_a = \underline{V}_s$$
(8.162)

Z_a is the auxiliary impedance (a capacitor in general) added in series to the auxiliary phase for better starting and (or) running.

The torque T_e expression has two terms: the feedforward and feedback components:

$$T_e = T_{e+} + T_{e-} = \frac{2p_1}{\omega_1}\left[I_{rm+}^2 \frac{R_{rm}}{S} - I_{rm-}^2 \frac{R_{rm}}{2-S}\right]$$
(8.163)

When the auxiliary winding is open $Z_a = \infty$, $I_{rm+} = I_{rm-}$ and thus, at zero speed (S=1) the torque is zero, as expected, and the machine does not start.

A typical mechanical characteristic $T_e(\omega_r)$ for a split phase capacitor run motor comprises the positive torque (with synchronism at $+ \omega_1$) and the negative torque (with synchronism at $- \omega_1$), Figure 8.23.

The backward (negative) torque is rather small but present at all speeds, less the rated speed, where, by design, symmetry conditions are met. There are two symmetrization conditions that stem from $I_{m-}=0$ which lead to a value of turns ratio a and of the capacitance C for given slip S [8, pp. 851].

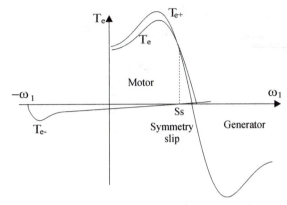

Figure 8.23. Typical mechanical characteristic of a split phase capacitor-run induction

motor;

For the case of motor with same copper quantity in both stator windings the symmetrization conditions are very simple [8]:

$$a = X_{m+} / R_{m+} = \tan \varphi_+$$

$$X_c = \frac{1}{\omega C} = Z_+ \cdot a\sqrt{a+1} \qquad (8.164)$$

So for an existing motor (with fixed turns ratio a), there may not be any capacitor to symmetrize the motor at no slip (speed). Fortunately perfect symmetrization is not necessary because the backward torque tends to be small for reasonably low rotor resistance (good efficiency) designed motors.The slip phase capacitor motor is widely used both at constant and variable speed for applications below 500W in general. More on the control of this motor in Chapter 9.

Example 8.4:
A capacitor run split-phase IM with $C_a = 8\mu F$, 2p = 6 pole, 230V, 50 Hz, $n_n = 960$ rpm has the following main and auxiliary winding parameters: $R_{sm} = 34\Omega$, $R_{sa} = 150\Omega$, a = 1.73, $X_{sm} = 35.9\Omega$, $X_{sa} = a^2 X_{sm}$, $X_{rm} = 29.32\Omega$, $R_{rm} = 23.25\Omega$, $X_m = 249\Omega$.

For the case of open auxiliary phase calculate:
a) The $\pm$ impedances $\underline{Z}_{m+}$, $\underline{Z}_{m\pm}$ at rated slip;
b) The equivalent circuit of the machine for this particular case;
c) The main winding current components I_{m+}, I_{m-} and the total current I_m.
d) The $\pm$ torque components, the resultant torque and the mechanical power, efficiency and power factor.
Solution:
a) The rated slip:

$$S_n = \frac{\omega_1 - \omega_r}{\omega_1} = \frac{2\pi50 - 2\pi3 \cdot 960/60}{2\pi50} = 0.06; 2 - S = 1.94; \quad (8.165)$$

From Figure 8.22. the $\pm$ impedances are:

$$\underline{Z}_{m+} = R_{sm} + jX_{sm} + \underline{Z}_{r+} = 34 + j35.9 + \frac{j249(23.25/0.06 + j29.32)}{23.25/0.06 + j(29.32 + 249)} =$$

$$= 34 + j35.9 + 105.4 + j174$$

$$(8.166)$$

$$\underline{Z}_{m-} = R_{sm} + jX_{sm} + \underline{Z}_{r-} = 34 + j35.9 + \frac{j249(23.25/1.94 + j29.32)}{23.25/1.94 + j(29.32 + 249)} =$$

$$= 34 + j35.9 + 9.6 + j27$$

$$(8.167)$$

Let us notice that because the auxiliary current $\underline{I}_a = 0$, from (8.158) – (8.159):

$$\underline{I}_{m+} = \underline{I}_{m-} = \frac{\underline{I}_m}{2}$$

Combining this with (8.160):

$$\underline{V}_m = \underline{V}_s = \underline{V}_{m+} + \underline{V}_{m-} = \underline{Z}_{m+}\underline{I}_{m+} + \underline{Z}_{m-}\underline{I}_{m-} = \underline{I}_m\left(\frac{\underline{Z}_{m+}}{2} + \frac{\underline{Z}_{m-}}{2}\right)$$

which means that we obtain a series equivalent circuit (Fig. A):

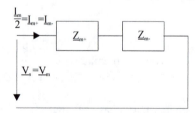

Figure A. Single phase ($\underline{I}_a$=0) IM equivalent circuit;

c) The main winding current components $\underline{I}_{m+} = \underline{I}_{m-}$ are then simply:

$$\underline{I}_{m+} = \underline{I}_{m-} = \frac{V_s}{|\underline{Z}_{m+} + \underline{Z}_{m-}|} = \frac{230}{2 \cdot 165.2} = 0.696A$$

The main winding current $I_m = 2I_{m+} = 1.392$ A.
The power factor for $S = 0.06$, $\cos\varphi_1$ is:

$$\cos\varphi_1 = \frac{\text{Real}(\underline{Z}_{m+} + \underline{Z}_{m-})}{|\underline{Z}_{m+} + \underline{Z}_{m-}|} = \frac{2 \cdot 92.8}{2 \cdot 165.2} = 0.5617!$$

d) The torque components in (8.163) are (Fig. A):

$$T_e = T_{e+} + T_{e-} = \frac{2p}{\omega_1}\left[I_{m+}^2 \frac{R_{rm}}{S} - I_{m-}^2 \frac{R_{rm}}{2-S}\right]$$

With $R_{rm+} = 105.4\ \Omega$ and $R_{rm-} = 9.6\ \Omega$ from $\underline{Z}_{m+}$ and $\underline{Z}_{m-}$ above:

$$T_{e+} = \frac{2 \cdot 3}{2\pi 50} 0.696^2 \cdot 105.4 = 0.9756\text{Nm}$$

$$T_{e-} = -\frac{2 \cdot 3}{2\pi 50} 0.696^2 \cdot 9.6 = -0.08886\text{Nm}$$

So the total torque:

$$T_e = T_{e+} + T_{e-} = 0.8687\text{Nm}$$

The mechanical power P_m is:

$$P_m = T_e\, 2\pi 50 = 0.8687 \cdot 2\pi \cdot \frac{940}{60} = 85.47\text{W}$$

Neglecting the core and mechanical loss the efficiency is:

$$\eta = \frac{P_m}{P_m + P_{Co}} = \frac{85.47}{85.47 + 1.392^2 \cdot 34} = 0.5647$$

The power factor $\cos\varphi_1$ is low, as calculated above because the capacitor is missing and the efficiency is low, mainly because the power is small.

The need to supply the auxiliary winding in series with a capacitor is evident.

Note: Even when the auxiliary phase winding is present, the performance computation process is as above, but a bit more tedious [8, pp. 847].

8.21. SUMMARY

- The three-phase induction motor is the workhorse of industry for powers from below 1 kW to 10 MW and over 100 MW in wound rotor configuration for pumped storage power plants.
- The a.c. stator and rotor windings are placed in slots. Skin effect in the rotor windings is put to work in constant speed high starting torque applications and is to be avoided in variable speed drives where variable voltage and frequency are provided by PECs.

- The inductances, between stator and rotor windings, vary sinusoidally with the rotor position and thus the voltage-current equations in phase coordinates are difficult to handle for transients even through numerical methods, in an 8th order nonlinear system.

- The space-phasor model, equivalent to the d-q (orthogonal axis, or Park's) model, is characterized by constant coefficients.

- For steady-state, the rotor current and flux space-phasors are spatially orthogonal to each other.

- At constant speed the space-phasor model is a second-order system with two complex eigenvalues, one for the stator, the other for the rotor. They depend on rotor speed and on the reference system speed ω_b, but their real part is negative for low slip values.

- Under steady-state the space-phasor model has the same equations irrespective of the reference system speed.

- Losses in an induction motor are distributed both in the stator and in the rotor: winding and core loss in the stator and winding, and additional and mechanical losses in the rotor. The equivalent circuit allows for calculating the steady-state performance conveniently.

- The induction motor operation modes are motoring, generating and a.c. braking (plugging). The torque/speed curve shows two peak values (one for motoring and the other for generating), where constant voltage and frequency fed.

- When the motor stator is d.c. current fed and the rotor speed is nonzero, motion induced voltages in the rotor produce rotor currents and as a consequence a braking torque. Its maximum is obtained at a small speed for cage rotor motors, as the rotor current frequency ω_2 is equal to rotor speed ω_r.

- Speed control methods include variable stator or rotor frequency and voltage through PECs. For limited speed control ($\pm$ 30% around ideal no-load speed) the variable rotor frequency supply method (through limited power PECs) is particularly suited for the scope, especially in high power drives.

- Stator frequency variation is followed along three methods of practical interest: V_1/f_1 method, constant rotor flux and constant stator flux methods.

- Only for constant rotor flux is the torque/speed curve linear (as for a PM d.c. brush motor) that is, ideal for electric drives control. This method is known as vector control.

- Split-phase capacitor induction motors are used in general below 500 W in single phase supply (home) applications.

8.22. PROBLEMS

8.1. Draw the complex root (eigenvalues) locus plot of a cage rotor induction motor with data as in example 8.2 for 8 different rotor speed values from zero to 2400 rpm, in stator coordinates ($\omega_b = 0$).

8.2. For the induction motor with resistances and inductances as in example 8.3, p = 1, $V_r^b = 0$, n = 900rpm, rotor flux $\lambda_r^b = 1$ Wb, torque level $T_e = 40$ Nm calculate: primary frequency ω_1, stator flux space-phasor amplitude in synchronous coordinates, stator voltage, stator current and power factor. Align the d axis along the rotor flux and draw the space-phasor diagram with the corresponding numbers on it (as in Figure 8.12).

8.3. An induction motor has the data ideal no-load (S = 0) phase current (rms) $I_{on} = 5$ A, p = 2, no-load losses $p_0 = 200$ W, stator phase voltage V = 120 V (rms); $f_1 = 60$ Hz; starting current per phase ($\omega_r = 0$, S = 1) $I_{start} = 15I_{on}$ and starting power factor $\cos\varphi_{1s} = 0.3$; $L_{ls} = L_{lr}$; $r_{start} = 3r_r$. Using the equivalent circuit of Figure 8.11 calculate: the stator and rotor resistances r_s, r_{start} and the leakage inductances L_{ls} and L_{lr}, the starting torque, core loss resistance r_m and core loss p_{iron}, and main inductance L_m. Observing that we have a deep bar (skin effect) rotor and the rotor resistance during normal operation is $r_r = 1.2r_s$, calculate all the electrical losses in the machine for S = 0.02.

8.4. For an induction motor with the data: $r_s = 0.2$ Ω, p = 2, $r_r = 0.2$ Ω, $L_{lr} = L_{ls} = 5\times10^{-3}$ H, ($L_m = 0.1$ H), $f_1 = 60$ Hz, $V_L = 220$ V (rms) operating in regenerative braking at S = -0.02, calculate electromagnetic torque T_e and the electric power retrieved if the core and mechanical losses are neglected.

8.5. For the induction motor with the data (resistances and inductances) of problem 8.4, calculate the d.c. braking peak torque (for the d.c. connection of Figure 8.15b at $I_o = 4$ A), for $\omega_1 = 2\pi10$ rad/s.

8.6. An induction motor works with a rotor flux level $\lambda_{r0} = 1$ Wb and has four poles (2p = 4) and the rotor resistance $r_r = 0.2$ Ω. For $\omega_1 = 2\pi20$rad/s calculate and draw the torque/speed curve. For $\tau_r' = 0.01s$, S = 0.1, $K_s = 0.95$ determine the stator flux level λ_{so} and, neglecting the stator resistance, the stator voltage V_{s0} required.

8.7. An induction motor works at constant stator flux and variable frequency and has the data as in problem 8.6. Determine the critical slip frequency and $(S\omega_1)_k$ and the critical (breakdown) torque for $\lambda_{s0} = 1.06$ Wb and draw the torque/speed curve for $f_1 = 20$ Hz. For $f_1' = 60$ Hz, calculate the maximum flux level available for the same voltage as in the case of 20Hz, and determine again the breakdown torque available, and the corresponding electromagnetic power.

8.23. SELECTED REFERENCES

1. **K.P. Kovacs, I. Racz**, Transient regimes of a.c. machines, Springer Verlag 1995 (the original edition, in German, in 1959).
2. **P. Vas,** Vector control of a.c. machines, O.U.P., 1990.
3. **I. Boldea, S.A. Nasar,** Vector control of a.c. drives, Chapter 2, CRC Press, Florida, USA, 1992.
4. **D.W. Novotny, T.A. Lipo**, Vector control and dynamics of a.c. drives, Oxford University Press, 1996.
5. **R.H. Park**, "Two reaction theory of synchronous machines: generalized method of analysis", AIEE Trans.vol.48, 1929, pp.716-730.
6. **J. Holtz**, "On the spatial propagation of transient magnetic fields in a.c. machines", IEEE Trans.vol.IA-32, no. 4, 1996, pp.927-937.
7. **I. Boldea, S.A. Nasar**, "Unified treatement of core losses and saturation in orthogonal axis model of electric machines", Proc.IEE, vol.134, PartB, 1987, pp.355-363.
8. **I. Boldea, S.A.Nasar,** Induction machine handbook, CRC Press 2001.

Chapter 9

PWM INVERTER-FED INDUCTION MOTOR DRIVES

9.1. INTRODUCTION

In this chapter we will deal only with cage-rotor induction motor drives of small and medium power ratings provided with voltage source PWM inverters for variable frequency and voltage amplitude.

Open-loop coordinated voltage-frequency V_1 / f_1 control or current-slip frequency control — called scalar control — do not include a speed (or position) closed-loop and have so far traditionally provided satisfactory variable speed with low dynamics and low speed control precision.

Fast torque dynamics and speed control require closed-loop speed control and even torque closed-loop control. Linearizing the torque/speed characteristic, or torque/current characteristic, as in d.c. brush PM motors, would be ideal for the scope.

There are three main schemes to produce this linearization by intelligent manipulation of IM equations in space - phasors:

- Vector current and voltage control - VC [1];
- Direct torque and flux control - DTFC [2-4];
- Feedback linearization control - FLC [5-7]

For closed-loop motion control, — speed and (or) position — sensors are to be used for wide speed control range (above 100 to 1), while below this value no motion sensors are provided, with speed (position) calculated through observers based on measured motor voltages and currents.

So we distinguish:

- Control with motion sensors;
- Control without motion sensors (sensorless).

The rapid development of powerful digital signal processors (DSPs) and ASICs led to a kind of universal induction motor drive that, on a menu basis, may work in various modes:

- V_1 / f_1 with slip compensation with motion sensors or sensorless;
- Vector control (VC) or direct torque and flux (DTFC) or feedback linearization control (FLC) with motion sensors or sensorless.

As V_1 / f_1 control is losing momentum we will present it at the end of the chapter though historically it was the first sensorless (low dynamics) induction motor drive. Consequently, we will deal first with the principle of VC, state observers implementation examples with motion sensors and sensorless. Then the same treatment will be applied to DTFC and FLC, and V_1 / f_1 schemes.

9.2. VECTOR CONTROL — GENERAL FLUX ORIENTATION

Vector control implies independent (decoupled) control of flux-current and torque/current components of stator current through a coordinated change in the supply voltage amplitude, phase and frequency. As the flux variation tends to be slow, especially with current control, constancy of flux should produce a fast torque/current (and torque) response and finally fast speed (position) response. On the other hand, flux level control is essential to avoid magnetic saturation (and heavy core losses) and reduce the core losses at light loads through flux weakening.

There are three distinct flux space-phasors in the induction machine (Chapter 8, Equations (8.92-8.93)): λ_m^b — airgap flux; λ_s^b — stator flux and λ_r^b — rotor flux. Their relationships with currents are

$$\overline{\lambda}_m = L_m \cdot (\overline{i}_s + \overline{i}_r); \quad \overline{\lambda}_s = L_s \overline{i}_s + L_m \overline{i}_r; \quad \overline{\lambda}_r = L_r \overline{i}_r + L_m \overline{i}_s \qquad (9.1)$$

Vector control could be performed with respect to any of these flux space-phasors by attaching the reference system d axis to the respective flux linkage space-phasor direction and by keeping its amplitude under surveillance. A general flux may be defined, with the above cases, as a particular situation [8-9].

In order to facilitate comparisons between the three strategies we introduce new, general, rotor variables $\overline{i}_{ra}$, $\overline{\lambda}_{ra}$

$$\overline{i}_{ra} = \overline{i}_r / a; \quad \overline{\lambda}_{ra} = \overline{\lambda}_r \cdot a; \quad r_{ra} = r_r \cdot a^2 \qquad (9.2)$$

and define a general flux

$$\overline{\lambda}_{ma} = aL_m \left(\overline{i}_s + \overline{i}_{ra} \right) \qquad (9.3)$$

From the complex variable equations (8.50-8.53) (with d/dt = s)

$$\overline{V}_s = r_s \overline{i}_s + s\overline{\lambda}_s + j\omega_1 \overline{\lambda}_s$$

$$\overline{V}_r = r_r \overline{i}_r + s\overline{\lambda}_r + j(\omega_1 - \omega_r)\overline{\lambda}_r \qquad (9.4)$$

$$T_e = \frac{3}{2} p \operatorname{Re}\left(j\overline{\lambda}_s \overline{i}_s^* \right)$$

with ω_1 the general flux space-phasor speed, $\overline{V}_r = 0$. Making use of (9.1)-(9.3) we obtain

$$\overline{V}_s = \left[r_s + (s + j\omega_1)(L_s - aL_m) \right] \cdot \overline{i}_s + (s + j\omega_1) \cdot \overline{\lambda}_{ma} \qquad (9.5)$$

$$0 = -\left[r_r + (s + jS_a\omega_1)(L_r - L_m / a) \right] \cdot \overline{i}_s + \left[r_r + (s + jS_a\omega_1)L_r \right] \cdot \frac{\overline{\lambda}_{ma}}{aL_m} \qquad (9.6)$$

$$T_e = \frac{3}{2}p\,Re\left(j\bar{\lambda}_{ma}\bar{i}_s^*\right) \tag{9.7}$$

with the slip S_a defined for the general flux $\bar{\lambda}_{ma}$

$$S_a = \frac{\omega_1 - \omega_r}{\omega_1} \tag{9.8}$$

The complex variables are the stator current $\bar{i}_s$ and the general flux $\bar{\lambda}_{ma}$.

The reference system is attached to axis d along the general flux (Figure 9.1).

$$\bar{\lambda}_{ma} = \lambda_{ma}; \quad \bar{i}_s = i_{da} + ji_{qa} \tag{9.9}$$

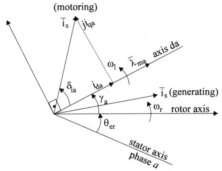

Figure 9.1. General flux orientation axes

According to (9.9), the torque expression (9.7) becomes

$$T_e = \frac{3}{2}p\lambda_{ma}i_{qa}; \quad i_{qa} = i_s \sin\delta_{ia} \tag{9.10}$$

For motoring $T_e>0$ (for $\omega_r>0$) for $\delta_{ia}>0$ and thus generating is obtained with $\delta_{ia}<0$. *In other words, for direct (trigonometric) motion, the stator current leads the general flux for motoring and lags it for generating.*

Also from (9.8) and Figure 9.1

$$\frac{d\gamma_a}{dt} = S_a \omega_a \tag{9.11}$$

We may consider that i_{qa} is the torque current component (9.10) while i_{da} is the flux current component of stator current (Figure 9.1). <u>Their decoupled control is the essence of vector control.</u> This is so because, for constant λ_{ma}, torque control means i_{qa} control and thus flux control means i_{da} control.

9.3. GENERAL CURRENT DECOUPLING

The input (command) variables in vector control are the reference flux $\lambda_{ma}{}^*$ and the reference torque $T_e{}^*$. Consequently, the IM is a two-input system. The value of $\lambda_{ma}{}^*$ is either constant or it depends on speed or on torque, according to an *a priori* criterion of optimization heavily related to application.

General current decoupling means to determine the reference current space-phasor $\bar{i}_s{}^*$ ($i_{da}{}^*$, $i_{qa}{}^*$) based on reference flux $\lambda_{ma}{}^*$ and torque $T_e{}^*$. The reference torque may be the output of a position or speed closed-loop controller in speed (position) control drives or the required torque (speed or time dependent) in torque-control-only drives.

For general current decoupling we make use purely of (9.6)-(9.11), skipping the stator voltage equation, to obtain

$$S_a\omega_1 = \frac{i_{qa}\left(1+s\sigma_a\tau_r\right)}{\left(\dfrac{\lambda_{ma}}{aL_m}-\sigma_a i_{da}\right)\tau_r}; \quad \sigma_a = 1-\frac{L_m}{aL_r}; \quad \tau_r = L_r/r_r \qquad (9.12)$$

$$\left(1+s\sigma_a\tau_r\right)i_{da} = \left(1+s\tau_r\right)\frac{\lambda_{ma}}{aL_m}+S_a\omega_1\sigma_a\tau_r i_{qa} \qquad (9.13)$$

Equations ((9.12)-(9.13)) are illustrated in the general current decoupling network shown in Figure 9.2.

It becomes now possible, through PWM inverter control, to force the motor currents to follow $i_{da}{}^*$ and $i_{qa}{}^*$, in any flux orientation.

It is interesting to see that for $\sigma_a = 0$

$$a = \frac{L_m}{L_r} \qquad (9.14)$$

For $a = \dfrac{L_m}{L_r}$ from ((9.1)-(9.3)) and ((9.12)-(9.13)) we obtain

$$\left(\lambda_{ma}\right)_{a=\frac{L_m}{L_r}} = \frac{L_m}{L_r}\lambda_r; \quad S_a\omega_1 = \frac{i_q}{i_d\tau_r}$$

$$i_{da} = \left(1+s\tau_r\right)\frac{\lambda_r}{L_m}; \quad T_e = \frac{3}{2}p\frac{L_m}{L_r}\lambda_r i_{qa} = \frac{3}{2}\lambda_r{}^2\frac{S_a\omega_1}{r_r}p \qquad (9.15)$$

So only for rotor flux (λ_r) orientation the current decoupling network is simplified, to the form in Figure 9.3.

For $a = \dfrac{L_s}{L_m}$, from ((9.1)-(9.3))

$$\left(\overline{\lambda}_{ma}\right)_{a=\frac{L_s}{L_m}} = \overline{\lambda}_s \tag{9.16}$$

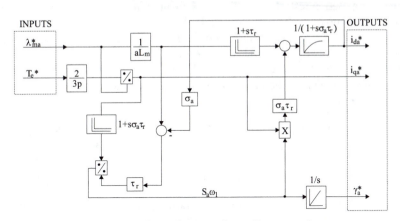

Figure 9.2. General current decoupling network

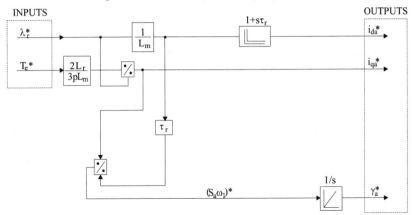

Figure 9.3. Current decoupling network in rotor flux orientation

So we end up with stator flux orientation but

$$\left(\sigma_a\right)_{a=\frac{L_s}{L_m}} = 1 - \frac{L_m^2}{L_s L_r} = \sigma \neq 0 \tag{9.17}$$

Consequently, in stator flux orientation, the current decoupling network retains the complicated form of Figure 9.2. A similar situation occurs with a=1.

$$\left(\overline{\lambda}_{ma}\right)_{a=1} = \overline{\lambda}_m; \quad \left(\sigma_a\right)_{a=1} = 1 - \frac{L_m}{L_r} \tag{9.18}$$

That is airgap flux orientation.

The current network is a feedforward (indirect) method to produce flux orientation. It presupposes a knowledge of machine parameters and an on-line computation effort (in DSP implementation) commensurable to the current decoupling network. The simpler it is, the better.

This is why indirect vector control is most adequate with rotor flux orientation. However simple it is (Figure 9.3), the current decoupling network in rotor flux orientation depends heavily on motor parameters L_m, L_r / L_m and τ_r which, in the end, depend on magnetic saturation level (L_m, τ_r) and rotor temperature (τ_r) [10].

9.4. PARAMETER DETUNING EFFECTS IN ROTOR FLUX ORIENTATION CURRENT DECOUPLING

Let us consider constant slip frequency

$$\left(S\omega_1\right) = \left(S\omega_1\right)^* \tag{9.19}$$

and constant stator current

$$i_s^* = i_s \tag{9.20}$$

But α and β, defined in ((9.21)-9.22)) vary with magnetic saturation and rotor temperature

$$\frac{\tau_r}{\tau_r^*} = \alpha \tag{9.21}$$

$$\frac{L_m}{L_m^*} = \frac{L_r}{L_r^*} = \beta \tag{9.22}$$

For steady-state, from (9.15) we obtain

$$i_{da} = \frac{\lambda_r}{L_m}; \quad i_{da}^* = \frac{\lambda_r^*}{L_m}$$

$$\left(S\omega_1\right) = \left(S\omega_1\right)^* = \frac{i_{qa}}{i_{da}\tau_r} = \frac{i_{qa}^*}{i_{da}^*\tau_r^*}$$

$$T_e = \frac{3}{2}p\frac{L_m}{L_r}\lambda_r i_{qa} = \frac{3}{2}p\frac{L_m^2}{L_r}i_{da}i_{qa} = \frac{3}{2}p\lambda_r^2\frac{S\omega_1}{r_r} \tag{9.23}$$

$$T_e^* = \frac{3}{2}p\frac{\left(L_m^*\right)^2}{L_r^*}i_{da}^*i_{qa}^*; \quad i_s^2 = \left(i_s^*\right)^2 = \sqrt{i_{da}^2 + i_{qa}^2} = \sqrt{\left(i_{da}^*\right)^2 + \left(i_{qa}^*\right)^2}$$

From (9.23) and ((9.21) (9.22))

$$\frac{T_e}{T_e^*} = \alpha\beta \left[\frac{1 + \left[(S\omega_1)^* \tau_r^* \right]^2}{1 + \left[(S\omega_1)^* \alpha\tau_r^* \right]^2} \right] \tag{9.24}$$

$$\frac{\lambda_r}{\lambda_r^*} = \frac{L_m}{L_m^*} \frac{i_{da}}{i_{da}^*} = \sqrt{\frac{\beta}{\alpha} \frac{T_e}{T_e^*}} \tag{9.25}$$

T_e and λ_r are the actual torque and rotor flux produced.

We may now illustrate the influence of α and β on T_e / T_e^* and λ_r / λ_r^*, for constant stator current i_s^* and slip frequency $(S\omega_1)^*$, by considering

$$(S\omega_1)^* \tau_r^* = K \tag{9.26}$$

In general, $K \le 1$; introducing K in (9.16) yields

$$\frac{T_e}{T_e^*} = \alpha\beta \left[\frac{1 + K^2}{1 + K^2\alpha^2} \right] \tag{9.27}$$

$$\frac{\lambda_r}{\lambda_r^*} = \beta \sqrt{\frac{1 + K^2}{1 + K^2\alpha^2}} \tag{9.28}$$

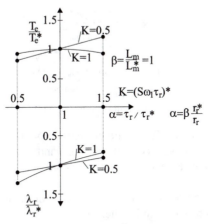

Figure 9.4. Actual / command torque (T_e / T_e^*) and rotor flux (λ_r / λ_r^*) versus rotor time constant detuning ratio α, for current decoupling in rotor flux orientation

It is now clear that both saturation and temperature (α, β) have a monotonous influence on the flux ratio while the influence of α on the torque ratio shows a maximum for $\alpha = 1 / K$

$$\left(\frac{T_e}{T_e^*}\right)_{max} = \beta\frac{1+K^2}{2K} \geq \beta \tag{9.29}$$

Results for $K = 0.5$ and 1.0, $\beta = 1$ an α variable from 0.5 to 1.5 are shown in Figure 9.4.

Being monotonous, the rotor flux detuning may be used to correct the rotor time constant in rotor flux orientation indirect vector control.

The torque detuning is higher when the slip frequency $S\omega_1$ is smaller while the converse is true for rotor flux detuning.

Correcting the rotor time constant τ_r, or for the slip frequency $(S\omega_1)^*$, to compensate for temperature and magnetic saturation detuning effects seems thus imperative in indirect vector current control, especially if reliable precise and quick torque control is required.

9.5. DIRECT VERSUS INDIRECT VECTOR CURRENT DECOUPLING

As seen above, the indirect current decoupling is either complicated (for stator flux orientation) or (and) strongly parameter dependent for rotor flux orientation.

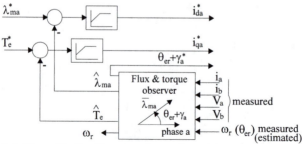

Figure 9.5. Direct (feedback) current decoupling with general flux orientation

To circumvent this difficulty the current decoupling may be performed through direct closed-loop control of flux and torque, based on their calculation through observers to produce the corresponding feedback (Figure 9.5).

In fact, parameter detuning problems are resurfacing into observers but here they proved to be easier to handle.

9.6. A.C. VERSUS D.C. CURRENT CONTROLLERS

Once the reference d-q currents i_{da}^*, i_{qa}^* and flux orientation angle $\theta_{er}+\gamma_a^*$ are known, we have to "translate" these commands into stator currents and to use current controllers to impose these currents through the power electronic converter (PEC).

There are two ways toward this scope:

- through a.c. current controllers;
- through d.c. (synchronous) current controllers [11].

In any case Park transformation is required (Chapter 8, Equation (8.57))

$$\left|P\left(\theta_{er}+\gamma_a\right)\right|=\frac{2}{3}\left|\begin{matrix}\cos\left(-\theta_{er}-\gamma_a\right) & \cos\left(-\theta_{er}-\gamma_a+\frac{2\pi}{3}\right) & \cos\left(-\theta_{er}-\gamma_a-\frac{2\pi}{3}\right)\\ \sin\left(-\theta_{er}-\gamma_a\right) & \sin\left(-\theta_{er}-\gamma_a+\frac{2\pi}{3}\right) & \sin\left(-\theta_{er}-\gamma_a-\frac{2\pi}{3}\right)\end{matrix}\right|$$

(9.30)

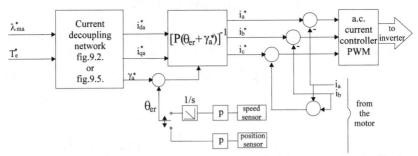

Figure 9.6. Indirect (or direct) vector current control with a.c. current controllers

The current decoupling network in Figure 9.6 may be replaced by the direct (feedback) current decoupling (Figure 9.5) to obtain direct vector current control with a.c. current controllers. A.c. current controllers presuppose, as shown later in this chapter, high switching frequency to produce adequate phase current waveforms and lead to a kind of closed-loop control PWM in the inverter. On the other hand, open-loop PWM to produce close to sinusoidal phase voltages may work well even at lower switching frequencies to produce acceptable phase current waveforms due to the low-pass filter action of the induction machine.

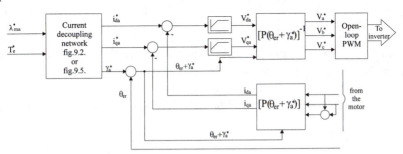

Figure 9.7. Indirect (or direct) vector current control with d.c. (synchronous) current

contollers

D.c. current controllers serve such a solution (Figure 9.7) and are load and frequency independent [11].

In presenting the vector current control we assumed that the IM response is rather instantaneous and speed independent. However, in general, with a voltage source PWM inverter, constant voltages are switched on and off and the current (torque) response depends on speed as the motion induced voltage opposes current variation.

Also, above base speed, the voltage is constant and thus besides current decoupling, voltage decoupling is required to avoid the saturation of current (torque) controllers and thus avoid sluggish response of the drive.

9.7. VOLTAGE DECOUPLING

Let us remember that so far we have not made use of the stator equation (9.5) summarized here for convenience

$$\overline{V}_{sa} = \left[r_s + (s + j\omega_1)(L_s - aL_m) \right] \cdot \overline{i}_{sa} + (s + j\omega_1) \cdot \overline{\lambda}_{ma} \qquad (9.31)$$

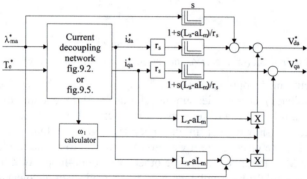

Figure 9.8. Voltage decoupling network for general flux orientation

In (9.31) ω_1 is the instantaneous speed of the general flux space-phasor $\overline{\lambda}_{ma}$ and $\overline{i}_{sa}$ and $\overline{V}_{sa}$ are also written in general flux coordinates. Now, voltage decoupling means to calculate the voltage space-phasor $\overline{V}_{sa}$ required to "produce" the $\overline{i}_{sa}^*$ just calculated through the current decoupling network (Figure 9.8). It should be noted that the voltage decoupling network gets simplified only in stator flux orientation, when $a = L_s / L_m$ and $\overline{\lambda}_{ma} = \overline{\lambda}_s$. This simplified form is shown in Figure 9.9. Equation (9.31) becomes

$$\overline{V}_s = r_s \overline{i}_s + (s + j\omega_1)\overline{\lambda}_s \qquad (9.32)$$

It may thus be considered that, in vector current control, rotor flux orientation is less complicated while, for voltage vector control, stator flux orientation is to be preferred.

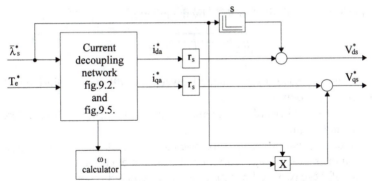

Figure 9.9. Voltage decoupling network for stator flux orientation

Note that even though in vector current control with d.c. current controllers (Figure 9.7.) voltage (open-loop) control (PWM) is performed, there is no e.m.f. (voltage) decoupling (compensation). Consequently, above base speed, at constant voltage, there is no guarantee that the drive will not lose control by requiring higher than available voltage levels.

To facilitate good control above base speed voltage decoupling is required. For implementation, in general, only the motion induced voltage, E, is considered (s = 0 in Figure 9.8.)

$$\overline{E} = j\omega_1\left[\overline{\lambda}_{ma} + \left(L_s - aL_m\right)\overline{i}_{sa}\right] = j\omega_1\overline{\lambda}_s \qquad (9.33)$$

This way combined voltage-current vector control is obtained. However only d.c. current controllers (Figure 9.7) allow for a practical solution (Figure 9.10).

Vector voltage and d.c. current control performed in rotor or stator flux orientation (preferred for wide speed range control), implies, in fact, d.c. current closed-loop control at low speeds and then open-loop voltage control takes over at high speeds (above base speed), in the flux weakening zone.

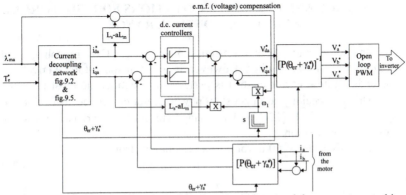

Figure 9.10. Indirect (or direct) combined vector voltage and d.c. current control in general flux orientation

The voltage is limited to V_{smax} by the inverter

$$\sqrt{\left(V_{da}^{*}\right)^{2}+\left(V_{qa}^{*}\right)^{2}} \leq V_{smax} \qquad (9.34)$$

By now we have discriminated between the following main methods to perform vector control:

- Indirect (feedforward) - with flux orientation calculated from reference flux and torque.
- Direct (feedback) - with flux orientation calculated through observers from the motor voltage, current and (sometimes) rotor speed.
  Any of these two methods may be implemented through:
- Vector current control, with a.c. or d.c. (synchronous) current controllers;
- Vector voltage-d.c. current control, with d.c. current controllers and e.m.f. (voltage) decoupling and open-loop (voltage) PWM.

Table 9.1. Summary of most appropriate strategies of vector control for induction motors.

Flux orientation	Current control		Voltage and d.c. current control	Indirect methods	Direct methods	Constant power speed range	
	a.c.	d.c				small	large
rotor flux	X	XX	XX	XX	XXX	XXX	XX
stator flux		X	XXX		XX		XXX
airgap flux	X				X	X	

X - satisfactory;   XX - good;   XXX - very good.

Vector voltage-d.c. current control is, due to the feedforward e.m.f. compensation, less dependent on rotor time constant τ_r or on magnetic saturation, above 10% of base speed. Below that speed, if indirect current decoupling is used, the dependence of response on τ_r and saturation stays, as the e.m.f. is rather low. *At least for torque control, τ_r (rotor time constant) has to be adapted (corrected) if indirect vector control methods are used.*

9.8. VOLTAGE AND CURRENT LIMITATIONS FOR THE TORQUE AND SPEED CONTROL RANGE

Both motors and power electronic converters (PECs) are voltage and current limited or kVA limited. However, in terms of speed-torque envelope, it depends on how vector control is performed to extract the most from the drive. This is especially so in constant power operation for speeds from ω_b to ω_{max} ($\omega_{max}/\omega_b = 2-4$) [12-13]. The voltage-current limits are easy trackable in rotor flux orientation.

For steady-state (s = 0 in rotor flux orientation) from (9.15);

$$\lambda_r = L_m i_d; \quad i_q = S\omega_1 \tau_r i_d \qquad (9.35)$$

$$T_e = \frac{3}{2} p \frac{L_m^{2}}{L_r} i_d i_q \qquad (9.36)$$

Equation (9.35) stresses the conjecture that i_d is the flux current. Using (9.33) we may obtain

$$\overline{\lambda}_s = \overline{\lambda}_{ma} + \left(L_s - aL_m\right)\overline{i}_{sa} \qquad (9.37)$$

For rotor flux orientation $a = L_m / L_r$ and thus,

$$L_s - \frac{L_m^2}{L_r} = \left(L_m + L_{se}\right) - \frac{L_m^2}{\left(L_m + L_{lr}\right)} \approx L_{ls} + L_{lr} = L_{sc} \qquad (9.38)$$

With (9.15) and (9.38), (9.37) becomes

$$\overline{\lambda}_s = \frac{L_m}{L_r}\lambda_r + L_{sc}\overline{i}_s = \frac{L_m^2}{L_r}i_d + \left(L_s - \frac{L_m^2}{L_r}\right)\left(i_d + ji_q\right) =$$

$$= L_s i_d + jL_{sc}i_q = \lambda_d + j\lambda_q \qquad (9.39)$$

The above mathematical manipulations are illustrated in Figure 9.11. Using (9.39) in (9.36) we obtain a new torque expression

$$T_e = \frac{3}{2}p\left(L_s - L_{sc}\right)i_d i_q \qquad (9.40)$$

We may read (9.40) as if the I.M., under constant rotor flux, is a reluctance synchronous motor with L_s as the d axis inductance (L_d) and L_{sc} as the q axis inductance (L_q) : $L_s \gg L_{sc}$.

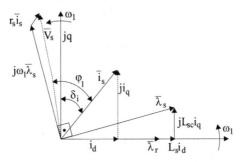

Figure 9.11. Stator flux and voltage in rotor flux coordinates

Constant rotor flux is implicitly maintained during steady-state. From (9.31) and (9.39) the stator voltage becomes

$$\overline{V}_s = r_s\left(i_d + ji_q\right) + j\omega_1\left(L_s i_d + jL_{sc}i_q\right) \qquad (9.41)$$

or

$$\begin{array}{l} V_d = r_s i_d - \omega_1 L_{sc}i_q \\ V_q = r_s i_q + \omega_1 L_s i_d \end{array} \qquad (9.42)$$

Note that $L_s = L_{ls} + L_m$ (the no-load inductance) is dependent on magnetic saturation through L_m for the main flux path and through L_{ls} for the leakage path.

The base speed ω_b is a matter of design options in the sense that the maximum flux level λ_{smax} is related to motor size while the same motor has to provide continuous duty power (torque and power) at rated motor temperature and maximum available voltage V_{smax}

$$\omega_b \approx \frac{V_{smax}}{\lambda_{smax}} \tag{9.43}$$

with

$$i_{dn}^2 + i_{qn}^2 = i_{sn}^2 \tag{9.44}$$

where i_{sn} is the rated current.

For a short time $1.5\ i_{sn}$ or even $2.0\ i_{sn}$ is available in most commercial drives. This, of course, means adequate motor and PEC rating. Well below the base speed the drive is current limited in the sense that there is enough voltage reserve in the PEC, while around and above base speed strict voltage limitations have to be observed for a reliable control.

One problem is how to produce, with V_{smax} at base speed, not only i_{sn} but $(1.5-2.0)i_{sn}$ and higher torque for short time intervals. To solve this problem we have to calculate the maximum torque available at V_{smax} and $\omega_1 = \omega_b$.

Neglecting r_s ($r_s = 0$) in (9.42) we obtain

$$\frac{V_{smax}^2}{\omega_b^2} = \lambda_{smax}^2 = \left(\left(L_s i_d \right)^2 + \left(L_{sc} i_q \right)^2 \right) \tag{9.45}$$

The maximum torque T_{ek} under these conditions is obtained from

$$\left(\frac{\partial T_e}{\partial i_d} \right) = 0 \tag{9.46}$$

with

$$i_q = \frac{1}{L_{sc}} \sqrt{V_{smax}^2 - \omega_b^2 \left(L_s i_d \right)^2} \tag{9.47}$$

Finally we obtain

$$i_{dk} = \frac{\lambda_{smax}}{\sqrt{2}L_s}; \quad i_{qk} = \frac{\lambda_{smax}}{\sqrt{2}L_{sc}} \tag{9.48}$$

$$T_{ek} = \frac{3}{2} p \left(1 - \frac{L_{sc}}{L_s} \right) \cdot \frac{\lambda_{smax}^2}{2L_{sc}}; \quad i_{sk} = \frac{\lambda_{smax}}{\sqrt{2\left(L_s^2 + L_{sc}^2 \right)}} \tag{9.49}$$

For high values of peak torques the short circuit (transient) inductance should be small by design. Note also that (9.48) means

$$\lambda_{dk} = \lambda_{qk} = \frac{\lambda_{smax}}{\sqrt{2}}; \quad \frac{i_{dk}}{i_{qk}} = \frac{L_{sc}}{L_s} \tag{9.50}$$

$$(S\omega_1)_k = \frac{i_q}{i_d \tau_r} = \frac{L_s}{L_{sc}} \frac{r_r}{L_r} \approx \frac{r_r}{L_{sc}} \tag{9.51}$$

The power factor angle φ_1 (Figure 9.11) is

$$\varphi_1 = \tan^{-1} \frac{i_d}{i_q} + \tan^{-1} \frac{L_{sc} i_q}{L_s i_d} \tag{9.52}$$

The peak current i_{sk} determines the motor rating and sizing. The i_d-i_q current components are limited by flux limitation as in (9.45) which represents an ellipse whose span diminishes with increasing speed.

On the other hand, i_d-i_q are limited by the current limitations i_{sn} and i_{sk} as in (9.44), which depicts a circle (Figure 9.12).

Point A_k corresponds to the peak torque T_{ek} and current i_{sk} at base speed ω_b and full voltage V_{smax} and

$$\tan \delta_{ik} = \frac{i_{dk}}{i_{qk}} = \frac{L_{sc}}{L_s} \tag{9.53}$$

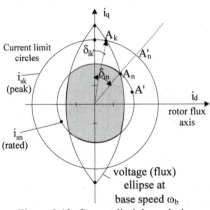

Figure 9.12. Current limit boundaries

The corresponding power factor angle φ_{1k} is

$$\varphi_{1k} = \tan^{-1}\left(\frac{i_{dk}}{i_{qk}}\right) + \tan^{-1}\left(\frac{L_{sc}}{L_s}\frac{i_{qk}}{i_{dk}}\right) = \tan^{-1}\left(\frac{L_{sc}}{L_s}\right) + 45° \tag{9.54}$$

So, for maximum torque per given flux, (given voltage and frequency), the power factor is slightly below 0.707 ($\cos \varphi_{lk} < 0.707$).

The current angle δ_{in} for rated (base) torque and current is to be obtained by observing (9.40) at rated (base) torque T_{eb} and (9.44) at rated current i_{sn}.

In an existing motor L_s and L_{sc} are given for a reasonable degree of saturation. If the values of i_{dn}, i_{qn} correspond to point A' in Figure 9.12 situated outside the hatched area, it means that they may not be met.

For a new motor it seems reasonable to choose the rated conditions for maximum power factor

$$\frac{\partial \varphi_1}{\partial \left(i_d / i_q \right)} = 0 \tag{9.55}$$

From (9.54)-(9.55) it follows that

$$\tan \delta_{im} = \frac{i_{dm}}{i_{qm}} = \sqrt{\frac{L_{sc}}{L_s}} \tag{9.56}$$

$$\left(\cos \varphi_1 \right)_{max} = \frac{1 - L_{sc} / L_s}{1 + L_{sc} / L_s} \tag{9.57}$$

The above rationale only introduces us to the induction motor design for variable speed drives. For constant power P_e, above base speed ω_b and for given stator current i_s, we should meet the conditions

$$P_{eb} = T_{eb} \cdot \frac{\omega_b}{p} = T_e \cdot \frac{\omega_1}{p} = \frac{3}{2} \omega_1 \left(L_s - L_{sc} \right) i_d i_q; \quad \omega_1 > \omega_b \tag{9.58}$$

and

$$i_s^2 \le i_d^2 + i_q^2; \quad S\omega_1 \le \frac{i_q}{i_d \tau_r} \tag{9.59}$$

Starting from the initial point (A_n on Figure 9.12) for base speed, according to (9.58)-(9.59) we have to calculate i_d, i_q, $S\omega_1$, for increasing frequency ω_1 and with voltage limitation

$$\left(L_s i_d \right)^2 + \left(L_{sc} i_q \right)^2 \le \left(V_{s\,max} / \omega_1 \right)^2; \quad \omega_1 > \omega_b \tag{9.60}$$

When ω_1 increases so does $S\omega_1$, the limit being the critical value corresponding to maximum torque per flux condition (9.51). Beyond that point, at higher speeds, $S\omega_1$ remains constant, at its critical value: $\left(S\omega_1 \right)_k \approx r_r / L_{sc}$. It is not advisable to start from A_k (δ_{ik}) because in that case δ_i remains equal to δ_{ik} and practically no constant power zone speed range is available. Consequently, placing ourselves at A_n' (Figure 9.12), for example, for peak (short duration) torque and current, leaves us a zone for constant speed. Thus the flux weakening zone ($\omega_b > \omega_1$) may be divided in two regions. The first zone of constant electromagnetic power P_{eb} $(\omega_b < \omega_1 < \omega_{max1})$, is characterized by variable current angle δ_i (δ_i decreasing from δ_{in} to δ_{ik}) or $S\omega_1$ increasing from $(S\omega_1)_b$ to $(S\omega_1)_k$.

In the second zone, $\omega_{max1} < \omega_1 < \omega_{max2}$, the motor works at constant current angle δ_{ik} ($(S\omega_1)_k = r_r / L_{sc}$) and is not able (anymore) to produce constant power (Figure 9.13).

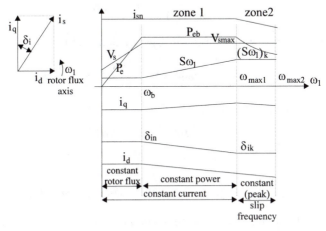

Figure 9.13. Flux weakening zone and the constant power subzone (region)

The lower the critical current angle δ_{ik}, that is, the higher the critical slip frequency $(S\omega_1)_k$ or the smaller the short-circuit inductance L_{sc} (for given rotor resistance), *the wider the constant power zone.*

Temperature and magnetic saturation variation introduce some departure from the above approximate analysis, but this is beyond our scope here. In terms of torque transients, rotor flux and stator flux vector control schemes produce similar performance in the flux weakening zone.

Example 9.1.
Let us consider a standard induction motor with the following parameters: $r_s = r_r = 0.2\Omega$, $L_{ls} = L_{lr} = 0.005H$, $L_m = 0.075H$, rated line voltage 220V (rms) (star), rated frequency $\omega_b = 2\pi 60$ rad/s, rated slip $S_n = 0.02$, p = 2 pole pairs. Determine the ideal no-load current i_{on}; for $i_{on} = i_{dn} / \sqrt{2}$, calculate i_{qn} for rated slip, the rotor flux, stator flux components and the rated electromagnetic torque T_{eb} and power P_{eb}; calculate the maximum frequency ω_{max1} for which constant electromagnetic power P_{eb} may be produced.
Solution:
From the equivalent circuit (Chapter 8, Figure 8.11), for S = 0,

$$\overline{V}_{sn} = \overline{i}_{s0}\left(r_s + j\omega_b L_s\right) \tag{9.61}$$

Neglecting r_s

$$i_{s0} = i_{dn} = \frac{V_{sn}}{\omega_b L_s} = \frac{\left(220/\sqrt{3}\right)\sqrt{2}}{2\pi60\cdot0.08} = 5.948 \text{ A} \qquad (9.62)$$

Consequently, the ideal no-load phase current i_{on} (rms) is

$$i_{on} = \frac{i_{dn}}{\sqrt{2}} = \frac{5.948}{1.41} = 4.218 \text{ A} \qquad (9.63)$$

The torque current i_{qn} - in rotor flux orientation - is (9.15)

$$i_{qn} = S_n \cdot \omega_1 \cdot \tau_r \cdot i_{dn} = 0.02 \cdot 2\pi60 \cdot \frac{0.08}{0.2} \cdot 5.948 = 17.929 \text{ A} \qquad (9.64)$$

The rotor flux, λ_{rn} (9.15), is

$$\lambda_{rn} = L_m i_{dn} = 0.075 \cdot 5.948 = 0.4461 \text{ Wb} \qquad (9.65)$$

The stator flux $\overline{\lambda}_s$ (9.39) is

$$\overline{\lambda}_{sn} = L_s i_{dn} + jL_{sc} i_{qn} = 0.080 \cdot 5.948 + j0.01 \cdot 17.929 = 0.4758 + j0.1793 \qquad (9.66)$$

The rated electromagnetic torque T_{eb} (9.40) is

$$T_{eb} = \frac{3}{2}p\left(L_s - L_{sc}\right)i_{dn}i_{qn} =$$
$$= \frac{3}{2}\cdot2\cdot(0.08-0.01)\cdot5.948\cdot17.929 = 22.3947 \text{ Nm} \qquad (9.67)$$

And the electromagnetic power P_{eb} (9.58) is

$$P_{eb} = T_{eb}\cdot\frac{\omega_b}{p} = 22.3947\cdot\frac{2\pi60}{2} \approx 4220 \text{ W} \qquad (9.68)$$

It is known that for ω_{max1} (Figure 9.13) the critical value of slip frequency is reached (9.51)

$$\left(S\omega_1\right)_k = \frac{r_r}{L_{sc}} = \frac{0.2}{0.01} = 20 \text{ rad/s} \qquad (9.69)$$

The corresponding electromagnetic power P_{eb} at ω_{max1} is (9.58)

$$P_{eb} = T_{ek}\cdot\frac{\omega_{max1}}{p} = \frac{3}{2}p\left(1 - \frac{L_{sc}}{L_m}\right)\cdot\frac{V_{sn}^{2}}{\omega_{max1}^{2}}\cdot\frac{1}{2L_{sc}}\cdot\frac{\omega_{max1}}{p} \qquad (9.70)$$

$$\omega_{max1} = V_{sn}^2 \cdot \frac{3\left(1-\dfrac{L_{sc}}{L_s}\right)}{4L_{sc}P_{eb}} = \left(\frac{220\sqrt{2}}{\sqrt{3}}\right)^2 \frac{3\left(1-\dfrac{0.01}{0.08}\right)}{4\cdot 0.01\cdot 4220} = 501.77 \text{ rad/s} \quad (9.71)$$

Note that the base frequency $\omega_b = 2\pi 60 = 376.8$ rad/s and, consequently, the constant power zone covers a speed ratio $\omega_{max1} / \omega_b = 501.77 / 376.8 = 1.33$. The corresponding currents, i_{dk}, i_{qk} are obtained from

$$i_{qk} / i_{dk} = 20 \cdot \frac{0.08}{0.2} = 8 \quad (9.72)$$

$$\lambda_{smax} = \frac{V_{sn}}{\omega_{max1}} = i_{dk} \sqrt{L_s^2 + \left(L_{sc} i_{qk} / i_{dk}\right)^2} = \qquad (9.73)$$
$$= i_{dk} \sqrt{0.08^2 + (0.01\cdot 8)^2} = 0.1128 \cdot i_{dk}$$

$$i_{dk} = \frac{220\sqrt{2}}{\sqrt{3}\cdot 501.77} \cdot \frac{1}{0.1128} = 3.1679 \text{ A} \quad (9.74)$$

$$i_{qk} = 8 \cdot i_{dk} = 8\cdot 3.1679 = 25.34 \text{ A} \quad (9.75)$$

For the base speed, the rated current i_{sn} is

$$i_{sn} = \sqrt{i_{dn}^2 + i_{qn}^2} = \sqrt{5.948^2 + 17.929^2} = 18.89 \text{ A} \quad (9.76)$$

$$i_{sk} = \sqrt{i_{dk}^2 + i_{qk}^2} = \sqrt{3.1679^2 + 25.34^2} = 25.537 \text{ A} \quad (9.77)$$

So even this narrow constant power speed range ratio $\omega_{max1} / \omega_b = 1.33$ is obtained at the price of higher stator current which implies lower power factor and, perhaps, efficiency. Reducing L_{sc} is a sure way to increase the value of ω_{max1} (9.71) and thus a wider constant power speed range is obtained. Leaving a voltage reserve at base speed or switching from a star to delta winding connection in the motor are two practical methods to widen the constant power zone.

9.9. IMPRESSING VOLTAGE AND CURRENT WAVEFORMS THROUGH PWM

As already mentioned, for vector control, PWM is either closed-loop, that is a.c. current control driven (Figure 9.6), or open-loop, that is voltage waveform driven (Figure 9.10). PWM is thus a key part of vector control with voltage source inverters and has been given worldwide attention [15] with quite a few commercial implementation schemes.

9.9.1. Switching state voltage vectors

A PWM voltage source inverter (Figure 9.14) produces in the a.c. motor symmetrical rectangular voltage potentials V_{ap}, V_{bp}, V_{cp} provided that the conducting PES triplet is on for 60° electrical degrees. This means six pulses per period or only six switchings per period.

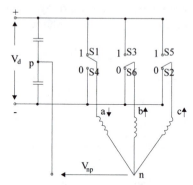

Figure 9.14. PWM voltage source inverter:

one switch per leg conducting at any time

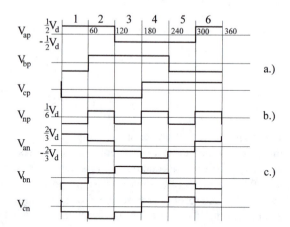

Figure 9.15. Voltage waveforms for six switchings per period:

a.) Voltage potentials at motor terminals, b.) neutral potential, c.) phase voltages

The neutral potential V_{np} is either positive or negative as two upper or lower P.E.Ss in the inverter lag are on. V_{np} has three times the fundamental frequency and thus contains the triple harmonics which do not appear in the phase voltages V_{an}, V_{bn}, V_{cn}. This is true also when the conduction angle of various PES triplets is less than 60° via pulse width modulation (PWM) performed to modify the fundamental of the motor phase voltages.

The maximum voltage fundamental $V_{1six-step}$ is obtained for six-pulse switching and the modulation index m is

$$m = \frac{V_1}{V_{1six-step}}; \quad 0 \le m \le 1 \tag{9.78}$$

where from Figure 9.15a

$$V_{1six-step} = \frac{2}{\pi} V_d \tag{9.79}$$

The ideal maximum modulation index is equal to unity. Various PWM schemes allow an $m_{max}<1$ which represents an important performance criterion as the inverter maximum kVA depends on the maximum voltage at motor terminals.

The PWM voltage source inverter allows for 6 SCR triplets that produce nonzero voltage space-vectors (Figure 9.14): (100, 110, 010, 011, 001, 101) and two zero voltage space-vectors (111, 000).

We may use space-phasors to describe the 6 nonzero switching situations as

$$\overline{V}_s(t) = \frac{2}{3}\left(V_{an}(t) + V_{bn}(t) \cdot e^{j\frac{2\pi}{3}} + V_{cn}(t) \cdot e^{-j\frac{2\pi}{3}} \right) \tag{9.80}$$

with V_{an}, V_{bn}, V_{cn} from Figure 9.15.c, we obtain six space-phasors, 60° apart (Figure 9.16).

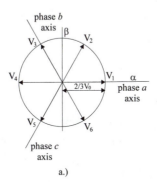

a.)

$\overline{V}_s$	V_1	V_2	V_3	V_4	V_5	V_6	V_0, V_7
	100	110	010	011	001	101	111, 000
V_{an}	$2/3V_0$	$1/3V_0$	$-1/3V_0$	$-2/3V_0$	$-1/3V_0$	$1/3V_0$	0
V_{bn}	$-1/3V_0$	$1/3V_0$	$2/3V_0$	$1/3V_0$	$-1/3V_0$	$-2/3V_0$	0
V_{cn}	$-1/3V_0$	$-2/3V_0$	$-1/3V_0$	$1/3V_0$	$2/3V_0$	$1/3V_0$	0

b.)

Figure 9.16. a.) Voltage space-vectors, b.) The corresponding phase voltages

Timing the eight voltage space-vectors V_1, ..., V_8 is, in fact, the art of PWM. *Impressing the voltage commands* required by the vector control strategies may be done directly by *open-loop PWM. Impressing the current commands* through the same voltage space-vectors is done through *closed-loop PWM.* Among various PWM methods — treated extensively in the power electronics literature [15-16] — we deal with the two open-loop and two closed-loop PWM strategies considered here most representative.

9.9.2. Open-loop space-vector PWM

In space-vector PWM the reference voltage space-vector of the motor is treated directly and not phase by phase. The reference voltage space-vector V_i^* is sampled at a fixed clock frequency $2f_s$ (Figure 9.17a) being constructed through adequate timing of adjacent nonzero inverter voltage space-vectors V_1 to V_6 and the zero voltage space-vectors V_0, V_7 (Figure 9.17b)

$$2f_s(t_1 V_i + t_2 V_{i+1}) = V_s^*(t) \tag{9.81}$$

$$t_0 = \frac{1}{2f_s} - t_1 - t_2 \tag{9.82}$$

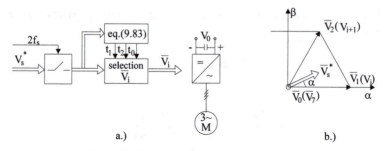

a.)   b.)

Figure 9.17. Open-loop space-vector PWM:

a.) structural diagram; b.) voltage space-vector in the first sector.

The respective timings t_1, t_2 are

$$t_1 = \frac{1}{2f_s} V_s^*(t) \frac{2\sqrt{3}}{\pi} \sin(60^0 - \alpha) \tag{9.83}$$

$$t_2 = \frac{1}{2f_s} V_s^*(t) \frac{2\sqrt{3}}{\pi} \sin \alpha \tag{9.84}$$

In fact, this technique produces an average of three voltage space-vectors V_i, V_{i+1} and V_0 (V_7) over a subcycle $T = 1/2f_s$.

For the minimum number of commutations, with V_1^* in the first sector, the switching sequence is

$$V_0(t_0/2)...V_1(t_1)...V_2(t_2)...V_7(t_0/2)... \qquad (9.85)$$

in all odd subcycles (of all 6 sectors) and

$$V_7(t_0/2)...V_2(t_2)...V_1(t_1)...V_0(t_0/2) \qquad (9.86)$$

for all even subcycles of all sectors.

Other sequences [17] might be imagined. For example,

$$V_0(t_0/3)...V_1(t_1/3)...V_2(t_2/3)...V_2(t_2/3)...V_1(t_1/3)...V_0(t_0/3) \qquad (9.87)$$

The loss factor is smaller in sequence (9.87) modified space-vector PWM than in ((9.85)-(9.86)).

The space-vector PWM produces high performance but requires prediction (online computation) of the reference voltage space-vector $V^*(t)$.

Carrier-based (or fixed frequency) PWM exhibits selected harmonics especially around the carrier frequency and thus produces increased noise in the machine. Distributing the harmonics energy over a large frequency band reduces noise. Random PWM does just that [18] — Figure 9.18.

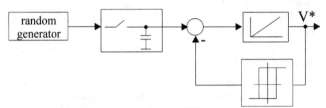

Figure 9.18. Random PWM principle

When the carrier signal reaches one of its peak values, its slope is reversed by a hysteresis block and a sample is taken from the random generator (Figure 9.18) which triggers an additional variation on the slope. This way the duration of subcycles is obtained while only the average switching frequency remains constant.

Overmodulation

The zero voltage timing decreases with the increase of reference voltage. This means that the reference voltage $\overline{V}_s^*$ touches the outer hexagon of the inverter space-vectors $\overline{V}_{1,...,6}$ (Figure 9.17b).

At this point the modulation index reaches $m_{max1} \approx 0.9$, and the next step would be six-pulse operation. Special overmodulation techniques are required for a smooth transition up to $m_{max2} = 1$ [15]. To further improve performance, optimized open-loop PWM methods have been introduced, especially for low switching frequency (GTOs) [15].

Dead time: effect and compensation

In order to prevent short-circuiting an inverter leg, there should be a lock-out time T_d between the turn-off of one PES and the turn-on of the next. T_d should be larger than the maximum particle storage time of the PES, T_{st}. The effect of the lock-out time T_d is a distortion ΔV on the reference voltage V_s^*.

$$V_{av} = V_s^* - \Delta V; \quad \Delta V = \frac{T_d - T_{st}}{T_d} \overline{sigi}_s \qquad (9.88)$$

The voltage distortion ΔV changes sign with current space-vector $\overline{sigi}_s$ function

$$\overline{sigi}_s = \frac{2}{3}\left[sign(i_a) + e^{j\frac{2\pi}{3}} sign(i_b) + e^{-j\frac{2\pi}{3}} sign(i_c) \right] \qquad (9.89)$$

and is proportional to safety time $T_d - T_{st}$.

This voltage distortion is considered by the fact that the on-time of the upper bridge arm is shortened by $T_d - T_{st}$ for positive current and is increased by the same amount for negative sign of current.

If closed-loop (current controller) PWM is used, a compensator of dead time may not be required.

For open-loop PWM, dead time compensation is mandatory to avoid voltage distortion caused by electromechanical instabilities at low frequency (speed). As the delay is present every PWM cycle, the errors increases with speed (frequency) so dead time compensation is required at high speed.

Hardware or software compensation is used [16]. This is added to the pulse before the dead time generation for positive current and subtracted from the pulse for negative current.

9.9.3. Closed-loop PWM

Closed-loop PWM involves, in general, current or flux closed-loop control and may be left nonoptimal or real-time optimized.

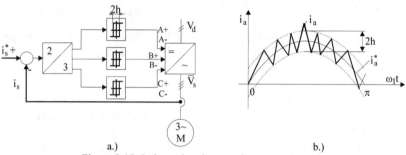

a.)   b.)

Figure 9.19. Independent hysteresis current control:
a.) signal flow diagram, b.) phase current waveform

Hysteresis current control may be performed either with three independent (phase) current (a.c.) controllers (Figure 9.19) or through coordinated control of the current space-phasor error $\overline{\Delta i}_s$ in stator coordinates (a.c.).

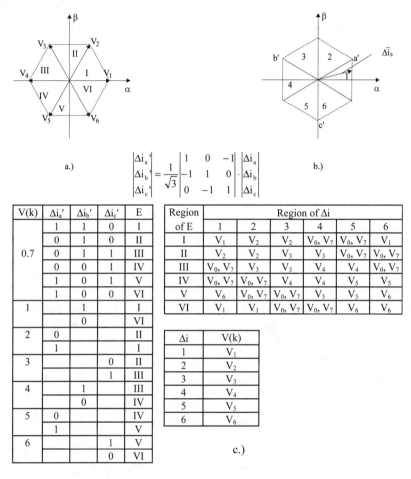

$$\begin{vmatrix} \Delta i_a{}' \\ \Delta i_b{}' \\ \Delta i_c{}' \end{vmatrix} = \frac{1}{\sqrt{3}} \begin{vmatrix} 1 & 0 & -1 \\ -1 & 1 & 0 \\ 0 & -1 & 1 \end{vmatrix} \cdot \begin{vmatrix} \Delta i_a \\ \Delta i_b \\ \Delta i_c \end{vmatrix}$$

V(k)	$\Delta i_a{}'$	$\Delta i_b{}'$	$\Delta i_c{}'$	E
0.7	1	1	0	I
	0	1	0	II
	0	1	1	III
	0	0	1	IV
	1	0	1	V
	1	0	0	VI
1		1		I
		0		VI
2	0			II
	1			I
3			0	II
			1	III
4		1		III
		0		IV
5	0			IV
	1			V
6			1	V
			0	VI

Region	Region of Δi					
of E	1	2	3	4	5	6
I	V_1	V_2	V_2	V_0, V_7	V_0, V_7	V_1
II	V_2	V_2	V_3	V_3	V_0, V_7	V_0, V_7
III	V_0, V_7	V_3	V_3	V_4	V_4	V_0, V_7
IV	V_0, V_7	V_0, V_7	V_4	V_4	V_5	V_5
V	V_6	V_0, V_7	V_0, V_7	V_5	V_5	V_6
VI	V_1	V_1	V_0, V_7	V_0, V_7	V_6	V_6

Δi	V(k)
1	V_1
2	V_2
3	V_3
4	V_4
5	V_5
6	V_6

Figure 9.20. Current vector hysteresis control
a., b.) $\overline{E}$ zone detection; c.) Switching tables after Reference [19].

When the phase current error $\Delta i_a = i_a{}^* - i_a > +h$, the upper inverter leg PES, (A+), is turned on, while, when $\Delta i_a < -h$, the lower leg PES (A–) is turned on. The same procedure is followed independently in phases b and c. Evidently, no zero voltage space-vector $V_0(V_7)$ may be applied. Also, the switching frequency f_s is variable, depending on the hysteresis band 2h, motor parameters and speed. The value of h may be adjusted to maintain the average switching frequency between acceptable limits, to limit current harmonics, noise, etc. The absence of zero voltage vector requires high switching frequency at low fundamental frequency (speed) — low voltage amplitude — that is, low motor speeds; subharmonics may also occur.

In order to reduce the switching frequency and decrease current harmonics, an appropriate nonzero or zero voltage vector of the inverter may be applied, based on current phasor error $\Delta \bar{i}_s$ and its derivative $d\,\Delta\bar{i}_s/dt$ position corroborated with the e.m.f. vector $\bar{E}$ and the existing voltage vector $\bar{V}_i$ position in one of the six 60°-wide sectors.

A table of optimal switchings may thus be defined [19] based on the machine equation in stator coordinates

$$L_{sc}\frac{d\Delta\bar{i}_s}{dt} \approx \bar{E}-\bar{V}_i; \quad \Delta\bar{i}_s = i_s^* - i_s \tag{9.90}$$

$$\bar{E} = \bar{E}_0 + L_{sc}\frac{d\bar{i}^*}{dt} + r_s\bar{i}^* \tag{9.91}$$

$\bar{E}_0$ is $\bar{E}$ when $\bar{i}_s = i_s^*$.

Equation (9.90) shows that $d\,\Delta\bar{i}_s/dt$ is determined solely by the choice of $\bar{V}_i$. The position of $\bar{E}$ is in one of the 6 sectors (Figure 9.20), which may be found knowing only the position of $\Delta\bar{i}_s$ and of the applied $\bar{V}_i$ (Figure 9.20a, b).

The new voltage vector to be applied to the inverter should provide the smallest $\Delta\bar{i}_s$ for steady-state (to reduce current harmonics) or the largest $\Delta\bar{i}_s$ for transients (to produce fastest current (torque) response) — Figure 9.20c. Results [19] obtained on a 1.5kW motor compare favorably with the case of independent hysteresis controllers [19]. Fast response is preserved while harmonics content and noise level are reduced.

However, a.c. controllers are shown to be load, motor parameter and frequency (speed) dependent while d.c. (d-q) current controllers are rather independent of frequency and crosscoupling effects [11]. D.c. (synchronous or d-q) current controllers are better and may be implemented in stator (a.c.) coordinates by transforming their equation (in PI form) in flux coordinates [11]

$$\bar{V}_s^e = K\left(\tau_i + \frac{1}{s}\right)\left(\bar{i}_s^{e*} - \bar{i}_s^e\right) \tag{9.92}$$

through Park transformation $e^{-j\theta_1}$, to the final form

$$\bar{V}_s = \bar{x}_s + K\tau_i\left(\bar{i}_s^* - \bar{i}_s\right); \quad \frac{d\theta_1}{dt} = \omega_1 \tag{9.93}$$

$$s\bar{x}_s = K\left(\bar{i}_s^* - \bar{i}_s\right) + j\omega_1 \bar{x}_s \qquad (9.94)$$

These equations may be conveniently implemented, though for crosscoupling compensation, the value of the flux speed ω_1 has to be calculated. [11, 17].

Note: As shown later in this chapter direct torque and flux control (DFTC) may be assimilated with d-q (synchronous) current and voltage closed-loop random PWM thus yielding high-grade performance both at low speeds and at high speeds (flux weakening zone, $\omega_1 > \omega_b$).

Now that most knowledge for <u>indirect</u> vector control has been presented, a complete numerical example is given below.

9.10. INDIRECT VECTOR A.C. CURRENT CONTROL — A CASE STUDY

We will present here the simulation of a feedforward (indirect) vector current control system for induction motors. The example was implemented in MATLAB/SIMULINK simulation program. The motor model was integrated in two blocks, the first represents the *current and flux calculation module* in d-q axis, the second represents the *torque, speed and position computing module*.

Changing of motor parameters for different simulations is as simple as possible; by clicking on their blocks, a dialog box appears and you can change them by modifying their default values.

The current decoupling network was integrated in a single block. At its first input you have to set the value of the flux psim*. You can modify the speed regulator (*PI regulator*) parameters (amplification and integration time constant, default values are $K_i = 10$, $T_i = 0,8s$) to study the behavior of the system. We have included a *Commutation Frequency* block to satisfy the sampling criteria (to control the commutation frequency). The integration step can be modified from the Simulink's *Simulation / Parameters*.

To find out the structure of each block presented above, unmask them (*Options/Unmask*). The *Park axis transformation* has been integrated in the Park transform block.

The block structure of the electric drive system is presented in Figures 9.21-9.23.

The motor used for this simulation has the following parameters: $P_n = 1100W$, $V_{nf} = 220V$, $2p = 4$, $r_s = 9.53\Omega$, $r_r = 5.619\Omega$, $L_{sc} = 0.136H$, $L_r = 0.505H$, $L_m = 0.447H$, $J = 0.0026 kgfm^2$.

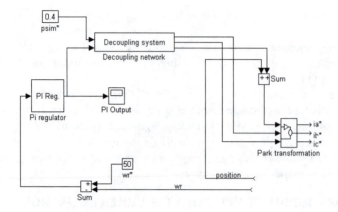

Figure 9.21. The indirect vector a.c. current control system for IMs

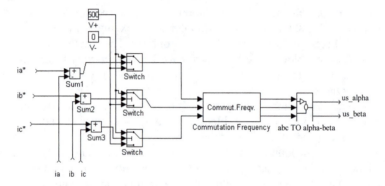

Figure 9.22. The a.c. current controllers

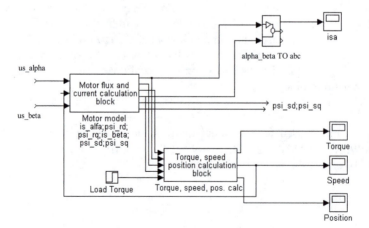

Figure 9.23. The motor space-phasor model

The following figures represent the speed, torque, current and flux responses for *starting transients and a load torque applied at 0.4s*.

Speed transients (Figure 9.24) are quick and smooth even immediately after load torque application at t = 0.4s. Torque response (Figure 9.25) is fast and the torque pulsations are moderate for a chopping frequency f_0 = 8kHz. Phase current waveform (Figure 9.26) also looks smooth with moderate harmonics content. The stator flux (Figure 9.27) and rotor flux (Figure 9.28) responses are somewhat similar, but the rotor flux transients are slower, as expected.

After starting from zero current and speed, the rotor flux settles to the reference value with very small transients when load torque occurs at t = 0.4s.

We should mention that the parameters were fully tuned. The influence of parameter detuning could be investigated also. This is beyond our scope here.

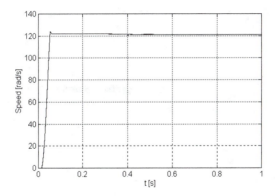

Figure 9.24. Speed transient response

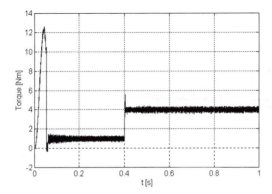

Figure 9.25. Torque response

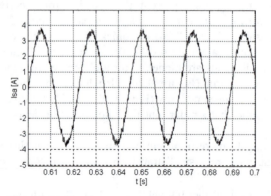

Figure 9.26. Phase current waveform under steady-state

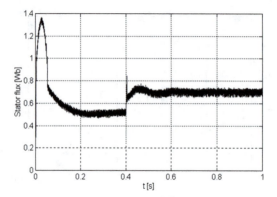

Figure 9.27. Stator flux amplitude

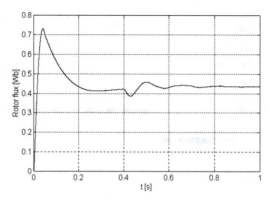

Figure 9.28. Rotor flux amplitude

9.11. FLUX OBSERVERS FOR DIRECT VECTOR CONTROL WITH MOTION SENSORS

The motor stator or airgap flux space-phasor amplitude λ_{ma} and its instantaneous position — $\theta_{er}+\gamma_a^*$ — with respect to the stator phase a axis have to be computed on-line, based on measured motor voltages, currents and, when available, rotor speed. The torque may be calculated from flux and current space-phasors and thus once the flux is computed and stator currents measured, the torque problem is solved.

So flux observation is paramount for direct vector control. We will present a few implementations of flux observers from simple to complex solutions.

Flux observers may be open-loop or close-loop.

9.11.1. Open-loop flux observers

Open-loop flux observers are based on the voltage model or on the current model. Voltage model makes use of stator voltage equation in stator coordinates (from (9.32) with $\omega_1 = 0$)

$$\overline{V}_s^{\,s} = r_s \overline{i}_s^{\,s} + s\overline{\lambda}_s^{\,s} \tag{9.95}$$

From (9.37) with $a = L_m / L_r$

$$\overline{\lambda}_r^{\,s} = \frac{L_r}{L_m}\left(\overline{\lambda}_s - L_{sc}\overline{i}_s\right) \tag{9.96}$$

Both stator flux, $\overline{\lambda}_s$, and rotor flux, $\overline{\lambda}_r$, space-phasors, may thus, in principle, be calculated based on $\overline{V}_s$ and $\overline{i}_s$ measured. The corresponding signal flow diagram is shown in Figure 9.29.

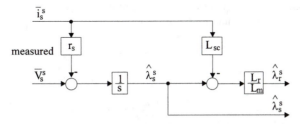

Figure 9.29. Voltage-model open-loop flux observer (stator coordinates)

Unfortunately the integration term and the variation of stator resistance with temperature make this observer hardly practical under 2Hz without special corrections.

On the other hand, the current model for the rotor flux space-phasor is based on the rotor equation in *rotor coordinates* ($\omega_b = \omega_r$)

$$-r_r \bar{i}_s^r + (r_r + sL_r)\frac{\bar{\lambda}_r^r}{L_m} = 0 \tag{9.97}$$

Two coordinate transformations — one for current and the other for rotor flux — are required to produce results in stator coordinates. This time, the observer works even at *zero frequency* but is very sensitive to the detuning of parameters L_m and τ_r due to temperature and magnetic saturation variation. Besides, it requires a rotor speed or position sensor. Parameter adaptation is a solution.

The corresponding signal flow diagram is shown in Figure 9.30.

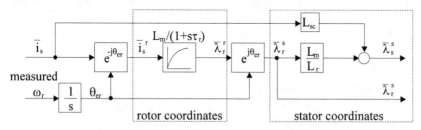

Figure 9.30. Current control open-loop flux observer

More profitable, however, it seems to use closed-loop flux observers.

9.11.2. Closed-loop flux observers

We start by combining the voltage and current models to put together their advantages. At high speeds the voltage model is better as the stator resistance influence is small while, at low speeds, the current model is better as it works even at zero frequency, during d.c. braking (Figure 9.31). The transition between the two models is decided by the two coefficients K1 and K2 such that at zero frequency only the current model remains active while at high frequency the voltage model prevails.

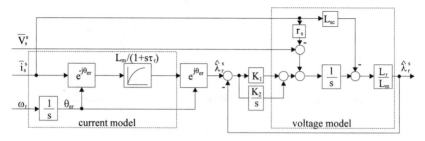

Figure 9.31. Close loop voltage and current model rotor flux observer

Stator flux may simply be calculated from rotor flux using (9.96). Values $K_1 = 33 \text{rad/s}$ and $K_2 = 90(\text{rad/s})^2$ are practical. More on the above flux observers are to be found in [20]. Many other flux observers have been proposed. Among them, the third flux (voltage) harmonic estimator [21] and Gopinath observer [22], model reference adaptive and Kalman filter observers.

They all require notable on-line computation effort and knowledge of induction motor parameters. Consequently, they seem more appropriate when used together with speed observers for sensorless induction motor drives as shown in the corresponding paragraph, to follow after the next case study.

9.12. INDIRECT VECTOR SYNCHRONOUS CURRENT CONTROL WITH SPEED SENSOR - A CASE STUDY

The simulation results of a vector control system with induction motor based on d.c. current control (Fig. 9.7) are now given. The simulation of this drive is implemented in MATLAB-SIMULINK. The motor model was integrated in two blocks, the first represents the *current and flux calculation module* in d-q axis (Figure 9.32), the second represents the *torque, speed and position computing module* (Figure 9.33).

The commutation table was implemented in MATLAB as a function. Changing of motor parameters for different simulations is simple. By clicking on these blocks, a dialog box appears and you can change them by modifying their default values.

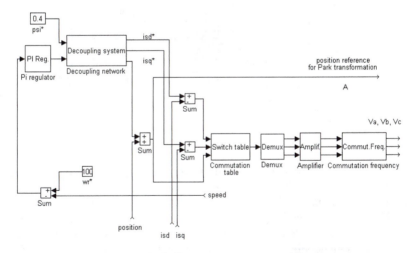

Figure 9.32. The indirect vector current control system

The current decoupling network was integrated in a single block. At its first input, you have to set the value of the flux λ_r^*. You can modify the speed regulator (*PI regulator*) parameters (amplification and integration time constant, default values are: $K_i = 10$, $T_i = 0.8s$) to study the behavior of the system. We included a *Commutation Frequency* block to satisfy the sampling criteria (to control the commutation frequency). The integration step (default: 50μs) can be modified from the Simulink's *Simulation / Parameters*.

To find out the structure of each block presented above, unmask it (*Options/Unmask*).

The errors in d-q currents id*–id, iq*–iq are translated directly into voltage vectors in the inverter in a kind of space-vector voltage PWM implemented through a commutation table.

The motor used for this simulation has the following parameters: P_n = 1100W, V_{nf} = 220V, 2p = 4, r_s = 9.53Ω, r_r = 5.619Ω, L_{sc} = 0.136H, L_r = 0.505H, L_m = 0.447H, J = 0.0026kgfm².

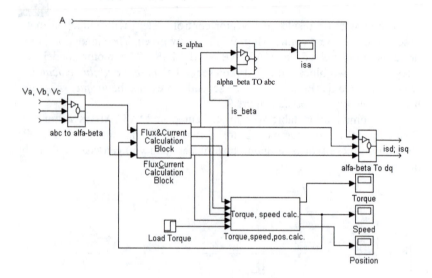

Figure 9.33. The motor space-phasor (d-q) model

The following figures represent the speed, torque, current and flux responses, for the *starting process and with load torque applied at 0.4s*. The value of load torque is 4Nm.

The speed response (Figure 9.34) is fast and smooth and so is the torque response with moderate torque pulsations. The current waveform has a small harmonics content (Figure 9.36) and the application of the load torque does not produce strong current transients, as expected. The stator and rotor flux linkage responses (Figures 9.37-9.38) are very similar to those of Section 9.10.

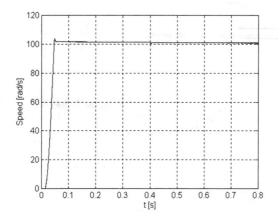

Figure 9.34. Speed transient response

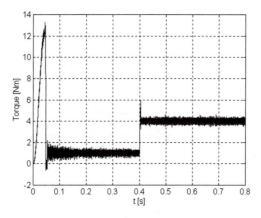

Figure 9.35. Torque response

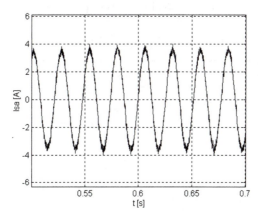

Figure 9.36. Phase current waveform

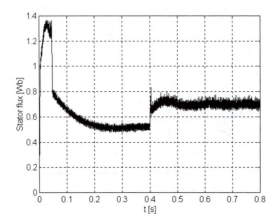

Figure 9.37. Stator flux amplitude

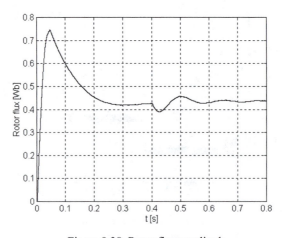

Figure 9.38. Rotor flux amplitude

The listing of MATLAB file for switching table follows:
```
function [sys, x0] = tab3(t,x,u,flag,zindex,table)
%*****************************************
%Here starts the switching table lookup routine...
%*****************************************
if flag == 3, Perform the desired table lookup
V0=[0 0 0];
V1=[1 0 0];
V2=[1 1 0];
V3=[0 1 0];
V4=[0 1 1];
V5=[0 0 1];
V6=[1 0 1];
V7=[1 1 1];

sq=0.866025404;
```

```
x0 = u(1);
y0 = u(2);
z0 = u(3);
if (x0>0), x0=1;end;
if (x0<0), x0=-1;end;
if (x0==0), x0=0;end;
if (y0>=0), y0=1;end;
if (y0<0), y0=-1;end;
if (x0==1)&(y0==1), xf=1;end;
if (x0==1)&(y0==-1), xf=2;end;
if (x0==-1)&(y0==1), xf=3;end;
if (x0==-1)&(y0==-1), xf=4;end;
if (x0==0),              xf=5;end;
if (cos(z0)>sq), zf=1;end;
if (cos(z0)<=sq)&(cos(z0)>0)&(sin(z0)>=0), zf=2;end;
if (cos(z0)>-sq)&(cos(z0)<=0)&(sin(z0)>=0), zf=3;end;
if (cos(z0)<-sq), zf=4;end;
if (cos(z0)>=-sq)&(cos(z0)<0)&(sin(z0)<0), zf=5;end;
if (cos(z0)<=sq)&(cos(z0)>=0)&(sin(z0)<0), zf=6;end;
r = table(xf, zf);
if r==0, sys=V0;end;
if r==1, sys=V1;end;
if r==2, sys=V2;end;
if r==3, sys=V3;end;
if r==4, sys=V4;end;
if r==5, sys=V5;end;
if r==6, sys=V6;end;
if r==7, sys=V7;end;
%**************************************
%Here ends the switching table lookup routine...
%**************************************
elseif flag == 0,
       % This part takes care of all initialization; it is used only once.
       x0 = [];
       % The system has no states, three outputs, and three inputs.
       sys = [0 0 3 3 0 1]';
else
       % Flags not considered here are treated as unimportant.
       % Output is set to [].
       sys = [];
end.
```

9.13. FLUX AND SPEED OBSERVERS IN MOTION SENSORLESS DRIVES

Sensorless drives are becoming predominant when up to 100 to 1 speed control range is required even in fast torque response applications (1-5 ms for step-rated torque response). There is a rich literature on the subject with quite a few solutions proposed [23, 24] and some are already on the markets worldwide. Essentially they are:

- Without signal injection;
- With signal injection

9.13.1. Performance criteria

To assess the performance of various flux and speed observers for sensorless drives, the following performance criteria have become widely accepted:
- steady-state error;
- torque response quickness;
- low speed behavior (speed range);
- sensitivity to noise and motor parameter detuning;
- dynamic robustness.
- complexity versus performance.

9.13.2. A classification of speed observers

The basic principles used for speed estimation (observation) may be classified as:
A. Speed estimators
B. Model reference adaptive systems
C. Luenberger speed observers
D. Kalman filters
E. Rotor slot ripple

With the exception of rotor slot ripple, all the other methods imply the presence of flux observers to calculate the motor speed.

9.13.3. Speed estimators

Speed estimators are in general based on the classical definition of rotor speed $\hat{\omega}_r$

$$\hat{\omega}_r = \hat{\omega}_1 - \left(\hat{S\omega}_1\right) \tag{9.98}$$

where ω_1 is the rotor flux vector instantaneous speed and $(S\omega_1)$ is the rotor flux slip speed. ω_1 may be calculated in stator coordinates based on the formula

$$\hat{\omega}_1 = \frac{d}{dt}\left[Arg\left(\overline{\lambda}_r^{\;s}\right)\right]; \quad \overline{\lambda}_r^{\;s} = \lambda_{dr} + j\lambda_{qr} \tag{9.99}$$

or

$$\hat{\omega}_1 = \frac{\dot{\lambda}_{qr}^{\;s}\lambda_{dr}^{\;s} - \dot{\lambda}_{dr}^{\;s}\lambda_{qr}^{\;s}}{\left(\lambda_r^{\;s}\right)^2} \tag{9.100}$$

$\dot{\lambda}_{qr}^{\;s}, \lambda_{dr}^{\;s}, \dot{\lambda}_{dr}^{\;s}, \lambda_{qr}^{\;s}$ are to be determined from a flux observer (see Figure 9.32, for example). On the other hand, the slip frequency $(S\omega_1)$, (9.23), is

$$\left(\hat{S\omega}_1\right) = \frac{\frac{3}{2}p\left(\lambda_{dr}i_{qs} - \lambda_{qr}i_{ds}\right)L_m}{\frac{3}{2}p\left(\lambda_r^2 / r_r\right)L_r} = \frac{\left(\lambda_{dr}i_{qs} - \lambda_{qr}i_{ds}\right)L_m}{\lambda_r^2\tau_r} \qquad (9.101)$$

Note that $\left(\hat{S\omega}_1\right)$ is strongly dependent only on rotor resistance r_r as L_m / L_r is rather independent of magnetic saturation. Still, rotor resistance is to be corrected if good precision at low speed is required. This slip frequency value is valid both for steady-state and transients and thus $\hat{\omega}_r$ is estimated quickly to allow fast torque response.

Such speed estimators may work even at 20 rpm although dynamic capacity of torque disturbance rejection at low speeds is limited.

This seems to be a problem with most speed observers. Asymmetrical Luenberger observers, with mechanical model, robust flux observer angle and rotor resistance correction, have been proven dynamically robust to 3 rpm with no signal injection.

9.13.4. Model reference adaptive systems (MRAS)

MRASs are based on comparision of two estimators. One of them does not include speed and is called the reference model. The other, which contains speed, is the adjustable model. The error between the two is used to derive an adaption model that produces the estimated speed $\hat{\omega}_r$ for the adjustable model.

To eliminate the stator resistance influence, the airgap reactive power q_m [25] is the output of both models

$$q_m = \bar{i}_s \otimes \left(\overline{V}_s - L_{sc}\frac{d\bar{i}_s^*}{dt}\right) \qquad (9.102)$$

$$\hat{q}_m = L_m\left[\hat{i}_m\,\hat{i}_s\,\hat{\omega}_r + \frac{1}{\tau_r}\left(\hat{i}_m \times \hat{i}_s\right)\right] \qquad (9.103)$$

The rotor flux magnetization current $\hat{i}_m$ equation in stator coordinates is ((9.15) with $\omega_1 = 0$)

$$\frac{d\hat{i}_m}{dt} = \hat{\omega}_r \otimes \hat{i}_m + \left(\hat{i}_s - \hat{i}_m\right)\frac{1}{\tau_r} \qquad (9.104)$$

Now the speed adaptation mechanism is

$$\hat{\omega}_r = \left(K_p + \frac{K_i}{s}\right)\left(q_m - \hat{q}_m\right) \qquad (9.105)$$

The signal flow diagram of the MRAS obtained is shown in Figure 9.39.

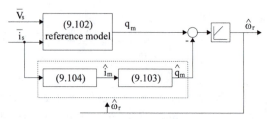

Figure 9.39. MRAS speed estimator based on airgap reactive power error

The effect of the rotor time constant τ_r variation persists and influences the speed estimation. However, if the speed estimator is used in conjunction with indirect vector current control, at least rotor field orientation is maintained as the same (wrong!) τ_r also enters the slip frequency calculator.

The MRAS speed estimator does not contain integrals and thus works even at zero stator frequency (d.c. braking) (Figure 9.40a) and does not depend on stator resistance r_s. It works even at 20 rpm (Figure 9.40b) [25].

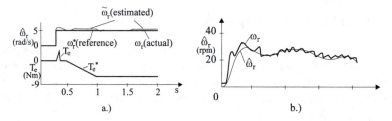

Figure 9.40. a.) Zero frequency, b.) low speed operation of MRAS speed estimator

9.13.5. Luenberger flux and speed observers

First the stator current and the rotor flux are calculated through a full order Luenberger observer based on stator and rotor equations in stator coordinates

$$\frac{d}{dt}\begin{vmatrix} \bar{i}_s^s \\ \bar{\lambda}_r^s \end{vmatrix} = \begin{vmatrix} A_{11} & A_{12} \\ A_{21} & A_{22} \end{vmatrix} \cdot \begin{vmatrix} \bar{i}_s^s \\ \bar{\lambda}_r^s \end{vmatrix} + \begin{vmatrix} B_1 \\ 0 \end{vmatrix} \cdot \begin{vmatrix} \bar{V}_s \end{vmatrix} \qquad (9.106)$$

with $\bar{i}_s^s = \begin{bmatrix} i_d & i_q \end{bmatrix}^T; \quad \bar{V}_s^s = \begin{bmatrix} V_d & V_q \end{bmatrix}^T; \quad \bar{\lambda}_r^s = \begin{bmatrix} \lambda_{dr} & \lambda_{qr} \end{bmatrix}^T; \qquad (9.107)$

$A_{11}, A_{12}, A_{21}, A_{22}$ are defined in (9.5) with $\omega_1 = 0$ and $a = L_m / L_r$

$$A_{11} = -\frac{r_r}{L_{sc}}\left(1+\frac{L_m^2}{L_r^2}\right) \cdot I = a_{r11} \cdot I; \quad B_1 = \frac{1}{L_{sc}} \cdot I$$

$$A_{12} = \frac{L_m}{L_r L_{sc}}\left(\frac{1}{\tau_r}I - \hat{\omega}_r I'\right) = a_{r12} \cdot I + a_{i12} \cdot I'$$

$$A_{21} = \frac{L_m}{\tau_r}I = a_{r21}I; \quad A_{22} = \frac{1}{\tau_r}I + \hat{\omega}_r I'$$

$$I = \begin{vmatrix} 1 & 0 \\ 0 & 1 \end{vmatrix}; \quad I' = \begin{vmatrix} 0 & -1 \\ 1 & 0 \end{vmatrix}$$

(9.108)

with

$$\hat{x} = \begin{bmatrix} i_{ds} & i_{qs} & \lambda_{dr} & \lambda_{qr} \end{bmatrix}^T$$

(9.109)

The full-order Luenberger observer writes

$$\frac{d\hat{x}}{dt} = \hat{A}\,\hat{x} + B\,\hat{V}_s + G\left(\hat{i}_s - i_s\right)$$

(9.110)

The matrix G is chosen such that the observer is stable.

$$G = \begin{bmatrix} g_1 & g_2 & g_3 & g_4 \\ -g_2 & g_1 & -g_4 & g_3 \end{bmatrix}^T$$

(9.111)

with

$$g_1 = (K-1)(a_{r11} + a_{r22}); \quad g_2 = (K-1)a_{i22}; \quad g_4 = -c(K-1)a_{i12}$$
$$g_3 = (K^2 - 1)(ca_{r11} + a_{r21}) - c(K-1)(a_{r11} + a_{r22}); \quad c = L_{sc}$$

(9.112)

$$a_{r22} = 1/\tau_r; \quad a_{i22} = \hat{\omega}_r$$

The speed estimator is based on rotor flux $\overline{\hat{\lambda}}_r{}^s$ and $\overline{\hat{i}}_s$ estimators

$$\hat{\omega}_r = \left(K_p + \frac{K_i}{s}\right) \text{Im ag}\left[\left(\overline{\hat{i}}_s - \hat{i}_s\right)\hat{\lambda}_r{}^s\right]$$

(9.113)

In essence the speed estimator is based on some kind of torque error.

If the rotor resistance r_r has to be estimated, an additional high reference current $i_{da}{}^*$ is added to the reference flux current $i_{ds}{}^*$. Then the rotor resistance may be estimated [26] as

$$\frac{d}{dt}\left(\frac{1}{\tau_r}\right) = -\lambda_r\left[\left(\hat{i}_{ds} - i_{ds}{}^*\right) \cdot i_{da}{}^*\right]$$

(9.114)

Remarkable results have been obtained this way with minimum speed down to 30 rpm.

The idea of an additional high frequency (10 times rated frequency) flux current may be used to determine both the rotor speed and rotor time constant τ_r [27].

Extended Kalman filters for speed and flux observers [28] also claim speed estimation at 20-25rpm though they require considerable on-line computation time.

9.13.6. Rotor slots ripple speed estimators

The rotor slots ripple speed estimators are based on the fact that the rotor slotting openings cause stator voltage and current harmonics $\omega_{s1,2}$ related to rotor speed $\hat{\omega}_r$, the number of rotor slot N_r and synchronous speed ω_1

$$\hat{\omega}_{s1,2} = N_r \hat{\omega}_r \pm \omega_1 \qquad (9.115)$$

Band pass filters centered on the rotor slot harmonics $\hat{\omega}_{s1,2}$ are used to separate $\hat{\omega}_{s1,2}$ and thus calculate $\hat{\omega}_r$ from (9.115). Various other methods have been proposed to obtain $\hat{\omega}_{s1,2}$ and improve the transient performance. The response tends to be rather slow and thus the method, though immune to machine parameters, is mostly favorable for a wide speed range, but for low dynamics (medium-high powers) applications [27].

For an update on sensorless IM control with and without signal injection refer to [30].

9.14. DIRECT TORQUE AND FLUX CONTROL (DTFC)

DTFC is a commercial abbreviation for the so-called direct self control proposed initially [32-33] for induction motors fed from PWM voltage source inverters and later generalized as torque vector control (TVC) in [4] for all a.c. motor drives with voltage or current source inverters.

In fact, based on the stator flux vector amplitude and torque errors sign and relative value and the position of the stator flux vector in one of the 6 (12) sectors of a period, a certain voltage vector (or a combination of voltage vectors) is directly applied to the inverter with a certain average timing.

To sense the stator flux space-phasor and torque errors we need to estimate the respective variables. So all types of flux (torque) estimators or speed observers good for direct vector control are also good for DTFC. The basic configurations for direct vector control and DTFC are shown on Figure 9.41.

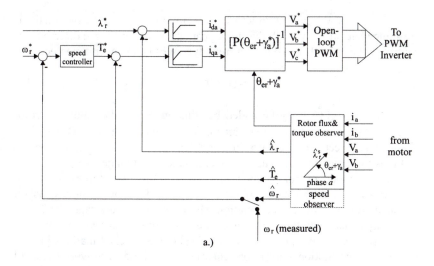

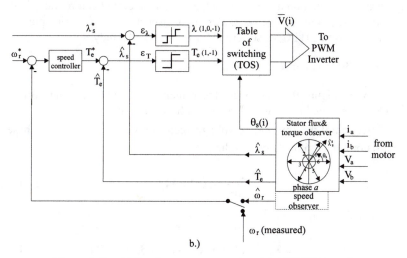

Figure 9.41. a.) Direct vector current control, b.) DTFC control

As seen from Figure 9.41, DTFC is a kind of direct vector d.c. (synchronous) current control.

Both strategies require flux and torque observers, (also, speed observers for motion sensorless control). However, direct torque and stator flux control (DTFC) have the right scales and thus no design of the PI d.c. current controllers (i_{da}, i_{qa}) is necessary. Also the vector rotator and the open-loop PWM are replaced with a table of switchings (TOS). Both these simplifications mean that DTFC is related to motor parameters only through the flux-torque-speed observers. Finally, DTFC works with stator flux and not with rotor flux.

It is true, however, that, in terms of dynamic properties, the two methods are somewhat similar though the flux dynamics is slower for the direct vector control as the rotor flux to d axis current dynamics is, in general, neglected.

$$i_d = \left(1 + s\tau_r\right)\frac{\lambda_r}{L_m} \qquad (9.116)$$

Equation (9.116) may be added but this operation also implies a rotor time constant τ_r adaptation mechanism. So it seems clear that DTFC does "indirectly" direct vector control in a simpler, more robust configuration but still preserves high torque dynamics over a wide speed range.

Sliding mode or fuzzy logic may replace the two (three) positional hysteresis flux and torque controllers in Figure 9.41. Knowing at any instant the stator flux amplitude and position, short source voltage sags (10-20 ms long) may be handled, thus providing for a ride through inherent capability.

These merits have prevailed so that DTFC has reached markets [33] in IM drives with motion sensors or sensorless, as a kind of universal (for all applications) induction motor drive. One drive with digital control does it all.

In what follows, we consider two aspects: DTFC principle and performance and DTFC sensorless IM drive — a case study.

9.14.1. DTFC principle

Though Figure 9.41 uncovers the principle of DTFC, finding how the T.O.S. is generated is the way to a successful operation.

Selection of the appropriate voltage vector in the inverter $\overline{V}_s(i)$ is based on a stator equation in stator coordinates

$$\frac{d\overline{\lambda}_s{}^s}{dt} = \overline{V}_s{}^s - r_s\overline{i}_s{}^s \qquad (9.117)$$

By integration

$$\Delta\overline{\lambda}_s{}^s = \overline{\lambda}_s{}^s - \overline{\lambda}_{s0}{}^s = \int_0^{T_i}\left(\overline{V}_s{}^s - r_s\overline{i}_s{}^s\right) \cdot dt \approx \overline{V}_s(i)T_i \qquad (9.118)$$

In essence the torque error ε_T may be canceled by stator flux acceleration or deceleration. To reduce the flux errors, the flux trajectories will be driven along appropriate voltage vectors (9.118) that increase or decrease the flux amplitude.

When the flux amplitude has to be increased, a voltage vector phase shifted by an angle larger than 90° with respect to existing flux vector $\overline{\lambda}_{s0}{}^s$ is applied. In contrast, if flux amplitude has to be reduced, an angle less than 90° will be observed.

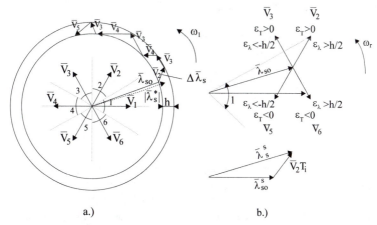

Figure 9.42. a.) Stator flux space-phasor trajectory, b.) Selecting the adequate voltage vector in the first sector (−30° to +30°)

For the flux space - phasor in the first sector, the voltage vector selection is shown in Figure 9.42b.

The complete table of optimal switching, TOS, is shown in Table 9.2.

Table 9.2. Basic voltage vector selection for DTFC

$\theta_s(i)$		$\theta_s(1)$	$\theta_s(2)$	$\theta_s(3)$	$\theta_s(4)$	$\theta_s(5)$	$\theta_s(6)$
λ	T						
1	1	V_2	V_3	V_4	V_5	V_6	V_1
1	−1	V_6	V_1	V_2	V_3	V_4	V_5
0	1	V_0	V_7	V_0	V_7	V_0	V_7
0	−1	V_0	V_7	V_0	V_7	V_0	V_7
−1	1	V_3	V_4	V_5	V_6	V_1	V_2
−1	−1	V_5	V_6	V_1	V_2	V_3	V_4

Either the hysteresis band will decide the timing of each voltage vector or, if the timing is constant, the switching frequency will be constant. The hysteresis band of the two controllers may be adapted to keep constant the average switching frequency.

DTFC looks thus deceptively simple. Still, the stator flux has to be estimated and at least a combined voltage-current model (Figure 9.31) is necessary if acceptable performance down to 0.5 Hz is required.

Once the stator flux is known, the torque is simply

$$\hat{T}_e = \frac{3}{2} p \operatorname{Real}\left(j\bar{\lambda}_s^{\,s} \bar{i}_s^{\,s*} \right) \tag{9.119}$$

As expected, the torque response is quick (as in vector control), but it is also rotor resistance independent above 1-2 Hz (Figure 9.43) [2].

Note also that the torque pulsations are directly controlled and the core losses and noise may be controlled through the stator flux level selection.

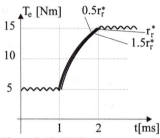

Figure 9.43. TVC torque response

For sensorless control both flux and speed observers, as presented in Section 9.13, are required; torque responses times of 1-5ms (as fast as for control with motion sensors) have been obtained for 100 to 1 speed control range and 0.1% of rated speed error. Such sensorless DTFC drives are now commercially available [31].

9.15. DTFC SENSORLESS: A CASE STUDY

The simulation results of a direct torque and flux control drive system for induction motors are presented. The example was implemented in MATLAB/SIMULINK. The motor model was integrated in two blocks, the first represents the current and flux calculation module in d-q axis; the second represents the torque, speed and position computing module (Figure 9.44).

The commutation table was implemented as a Matlab function to select the right voltage vector for the drive. The input values for this table are: the two stator flux values from the motor model (λ_{sd} & λ_{sq}), necessary to find out the position of the flux, the error values from comparing estimated torque (from the torque estimation block) with its reference value (the output of speed regulator), and, respectively, the calculated stator flux and its (constant) reference value.

The changing of motor parameters for different simulations is as simple as possible. By clicking on these blocks, a dialog box appears and you can change them by modifying their default values.

You can modify the speed regulator (*PI regulator*) parameters (amplification and integration time constant, default values are: $K_i = 10$, $T_i = 0.8s$) to study the behavior of the system. We included a *Commutation Frequency* block to satisfy the sampling criteria (to control the commutation frequency). The integration step (default: 25μs) can be modified from the Simulink's *Simulation / Parameters*.

To find out the structure of each block presented above, unmask it (*Options/Unmask*).

The motor used for this simulation has the following parameters: $P_n = 1100W$, $U_{nf} = 220V$, $2p = 4$, $r_s = 9.53\Omega$, $r_r = 5.619\Omega$, $L_{sc} = 0.136H$, $L_r = 0.505H$, $L_m = 0.447H$, $J = 0.0026kgfm^2$.

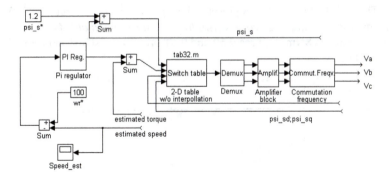

Figure 9.44.The DTFC system

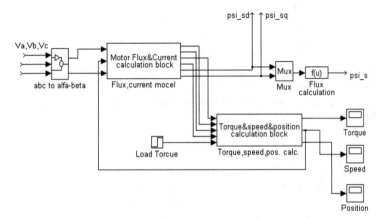

Figure 9.45. The I.M. model

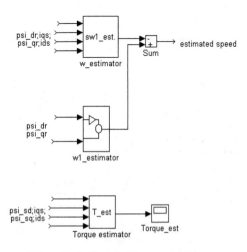

Figure 9.46. Speed and torque estimators

The following figures represent the speed, torque, current and flux responses for the *starting process with load torque applied at 0.4s*. The value of load torque is 8Nm. Figures 9.50 and 9.51 present the stator and rotor flux amplitudes.

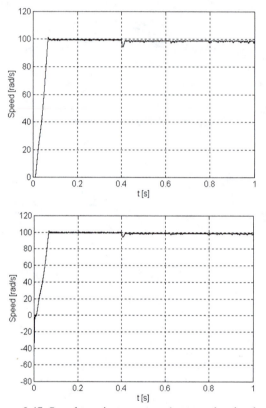

Figure 9.47. Speed transient response (measured and estimated)

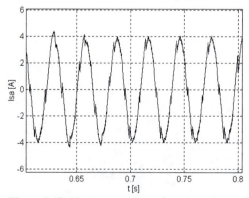

Figure 9.48. Phase current waveform (steady-state)

The measured (Figure 9.47a.) and estimated (Figure 9.47b) speed are very close to each other.

The current waveform is slightly different from that obtained with indirect vector a.c. or d.c. current control.

The speed response is fast as the speed observer.

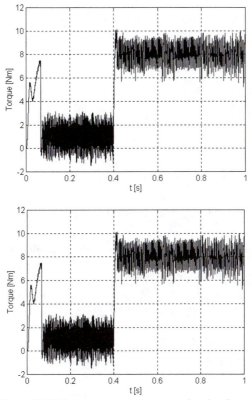

Figure 9.49. Torque response (measured and estimated)

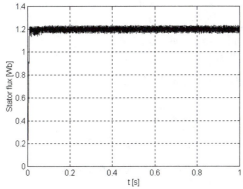

Figure 9.50. Stator flux amplitude

The torque calculator (Figure 9.49) also proves to be satisfactory. The stator flux response (Figure 9.50) is quick and stable. The rotor flux (Figure 9.51) varies only during transients. Please remember that this time the stator flux amplitude is directly controlled.

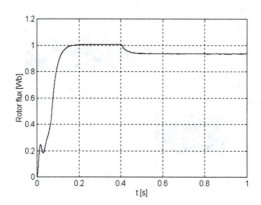

Figure 9.51. Rotor flux amplitude

The listing of the MATLAB file for switching table follows:

```
function [sys, x0] = tab3(t,x,u,flag,zindex,table)
% *****************************************
% Here starts the switching table lookup routine...
% *****************************************
if flag == 3,
% Perform the desired 2D table lookup

V0=[0 0 0];V1=[1 0 0];V2=[1 1 0];V3=[0 1 0];
V4=[0 1 1];V5=[0 0 1];V6=[1 0 1];V7=[1 1 1];
    x0 = u(1);
    y0 = u(2);
    z0 = u(3);
    z1 = u(4);
    if (x0>0.01), x0=1;end;
    if (x0<-0.01), x0=-1;end;
    if (x0<0.01)&(x0>-0.01), x0=0;end;
    if (y0>=0), y0=1;end;
    if (y0<0), y0=-1;end;
    if (x0==1)&(y0==1), xf=1;end;
    if (x0==1)&(y0==-1), xf=2;end;
    if (x0==-1)&(y0==1), xf=3;end;
    if (x0==-1)&(y0==-1), xf=4;end;
    if (x0==0),           xf=5;end;
    if  (sign(z0)>=0)&(sign(z1)>=0)&(sign(sqrt(3)*abs(z1)-abs(z0))==-1)&(sign(abs(z1)-
sqrt(3)*abs(z0))==-1), zf=1;end;
    if  (sign(z0)>=0)&(sign(z1)>=0)&(sign(sqrt(3)*abs(z1)-abs(z0))>=0)&(sign(abs(z1)-
sqrt(3)*abs(z0))==-1), zf=2;end;
    if  (sign(z0)>=0)&(sign(z1)>=0)&(sign(sqrt(3)*abs(z1)-abs(z0))>=0)&(sign(abs(z1)-
sqrt(3)*abs(z0))>=0), zf=2;end;
```

```
        if (sign(z0)==-1)&(sign(z1)>=0)&(sign(sqrt(3)*abs(z1)-abs(z0))>=0)&(sign(abs(z1)-
sqrt(3)*abs(z0))>=0), zf=3;end;
        if (sign(z0)==-1)&(sign(z1)>=0)&(sign(sqrt(3)*abs(z1)-abs(z0))>=0)&(sign(abs(z1)-
sqrt(3)*abs(z0))==-1), zf=3;end;
        if (sign(z0)==-1)&(sign(z1)>=0)&(sign(sqrt(3)*abs(z1)-abs(z0))==-1)&(sign(abs(z1)-
sqrt(3)*abs(z0))==-1),zf=4;end;
        if              (sign(z0)==-1)&(sign(z1)==-1)&(sign(sqrt(3)*abs(z1)-abs(z0))==-
1)&(sign(abs(z1)-sqrt(3)*abs(z0))==-1),zf=4;end;
        if (sign(z0)==-1)&(sign(z1)==-1)&(sign(sqrt(3)*abs(z1)-abs(z0))>=0)&(sign(abs(z1)-
sqrt(3)*abs(z0))==-1),zf=5;end;
        if (sign(z0)==-1)&(sign(z1)==-1)&(sign(sqrt(3)*abs(z1)-abs(z0))>=0)&(sign(abs(z1)-
sqrt(3)*abs(z0))>=0), zf=5;end;
        if (sign(z0)>=0)&(sign(z1)==-1)&(sign(sqrt(3)*abs(z1)-abs(z0))>=0)&(sign(abs(z1)-
sqrt(3)*abs(z0))>=0), zf=6;end;
        if (sign(z0)>=0)&(sign(z1)==-1)&(sign(sqrt(3)*abs(z1)-abs(z0))>=0)&(sign(abs(z1)-
sqrt(3)*abs(z0))==-1), zf=6;end;
        if (sign(z0)>=0)&(sign(z1)==-1)&(sign(sqrt(3)*abs(z1)-abs(z0))==-1)&(sign(abs(z1)-
sqrt(3)*abs(z0))==-1),zf=1;end;
        r = table(xf, zf);
        if r==0, sys=V0;end;if r==1, sys=V1;end;
        if r==2, sys=V2;end;if r==3, sys=V3;end;
        if r==4, sys=V4;end;if r==5, sys=V5;end;
        if r==6, sys=V6;end;if r==7, sys=V7;end;
    % Here ends the switching table lookup routine...
    % *************************************
    elseif flag == 0,
        % This part takes care of all initialization; it is used only once.
        x0 = [];
        % The system has no states, three outputs, and four inputs.
        sys = [0 0 3 4 0 1]';
    else
        % Flags not considered here are treated as unimportant.
        % Output is set to [].
        sys = [];
    end
```

9.15.1. DTFC with space vector modulation (SVM)

Notable torque and current pulsations are visible in Figures 9.48 - 9.49.
There is now a rich literature dealing with this problem [39 - 43].

In essence there are two main ways to reduce the torque ripple and current harmonics in DTFC – IM drives:

• Increasing the switching frequency in the PWM converter from 5 - 15 kHz to 20 – 40 kHz.

• With limited switching frequency, the number of voltage vectors applied within 60° of stator flux vector evolution is increased.

The latter type of method seems more practical with existing IGBT PWM voltage – source converters.

Open loop space vector modulation (paragraph 9.9.2) may be used for limited switching frequency, to produce low torque ripple.

To apply open loop SVM the reference (desired) voltage vector $\overline{V}_s^*$ in the stator has to be computed on line.

Apart from the so called stator flux vector control [4] quite a variety of flux and torque controllers may be used to produce the reference voltage $\overline{V}_s^*$ required for open loop SVM.

Linear variable structure [43] or hybrid (linear plus variable structure) torque and flux controllers have been proven practical for the scope [40, 42].

The sliding mode (variable structure) surfaces $S_{\lambda s}$, S_{Te} are defined based on stator flux and torque errors ($\varepsilon_{\lambda s}$, ε_{Te}):

$$S_{\lambda s} = \varepsilon_{\lambda s} + C_{\psi s} \frac{d\varepsilon_{\lambda s}}{dt}$$
$$S_{Te} = \varepsilon_{Te} + C_{Te} \frac{d\varepsilon_{Te}}{dt} \tag{9.120}$$

The reference (command) d-q voltages in stator flux coordinates ($\theta_{\lambda s}$), $\overline{V}_{sd}^*$, $\overline{V}_{sq}^*$ are:

$$\overline{V}_{sd}^* = \left(k_{p\lambda} + k_{i\lambda} \frac{1}{s} \right) sgn(S_{\lambda s})$$
$$\overline{V}_{sq}^* = \left(k_{pT} + k_{iT} \frac{1}{s} \right) sgn(S_{Te}) + \hat{\omega}_{\psi s} \hat{\lambda}_s \tag{9.121}$$

$\hat{\omega}_{\psi s}$, $\hat{\lambda}_s$ - estimated stator flux amplitude and speed.

Park transformation to stator coordinates is operated to obtain the so-called relay with constant gains DTFC (Fig. 9.52) [39]:

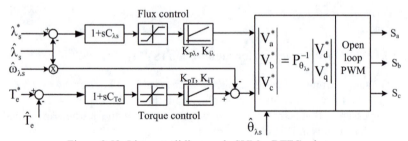

Figure 9.52. Linear + sliding mode SVM – DTFC scheme

As evident in eqn (9.121) and Fig. 9.52:

- A Park transformation from stator flux to stator coordinates is required.
- E.m.f. compensation is applied also to improve torque response at high speeds.
- The low torque and current ripple are secured by the linear (PI) controller section.
- Robustness to torque, inertia and machine parameter detuning is provided by the sliding mode controller section.

- To reduce chattering around the targeted variable values – under steady state – a boundary layer 2h may be used to replace the sign function with a saturation function:

$$\text{sat}(x) = \begin{cases} \text{sgn x for } |x| > h \\ \dfrac{x}{h} \text{ for } |x| < h \end{cases} \qquad (9.122)$$

Inside this layer the controller becomes high gain PI type and drives the error to zero. Outside the boundary layer sliding mode robust and fast response of original DTFC is preserved.

- The state observer has to estimate the stator flux amplitude $\hat{\lambda}_s$, angle $\hat{\theta}_{\lambda s}$, electromagnetic torque $\hat{T}_e$, and stator flux vector speed $\hat{\omega}_{\lambda s}$.

- To perform speed close loop control a rotor speed estimator is required.

A general voltage /current model inherently motion sersorless stator and rotor flux observer making use again of sliding mode concept is obtained simply by the stator equation in rotor flux coordinates:

$$\frac{d\bar{\hat{\lambda}}_s}{dt} = -R_s \bar{i}_s + \bar{V}_s + k_1 \text{sgn}(\bar{i}_s - \bar{\hat{i}}_s) + k_1'(\bar{i}_s - \bar{\hat{i}}_s) \qquad (9.123)$$

$$\frac{d\bar{\hat{\lambda}}_r^r}{dt} \approx \frac{L_m R_r \bar{\hat{\lambda}}_s^r}{L_s L_{sc}} - \frac{R_r \bar{\hat{\lambda}}_r^r}{L_{sc}} + k_2 \text{sgn}(\bar{i}_s^r - \bar{\hat{i}}_s^r) + k_2'(\bar{i}_s^r - \bar{\hat{i}}_s^r) \qquad (9.124)$$

$$L_{sc} = L_s - L_m^2 / L_r; \quad T_r = L_r / R_r$$

L_s, L_r – stator and rotor inductances;
R_s, R_r – stator and rotor resistances;

$(\bar{i}_s, \bar{\hat{i}}_s)$ and $(\bar{i}_s^r, \bar{\hat{i}}_s^r)$ are the actual and, respectively, estimated stator current vectors in stator and, respectively, rotor flux coordinates. As the stator and rotor resistances R_s, R_r depend on temperature, they have to be corrected online. In general, $R_s = k_{sr} R_r$, so only the stator resistance has to be corrected.

The nonlinear term in sgn reflects the sliding mode contribution while also a linear (proportional, P) loop is added.

The current estimator is obtained from the stator and rotor flux definitions in stator coordinates as:

$$\bar{\hat{i}}_s = \frac{L_r}{L_s L_{sc}} \bar{\hat{\lambda}}_s - \frac{L_m}{L_s L_{sc}} \bar{\hat{\lambda}}_r \qquad (9.125)$$

The relationship between rotor flux and stator flux in stator coordinates writes:

$$\bar{\lambda}_r^{\wedge s} = \frac{L_r}{L_m}\hat{\lambda}_s - L_{sc}\bar{i}_s; L_{sc} = L_s - \frac{L_m^2}{L_r} \approx L_{sl} + L_{rl} \qquad (9.126)$$

Magnetic saturation plays a minor role in the stator current and rotor flux estimators above, because L_{sc} does vary only a few percent when the current is higher than $2-3$ times rated current (for semiclosed or open slots in the motor) and the ratios L_r/L_m and L_r/L_m and L_r/L_s are mildly dependent on magnetic saturation.

The electromagnetic torque estimator is straightforward:

$$\hat{T}_e = \frac{3}{2}p\,\mathrm{Re}(j\bar{\lambda}_s^{\wedge}\bar{i}_s^*) \qquad (9.127)$$

The structural diagram of the above hybrid (linear + SM) flux observer is shown in Figure 4.53.

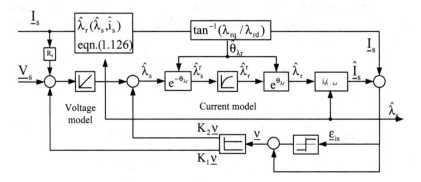

Figure 9.53. A generic hybrid (linear-SM) flux observer for IM

For the speed estimation an asymmetrical Luenberger (or equivalent PLL) speed observer may be adopted Fig. 9.54.

$$\frac{d}{dt}\begin{vmatrix}\hat{\alpha}\\ \hat{\omega}_r\\ \hat{T}_L\end{vmatrix} = \begin{vmatrix}0 & 1 & 0\\ 0 & 0 & -p/J\\ 0 & 0 & 0\end{vmatrix}\begin{vmatrix}\hat{\alpha}\\ \hat{\omega}_r\\ \hat{T}_L\end{vmatrix} + \begin{vmatrix}1 & 0\\ 0 & p/J\\ 0 & 0\end{vmatrix}\cdot\begin{vmatrix}\hat{\omega}_s\\ \hat{T}_e\end{vmatrix} + \begin{vmatrix}k_1\\ k_2\\ k_3\end{vmatrix}\cdot\varepsilon \qquad (9.128)$$

$$\varepsilon = \sin(\hat{\theta}_{\lambda r} - \hat{\alpha}); \quad \hat{\omega}_s = S\omega = \hat{\omega}_{\lambda r} - \hat{\omega}_r \qquad (9.129)$$

This observer roughly estimates the load torque $T_L^{\wedge}$ but estimates the rotor speed rather well. $\hat{\omega}_s$ is the slip (S) speed of the rotor flux vector. Also $\hat{\theta}_{\lambda r}$ is the estimated angle of rotor flux (Fig. 9.53).

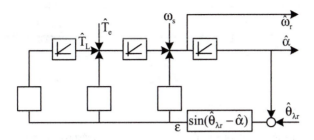

Figure 9.54. Asymmetrical Luenberger speed observer

The error dynamics for this observer is simply:

$$\frac{de}{dt} = |A| - |K||C| \; ; \; [C] = [1 \; , \; 0 \; 0] \; ; |K| = [K_1 \; K_2 \; K_3] \; ; \qquad (9.130)$$

|A| is the system matrix in eqn. (9.128).

The eigen values of |A|-|K||C| are the observer poles.

One negative real pole and two real negative but complex poles should be provided for good dynamics.

The speed observer good transient behavior and small error are shown in Fig. 9.55 for no parameters or inertia detuning [42].

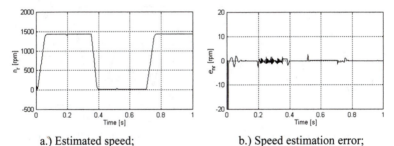

a.) Estimated speed;   b.) Speed estimation error;

Figure 9.55. Asymmetrical Luenberger speed observer

transient speed response (a) and error (b);

The speed error immunity to large inertia change is high but rotor resistance correction is mandatory to limit the speed error to 2 – 2.5 rpm at steady state.

The reduction of torque ripple at zero speed by SVM-DTFC with the above speed and flux observer is shown in Fig. 7.56.

Fast torque response from standstill and larger PI gains in the flux and torque controller are evident in Fig. 7.57 [42].

Steady 3 rpm speed control under load is shown in Fig. 9.58.

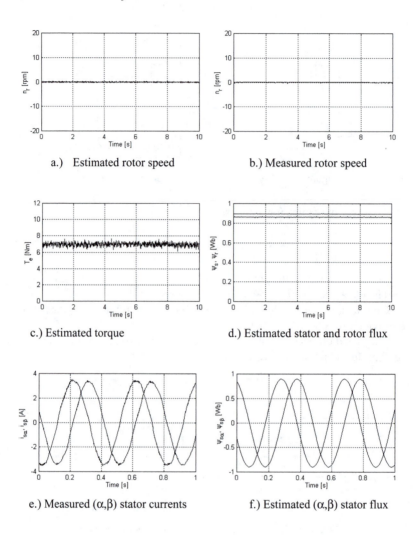

a.) Estimated rotor speed

b.) Measured rotor speed

c.) Estimated torque

d.) Estimated stator and rotor flux

e.) Measured (α,β) stator currents

f.) Estimated (α,β) stator flux

Figure 9.56. Measured current, estimated flux and speed at zero speed

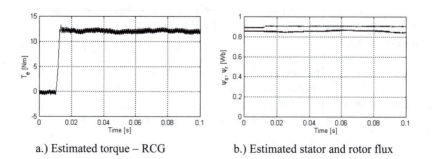

a.) Estimated torque – RCG

b.) Estimated stator and rotor flux

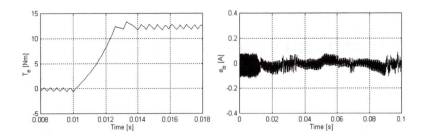

c.) Estimated torque – time zoom   d.) Current estimation error

Figure 9.57. Torque transients from standstill

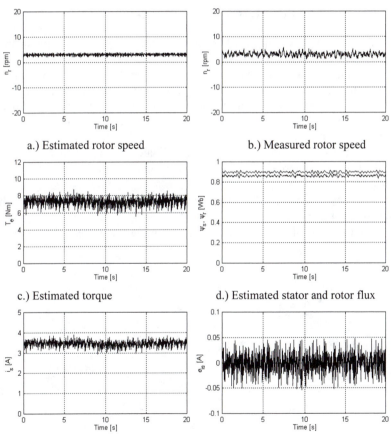

a.) Estimated rotor speed   b.) Measured rotor speed

c.) Estimated torque   d.) Estimated stator and rotor flux

e.) Estimated stator current magnitude   f.) Stator current estimation error

Figure 9.58 Very low speed (3 rpm) speed of DTFC – SVM IM drive with sliding
mode observers and torque and flux controllers.

The small current error (ripple) is also visible (Fig. 9.58f).

The still visible torque ripple is also due to the brush friction in the DC brush PM machine that loads the drive by operating as motor in the opposite direction.

The above method is by no means the only one to work well without signal injection at very low speed (see for example Ref. [44] for vector control at 5 rpm without signal injection).

The sensorless DTFC – SVM control of IMs proves to be capable of same performance as advanced vector control in all respects for slightly lower online computation effort.

Note: Below 3 rpm better speed and flux observers are required for motion sensorless control and perhaps the usage of signal injection may become mandatory [30].

9.16. FEEDBACK LINEARIZED CONTROL

Vector control was invented to produce separate flux and torque control as it is implicitly possible with d.c. brush motors.

It is also known that, with constant flux, the torque is proportional to torque current and linear torque/speed characteristics may be obtained. Such an artificial decoupling and linearization of induction motor equations may, in principle, be done with some other nonlinear transformations. Feedback linearization control [5-7] is such a method.

Still, ((9.106)-(9.108)) — in stator coordinates — with stator current and rotor flux d-q components, together with the equation of motion, are required.

The new output variables are the rotor flux squared $\Phi_1(x) = \lambda_r^2$ and the rotor speed $\Phi_2(x) = \omega_r$.

The basic signal flow diagram [7] is shown in Figure 9.59.

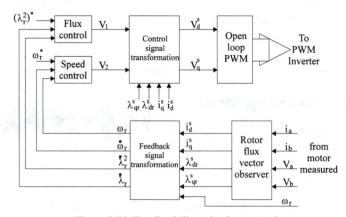

Figure 9.59. Feedback linearization control

It should be noted that the computation effort is greater than for vector control or DTFC and a thorough knowledge of motor parameters is necessary. Dynamic performance quite similar to both advanced vector control or DTFC has been obtained [32].

It seems too early to forecast the industrial prospects of feedback linearization control in competition with advanced vector control or DTFC.

9.17. SCALAR (V_1 / f_1) CONTROL

For pump and ventilator-like applications the speed control range is only from 3 to 1 up to 10 to 1. Motion (speed) sensors are avoided in such drives.

Traditionally scalar, V_1 / f_1, open-loop control has been used for such applications. In essence, the voltage amplitude V_1 and its frequency f_1 are related by

$$V_1^* = V_0^* + K_f f_1^*$$ (9.131)

V_0^* is called voltage boost and is required to run the motor properly at low speeds.

The primary frequency is ramped as desired and, based on (9.131), an open-loop PWM procedure is used to control the PWM inverter (Figure 9.60).

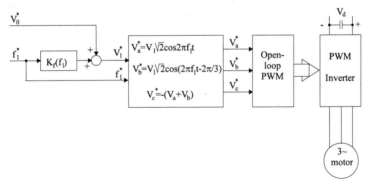

Figure 9.60. V_1 / f_1 open-loop (scalar) control

Ramping the frequency is performed slowly enough to maintain stability. Linearization of the IM equations around a steady-state point has backed experimental (industrial) findings of unstable open zones in the $i_{mr}(f_1)$ plane and ellipses of instability in the $V_1(f_1)$ plane [33] (Figure 9.61); i_{mr} is the rotor flux current.

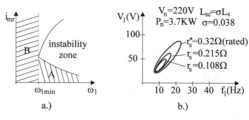

Figure 9.61. Instability zones for V_1 / f_1 open-loop scalar control:

a.) i_{mr} / ω_1 plane, b.) V_1 / f_1 plane

In general, ω_{1min} is [33]

$$\omega_{1min} > 2 / \tau_r \qquad (9.132)$$

where τ_r is the rotor time constant.

Also, it has been shown [33] that instabilities occur if the figure of merit f_m is

$$f_m = \frac{\tau_m}{\tau_r} < 0.5 \qquad (9.133)$$

where τ_m is the mechanical time constant.

The slope of the two asymptotes in Figure 9.61a are directly related to τ_r and τ_m. The values of f_m for a few 2, 4, 6 pole motors [33] are shown in Figure 9.62.

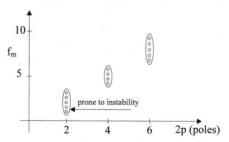

Figure 9.62. The figure of merit f_m (9.133)

Two-pole motors tend to be unstable ($f_m < 0.5$) while four- or six-pole motors are less susceptible to instabilities. It should be mentioned that a higher short-circuit inductance L_{sc} (in relative units) is beneficial from this point of view.

Besides the fact that open-loop scalar (V_1/f_1) drives are prone to instabilities, they are vulnerable to fast ramp accelerations or quick and (or) large torque perturbations. Moreover, increasing the load means a decrease in speed ω_r

$$\omega_r = \omega_1 - (S\omega_1) \qquad (9.134)$$

For low speed (frequency) the reduction of speed with load becomes important and may not be tolerable. Compensating for the slip frequency $(S\omega_1)$ with load is a practical solution to this problem.

Compensating the slip frequency is attempted for steady-state when, in fact, the rotor flux is constant in time and thus the torque T_e (9.15) is

$$T_e = H(\omega_1 - \omega_r) = HS\omega_1 \qquad (9.135)$$

with
$$H = \frac{3}{2}p\frac{\lambda_r^2}{r_r} \qquad (9.136)$$

The motion equation is

$$\frac{J}{p}\frac{d\omega_r}{dt} = T_e - T_{load} \qquad (9.137)$$

Assuming that the mechanical transients are slow, the rotor flux may be considered constant. The solution of ω_r for such slow transients is

$$\omega_r = \left(\omega_{r0} - \omega_1 + \frac{T_{load}}{H}\right)e^{-\frac{t}{\tau_m}} + \omega_1 - \frac{T_{load}}{H} \qquad (9.138)$$

$$\tau_m = \frac{J}{pH} \qquad (9.139)$$

For constant rotor flux the torque/speed curve is a straight line, so the stabilizing response in speed (9.138) is to be expected.

The variation of $r_r(\tau_r)$ with 25% produces, however, a 25% error in the slip frequency. For rated speed, this error is relatively low but at low speeds it remains a 25% speed error.

The principle of the slip frequency compensation method [34] consists of increasing the reference frequency ω_r^* by the estimated slip frequency $S\omega_1$ (Figure 9.63) to move, for given torque, from B to C and thus make ω_r independent of load.

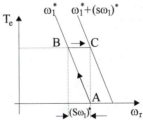

Figure 9.63. Principle of slip frequency compensation

The signal flow diagram (Figure 9.64) of such a scheme illustrates the estimation of the rotor flux from the voltage model.

The voltage model is adequate only above 2Hz but this is acceptable in V_1 / f_1 drives.

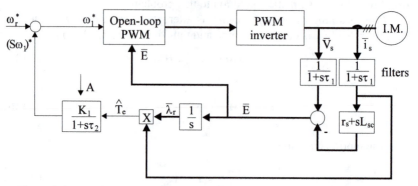

Figure 9.64. V_1 / f_1 scalar control with feedforward slip frequency compensation

Equations (9.135), (9.137) and Figure 9.65 suggest the signal flow diagram of Figure 9.65.

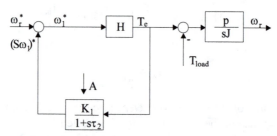

Figure 9.65. Signal flow diagram for slip frequency compensation

The ω_r / ω_r^* transfer function G_0 is simply (from Figure 9.65)

$$G_0 = \frac{sH(1+s\tau_2)}{J\tau_2^2 s^2 + (J + pH\tau_2 - JHK_1)s + pH} \qquad (9.140)$$

G_0 offers a stable response only if

$$HK_1 < 1 + pH\tau_2 / J \qquad (9.141)$$

On the other hand, the transfer function G_1 between ω_r and T_{load} is

$$G_1 = \frac{p(HK_1 - s\tau_2 - 1)}{J\tau_2^2 s^2 + (J + pH\tau_2 - JHK_1)s + pH} \qquad (9.142)$$

Equation (9.142) suggests that for steady-state (s = 0 in (9.142)) $\Delta\omega_r$ for unit step load torque is

$$\Delta\omega_r = \frac{HK_1 - 1}{H} \qquad (9.143)$$

So,

$$\Delta\omega_r < 0; \quad \text{for } HK_1 \leq 1$$
$$\Delta\omega_r = 0; \quad \text{for } HK_1 = 1 \qquad (9.144)$$
$$\Delta\omega_r > 0; \quad \text{for } HK_1 > 1$$

For full compensation $\Delta\omega_r = 0$ and thus $HK_1 = 1$ (point C in Figure 9.63). In this case, (9.141) is automatically met. In special cases, at low speeds, $\Delta\omega_r < 0$ ($HK_1 \leq 1$) in order to provide a stable operation in case of random friction torque perturbations. Good performance has been demonstrated down to 100 rpm [34].

Note: In case flux weakening is to be performed (in order to reduce core loss), for low load levels the gain K_1 in the filter in Figure 9.65 ($K_1 = 1 / H$) has to be adapted based on the *a priori* known rotor flux level.

A good example is represented by drives where the load torque is a function of speed only and thus the value of λ_r (and H) is solely a function of speed.

Again we mention that V_1/f_1 drives with slip compensation produces satisfactory precision in speed-sensorless control down to 100 rpm but they are adequate only for low dynamics applications (pumps, ventilators and the like).

9.18. SELF-COMMISSIONING

By self-commissioning we mean the parameter estimation and controller closed-loops calibration by the drive itself, when working in the initialization mode, at the site of application, before actual operation starts.

It is well understood that motor-converter parameters to be estimated depend on the particular control strategy adopted for the case.

As there are so many practical schemes for induction motor control we will treat here only a few basic principles of parameter estimation and controller calibration for a specific vector control system.

Let us consider a direct vector current control scheme (Figure 9.66) with d.c. (synchronous) current controllers and current model open-loop rotor flux estimator.

The apparent parameters are L_m (main inductance) and the rotor time constant τ_r.

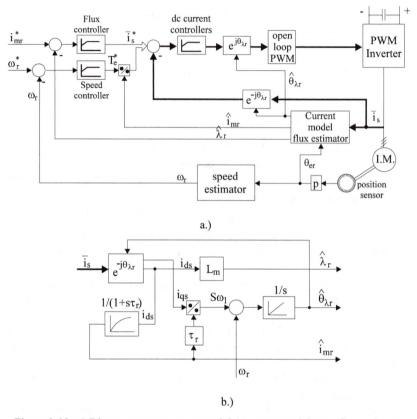

Figure 9.66. a.) Direct vector current control, b.) current model rotor flux estimator

However, in order to calibrate the current, flux and speed controllers:

- first, the stator resistance r_s and the transient time constant $\tau' = L_{sc} / (r_s+r_r)$ have to be estimated and, based on these values, the current controllers are to be calibrated;
- second, with the current controllers on, the rotor time constant τ_r is estimated and used in the rotor flux estimator (Figure 9.66.b). Further, the flux controller is calibrated.
- third, the reference rotor flux level and the corresponding magnetizing current $\hat{i}_{mr}$ are calculated;
- fourth, through a no-load acceleration test, the mechanical time constant τ_m (9.139) is calculated.

Various methods to estimate the induction motor parameters r_s, L_{sc}, L_m, τ_r, J may be classified into:

- step voltage response tests at standstill;
- frequency response tests at standstill;
- dynamic tests (nonzero speed).

Step voltage tests at standstill are rather easy to perform as the PWM may deliver voltage pulses for given time intervals. They still dominate the practice of self-commissioning programs.

However, recently it has been shown that step voltage response is influenced too much by the rotor cage skin effect, especially as the power of motor goes up. Consequently, the short-circuit inductance L_{sc} is underestimated in comparison with the case of actual operation conditions [35].

Standstill frequency tests are proven to produce better results for a wide power range (1kW-500kW) [35].

Here we still rely on step voltage responses at standstill [36]. The process of self-commissioning starts with introducing the motor specifications: rated voltage, rated current, rated frequency and the number of pole pairs p.

The stator resistance is first calculated from a d.c. test knowing the d.c. link voltage and the modulation index (no voltage sensors are used in our case m_d).

To account for the inverter dead time effect, measurements are taken for two different modulation indexes m_{d1} and m_{d2}

$$r_s = \frac{V_1\left(m_{d1} - m_{d2}\right)}{i_{s1} - i_{s2}}$$

(9.145)

Up to this point the current controllers are only grossly calibrated. The current levels are 50% and 100% of rated current and only a voltage vector is applied (V_1, for example). To estimate the motor transient time constant τ', the inverter is controlled through a binary port (with the modulator inhibited) for a few microseconds noting the time t_{peak} and the peak current reached, i_{peak} (Figure 9.67).

Due to the short time interval, the main flux does not occur and thus the motor equation is

$$\overline{V}_s = \left(r_s + r_r\right)\overline{i}_s + L_{sc}\frac{d\overline{i}_s}{dt}$$

(9.146)

$$\tau' = \frac{L_{sc}}{r_s + r_r} = \frac{V_d t_{peak}}{\left(r_s + r_r\right)i_{peak}}$$

(9.147)

The average value of τ' from a few tests is computed.

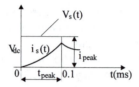

Figure 9.67. Step voltage response at standstill

Based on the above results, the current controllers (Figure 9.68) may be calibrated. T_s is the sampling time.

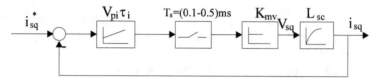

Figure 9.68. D.c. (synchronous) current controller

Based on the optimal gain method [36], the PI controller constants are

$$V_{Pi} = \frac{\tau'}{2T_s V_d K_{inv}}; \quad \tau_i = \tau' \qquad (9.148)$$

Further, we have to estimate the rotor time constant. We feed the inverter at standstill with d.c. current (applying V_1 with PWM) and then turn it off and record the stator voltage

$$\overline{V}_{s0}(t) = L_m \frac{d\overline{i}_r}{dt} \qquad (9.149)$$

$$0 = \overline{i}_r + \tau_r \frac{d\overline{i}_r}{dt} \qquad (9.150)$$

The rotor time constant τ_r is approximately

$$\tau_r \approx \frac{V_{s0}(t)}{V_{s0}(t_1) - V_{s0}(t_2)}(t_1 - t_2); \quad t_1 < t < t_2 \qquad (9.151)$$

V_{s0} has small values measured directly without attenuation. Once τ_r is known, the rated rotor flux magnetizing current i_{ms}^* can be determined.

To do so, we use the drive in the V_1 / f_1 open-loop mode at 10% of rated frequency and the motor no-load current is transformed into synchronous coordinates. The rotor time constant τ_r is changed until the i_q current in synchronous coordinates is zero. The d axis current will be the rated magnetizing current i_{mo}^* (provided V_1 / f_1 ratio is the rated one), [36].

The same test is performed above the rated frequency for rated voltage to get the value of $i_{mr}(\omega_r)$.

Also, $L_m = \lambda_r / i_{mr}$ has to be found in the process. The flux controller (Figure 9.66) may then be calibrated (Figure 9.69).

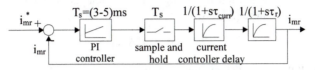

Figure 9.69. Rotor flux loop controller

The mechanical time constant τ_m may be calculated from constant torque (i_{mr}, i_q constant) rotor field orientation acceleration tests. Noting that the speed reached ω_a after the time t_a, the mechanical time constant is obtained from the motion equation

$$\frac{J d\omega_r}{p \, dt} = \frac{3}{2} p \frac{L_m^2}{L_r} i_{mr} i_q \tag{9.152}$$

$$J = \frac{3}{2} p^2 \frac{L_m^2}{L_r} i_{mr} i_q \frac{t_a}{\omega_a} \tag{9.153}$$

With τ_m from (9.139) and J from (9.153) we get

$$\tau_m = \frac{J}{pH} = \frac{1}{\tau_r} \frac{t_a}{\omega_a} \frac{i_q}{i_{mr}} \tag{9.144}$$

The speed controller may now be calibrated using standard methods.

The timing of self-commissioning operations may be summarized as in Figure 9.70 [36]

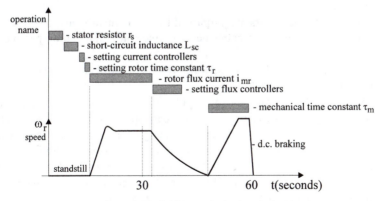

Figure 9.70. Self-commissioning timing

The whole process lasts about 60 seconds. After that the drive is ready for full performance operation. For sensorless operation slightly different techniques are required for self-commissioning.

In high performance drives the motor parameters are tuned through MRAC, on-line adaptation or through on-line estimators. These aspects are beyond our scope here [37, 38].

9.19. SUMMARY

- Induction motor drives may be used for torque, speed or position control. This control may use motion sensors for high performance response specifications or may be motion-sensorless.
- Historically, V_1/f_1-coordinated voltage/frequency open-loop (scalar) control without motion sensors - has been used for moderate performance (up to 10 / 1 speed range).
- Fast torque dynamics, however, requires linearization of the torque/speed curve and decoupled flux and torque control.
- Vector control (VC), direct torque and flux control (DTFC) or feedback linearization control (FLC) may all provide for fast torque dynamics high performance in variable speed induction motors.
- Vector control is performed with flux orientation and constant flux level so that fast torque control is obtained. Flux current and torque currents controls have to be decoupled for the scope.
- Though general flux orientation is, in principle, possible, rotor flux orientation leads to simplified current decoupling network and stator flux orientation leads to simple voltage decoupling network.
- Motor parameter detuning influence on rotor flux orientation is severe in vector current control and, at least for torque control, rotor time constant correction is required.
- Flux orientation may be achieved either indirectly (feedforward) or directly (feedback). In the latter case flux estimators (observers) are required.
- A.c. current controllers are surpassed by d.c. (synchronous, d-q) current controllers in terms of robustness to frequency and motor parameter detuning.
- Current control at low speeds and voltage control at high speeds are optimum in vector control.
- Extending the torque/speed envelope is performed through flux weakening and has two regions above base speed ω_b. One is the constant power zone when the d-q current (flux) angle (and slip frequency) varies up to the maximum torque / flux angle. In region two the slip frequency is constant at $(S\omega_1)_k = 1 / \tau_r$ but the power is no longer constant.
- Electric drives are current limited at low speeds and voltage (flux) limited at high speeds.

- Wide constant power zones (ω_{1max} / ω_b > 2) are obtained with low transient inductance L_{sc} motors.
- Voltage or current control according to their command values is performed through PWM techniques in the voltage source inverter. They may be open-loop or closed-loop.
- With open-loop PWM, the lockout time T_d between the turn-off of one PES and the turn-on of the next along the same inverter leg, has to be compensated to avoid voltage (current) distortion in low frequency (speed) operation.
- In direct vector control closed-loop flux observers are required. Full order or simplified observers are used. Up to 100 to 1 speed range, for fast torque response (1-5 ms for step rated torque at zero speed) has been achieved with recent commercial drives.
- Direct torque and flux control (DTFC) mean direct stator flux and torque error control to trigger a table of optimal switching in the PWM voltage source inverter.
- DTFC is a kind of combined current and voltage vector control in a simpler and more robust configuration but maintaining high steady-state and transient performance.
- Scalar open-loop (V_1/f_1) control is prone to instability regions in the V_1/f_1 plane, especially with two-pole motors.
- Compensating the slip frequency in (V_1/f_1) drives is the typical solution to improve speed control precision at low speeds. Still a less than 15 to 1 speed control ratio is available with low torque dynamics.
- Self-commissioning means parameter estimation and subsequent closed-loops-calibration by the drive itself at the site of application. It is an essential asset of modern electric drives. Typically self-commissioning takes 1-2 minutes only. After that the drive is ready to run at full performance.

9.20. PROBLEMS

9.1. In a rotor flux orientation current decoupling network, calculate the torque and flux detuning effects for cases a and b.
   - a.) L_m / L_m^* = β = 0.812. $S\omega_1\tau_r$ = 0.3 and 0.5 for rated rotor time constant τ_r^* = 40ms;
   - b.) for a ± 50% variation in the rotor resistance, $S\omega_1\tau_r$ = 0.5 and no saturation influence (β = 1).
   - c.) based on the rotor flux detuning monotonic function (Figure 9.4) and the rotor flux error, using a rotor flux estimator, introduce a PI controller to correct the rotor time constant τ_r in the rotor flux orientation current decoupling network (Figure 9.4) and draw the corresponding signal flow diagram for this case.

9.2. A vector control IM drive with rotor flux orientation operates in the flux weakening mode. The magnetizing inductance L_m of the motor varies linearly with the flux current i_d in rotor flux coordinates:

$L_m = 0.100H$, $0 < i_d \leq 3A$;

$L_m = 0.10 - 0.015i_d / 3$, for $i_d > 3A$

The leakage inductance $L_{ls} = L_{lr} = 0.005H$, $r_r = r_s = 0.35\Omega$, rated phase voltage $V_{1n} = 120V$ (rms per phase), $p = 2$, $\omega_{1b} = 2\pi60$ rad/s, $(S\omega_1)_{rated} = 3\pi$rad/s.

Calculate:

- the no-load phase rated current and the rated current;
- the rated stator, rotor and airgap flux motor amplitude;
- rated electromagnetic torque T_{en};
- maximum torque at $2\omega_b$.

9.3. As related to section 9.17 an induction motor has the data $p = 2$ pole pairs, $\omega_1 = 2\pi60$, $\lambda_r = 1Wb$, $r_r = 0.4\Omega$, $J = 0.1kgm^2$, $\tau_r = 0.1s$, $S_n = 0.01$. For constant rotor flux, and slow transients, calculate the speed response for a 10 Nm step increase in the load torque.

9.21. SELECTED REFERENCES

1.   **F. Blaschke**, The principle of field orientation as applied to the new transvector closed-loop control system for rotating field machines, Siemens Review, vol.34, 1972, pp.217-220 (in german).

2.   **I. Takahashi, T. Noguchi**, A new quick response and high efficiency control strategy of an induction motor, Record of IEEE-IAS-1985, Annual Meeting, pp.496-502.

3.   **M. Depenbrock**, Direct self control (DSC) of inverter-fed induction machine, IEEE Trans. vol.PE-3, 1988, pp.420-429.

4.   **I. Boldea, S.A. Nasar**, Torque vector control (TVC) a class of fast and robust torque, speed and position digital controllers for electric drives, EMPS-vol.15, 1988, pp.135-148.

5.   **R. Marino, R. Peresada, P. Valigi**, Adaptive partial feedback linearization of induction motors, Proceedings of the 29th Conference on Decision and Control, Honolulu, Hawaii, 1990, pp.3313-3318.

6.   **M. Bodson, J. Chiarsson, R. Novotnak**, High performance induction motor control via input-output linearization, IEEE Control Systems, 1994, pp.25-33.

7.   **D.L. Sobczuk**, Nonlinear control of PWM inverter fed induction motor drives, Record of ISIE '96, Warsaw, Poland, vol.2., pp.958-962.

8.   **R.De Doncker, D.W. Novotny**, The universal field oriented controller, Rec. IEEE-IAS, Annual Meeting, 1988, pp.450-456.

9.   **I. Boldea, S.A. Nasar**, Vector control of a.c. drives, Chapter 3, CRC Press, Florida, USA, 1992.

10.   **R. Krishnan, F.C. Doran**, Study of parameter sensitivity in high performance inverter-fed induction motor drives, Record of IEEE-IAS, Annual Meeting, 1984, pp.510-524.

11.   **T.R.Rowan, R.J. Kerkman**, A new synchronous current regulator and an analysis of current-regulated PWM inverters, IEEE Trans. Vol.IA-22, 1986, pp. 678-690.

12.   **S.H. Kim, S.K. Sul**, Maximum torque control of an induction machine in the field weakening region, IEEE Trans.vol.IA-31, no.4, 1995, pp.784-794.

13.   **R. Krishnan**, Review of flux weakening in high performance vector controlled induction motor drives, Record of ISIE'96, Warsaw Poland, vol.2, pp.917-922.

14.   **H. Grotstollen, A. Bünte**, Control of induction motor with orientation on rotor flux or on stator flux in a very wide field weakening region - experimental results, IBID, pp.911-916.

15. **J. Holtz**, Pulsewidth modulation for electronic power conversion, Proc. IEEE, vol.82, no.8, 1994, pp.1194-1214.
16. **D. Legatte, R.J. Kerkman**, Pulse-based dead-time compensation for PWM voltage inverters, IEEE Trans.vol.IE-44, no.2, 1997, pp.191-197.
17. **J. Holtz, E. Bube**, Field oriented asynchronous PWM for high performance a.c. machine drives operating at low switching frequency, IEEE Trans, vol.IA-27, no.3, 1991, pp.574-581.
18. **A.M. Trzynadlowski**, An overview of modern PWM tehniques for three phase-voltage-controlled voltage-source inverters, Record of ISIE-96, Warsaw, Poland, vol.1, pp25-39.
19. **A. Nabae, S.Ogasawara, H.Akagi**, A novel control scheme for current controlled PWM inverters, IEEE Trans, vol.IA-22, 1986, pp.697-701.
20. **P.L. Jansen, R.D. Lorenz, D.W. Novotny**, Observer based direct filed orientation and comparison of alternative methods, IEEE Trans, vol.IA-30, no.4, 1994, pp.945-953.
21. **L. Kwindler, J.C. Moreira, A. Testa, T.A. Lipo**, Direct field orientation controller using the stator phase voltage third harmonics, IEEE Trans., Vol.IA-30, no.2, 1994, pp.441-447.
22. **Y. Hori, T. Umeno**, Implementation of robust flux observer based field orientation controller for induction machines, Record of IEEE-IAS, Annual Meeting, 1989, pp.523-528.
23. **J. Holtz**, Methods for speed sensorless control of a.c. drives, Record of IEEE-PCC-Yokohama, 1993, pp.415-420.
24. **C. Has, A. Betini, L. Ferraris, G. Griva, F. Profumo**, Comparison of different schemes without shalf sensors for field oriented control drives, Record of IEEE-IECON, 1994, pp.1579-1588.
25. **F.Z. Peng, T. Fukao, J.S. Lai**, Low speed performance of robust speed identification using instantaneous reactive power for tacholess vector control of induction motors, Record of IEEE-IAS, Annual Meeting, 1994, pp.509-514.
26. **H. Kubota, K. Matsuse**, Speed sensorless, field oriented control of induction motor with rotor resistance adaption, IEEE Trans., vol.IA-30, no.5, 1994, pp.1219-1224.
27. **S.I. Yong, J.W. Choi, S.K. Sul**, Sensorless vector control of induction machine using high frequency current injection, Record of IEEE-IAS, Annual Meeting, 1994.
28. **Y.R. Kim, S.K. Sul, M.H. Park**, Speed sensorless control of an induction motor using an extended Kalman Filter, Record of IEEE-IAS, Annual Meeting, 1992, Part I, pp.594-599.
29. **K.D. Hurst, T.G. Habetler, G. Griva, F. Profumo**, Speed sensorless field-oriented control of induction machines using current harmonics spectral estimation, Record of IEEE-IAS, Annual Meeting, 1994.
30. **J. Holtz**, "Sensorless control of induction machines – with and without signal injection", Keynote address at OPTIM 2002, Poiana Brasov, Romania.
31. **P. Tiitinen, P. Pohyalainen, J. Lalu**, The next generation motor control method-direct torque control, DTC, in Proc. EPE Chapter Symp., Lausanne, Switzerland, 1994.
32. **M.P. Kazmierkowski**, Control philosophies of PWM inverter-fed induction motors, University of Aalborg lecture, Febr. 1997, DK.
33. **R. Ueda, T. Sonada, K. Koga, M. Ichikawa**, Stability analysis in induction motor driven V / f controlled general-purpose inverter, IEEE Trans. Vol.IA-28, no.2, 1992, pp.472-481.
34. **K. Koga, R. Ueda, T. Sonada**, Constitution of V / f control for reducing the steady-state error to zero in induction motor drive system, IEEE Trans., vol.IA-28, no.2, 1992, pp.463-471.
35. **R.J. Kerkman, J.D. Thunes, T.W. Rowan, D.W. Schlegel**, A frequency based determination of transient inductance and rotor resistance for field commissioning purposes. IEEE Trans.vol.IA-32, no.3, 1996, pp.577-584.
36. **A.M. Klambadkone, J. Holtz**, Vector controlled induction motor drive with a selfcommissioning scheme, IEEE Trans. Vol.IE-38, no.5, 1991, pp.322-327.
37. **M. Ruff, A. Bünte, H. Grotstolen**, A new selfcommissioning scheme for an asynchronous motor drive system, Record of IEEE-IAS, Annual Meeting, 1994, part.I., pp.616-623.
38. **R.J. Kerkman** et al., A new flux and stator resistance identifier for a.c. drives, IEEE Trans., vol.IA-32, no.3, 1996, pp.585-593.

39. **C. Lascu,** "Direct torque control of sensorless IM drives", Ph.D. Thesis, 2002, University Politehnica Timisoara, Romania.
40. **C. Lascu, I. Boldea, F. Blaabjerg,** "Variable-structure direct torque control – a class of fast and robust controllers for induction machine drives", IEEE Trans. Vol. IE – 51, 2004, pp. 785-792.
41. **D. Casadei, F. Profumo, G. Serra, A. Tani,** "FOC and DTC: two viable schemes for induction motors torque control", IEEE Trans. Vol PE – 17, 2002, pp. 779-787.
42. **C. Lascu, I. Boldea, F. Blaabjerg,** "Direct torque control of sensorless induction motor drives – a sliding mode approach", IEEE Trans. Vol. IA – 40, no. 2, 2004, pp. 582 – 590.
43. **I. Boldea, A. Trica,** "Torque vector control (TVC) voltage – fed induction motor drives – very low speed performance via sliding mode", Record of ICEM – 1990, vol. 3, pp. 1212-1217.
44. **J. Holtz, J. Quan,** "Sensorless vector control of induction motors at very low speed using a nonlinear inverter model and parameters identification", IEEE Trans., Vol. IA – 38, no. 4, 2002, pp. 1087-1095.

Chapter 10

SYNCHRONOUS MOTORS FOR DRIVES

10.1. INTRODUCTION

Synchronous motors (SMs) are, in general, three-phase a.c. fed in the stator and d.c. (or PM) excited in the rotor. As the stator currents produce an m.m.f. traveling at the electric speed ω_1

$$\omega_1 = 2\pi f_1, \tag{10.1}$$

the rotor m.m.f. (or PM) is fixed to the rotor. The rotor electrical speed ω_r is

$$\omega_r = \omega_1 = 2\pi np, \tag{10.2}$$

in order to obtain two m.m.f. waves at standstill with each other. It is a known fact that only in this situation is a nonzero average torque per revolution obtained.

Yet, in an alternative interpretation, nonzero average torque is produced when the magnetic coenergy in the machine W_{co} varies with rotor position

$$T_e = \left(\frac{\partial W_{co}}{\partial \theta_r}\right)_{i_i = ct} \quad ; \quad \frac{d\theta_r}{dt} = \frac{\omega_r}{p} \tag{10.3}$$

θ_r is the geometrical rotor position angle.

Thus a magnetically anisotropic (reluctance) — exciterless — rotor may also be used. In all cases the number of pole pairs is the same on the stator and on the rotor.

PM or reluctance rotors are preferred, in general, for low (even medium) power motors (up to 50 - 300 kW in general) and excited rotors are used for medium and high power (hundreds of kW, MW and tens of MWs) applications.

As the SM speed is rigidly related to the stator frequency only the development of PECs — variable voltage and frequency sources — has made the SM suitable for variable speed drives.

The higher efficiency, power density and power levels per unit have thus become the main assets of variable speed synchronous motor drives.

10.2. CONSTRUCTION ASPECTS

As with any electrical motor, the SM has a stator and a rotor. The stator is made of a laminated core with uniform slotting. The stator slots accommodate, in general, a three-phase, single- or double-layer winding made like for induction motors.

In general, the number of slots per pole and phase q ≥ 2 and thus the coupling inductances vary sinusoidally with rotor position. However, in small power PM-SMs, concentrated windings may be used (q = 1 slot per pole per phase) when the sinusoidality is lost [1,2].

On the other hand, the SM rotors may be:

- active: - excited
          - with PMs
- passive: - highly anisotropic (variable reluctance)

The excited and PM rotors may be with salient or nonsalient poles, that is magnetically anisotropic or isotropic.

Also excited-rotor SMs may have a squirrel cage on the rotor to reduce the commutation inductance in current source PEC variable speed drives.

Figure 10.1 shows salient pole SMs with excited and PM rotors while Figure 10.2 exhibits nonsalient pole SMs.

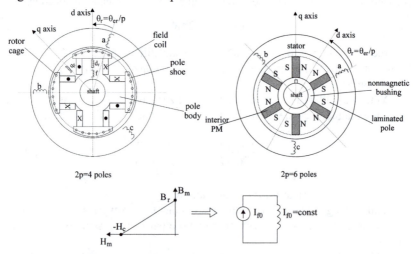

Figure 10.1. Salient pole active rotor SMs with a.) excited rotor, b.) interior PMs

The high energy PMs have a rather linear demagnetization characteristic in the second quadrant (Figure 10.1b) and thus such PMs may be considered as constant current ideal field coils or artificial superconducting coils at room temperature.

Once magnetized with a few high current pulses (milliseconds long), capable to produce (2 to 3) H_c and (2 to 3) B_r (see Figure 10.1), the magnetic energy stored in the PMs stays there for years if accidental demagnetization through ultrahigh stator currents is avoided.

The energy required to magnetize the PMs is much smaller than the power losses in an equivalent field winding over the active life of the PMs (4-5 years or more). However, the cost of PMs is notably higher than the field winding costs. If the capitalized cost of losses is considered however, the extra cost of PMs is paid back in less than 2-3 years.

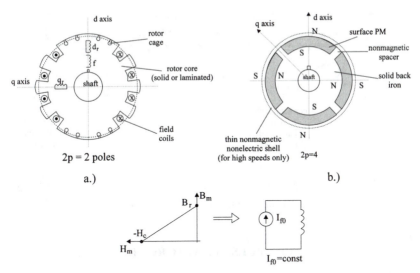

Figure 10.2. Nonsalient pole active rotors of SMs: a.) excited, b.) with surface PM

While the nonsalient pole configuration is more suitable for high peripheral speeds for excited rotors, with PM rotors the opposite is true; that is, interior PMs are, in general, preferred for high speed as PM "protection" against large centrifugal and radial electromagnetic forces are implicitly provided. Radial — attraction — forces between the PM rotor and the stator core are exerted through the surface PMs and through the laminated poles (for the interior PM rotors). Note that all PMs are rather severely temperature limited (100°C for NdFeB, 150°C for SmCo₅, etc) and thus, in adverse environments, the PMs are a liability.

Passive rotors are characterized by high magnetic anisotropy (or saliency). They may be made of conventional or of axial laminations (Figure 10.3a, b). Multiple flux barriers are provided to increase the saliency for the conventional rotor (Figure 10.3a). The same scope is served easily by alternating axial laminations with insulation layers (Figure 10.3b).

Special measures need to be taken to reduce the rotor core harmonics losses in the axially laminated anisotropic rotor [3].

Saliency ratios (for rated conditions) above 10 to 1 have been obtained, thus pushing the performance of reluctance synchronous motors (RSMs) to the level of IMs [3]. Obviously the absence of the rotor cage makes both PM-SMs and RSMs fully dependent on power electronic converters (PECs).

The insulation layers may be replaced by PMs along q axis (Figure 10.3b), in order to improve the torque density. Also, in the presence of a rotor cage on both PM rotor or variable reluctance rotor, the same motors may be line-started and could operate at constant speed for constant voltage-frequency conditions.

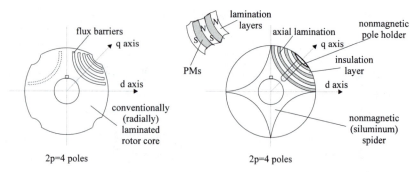

Figure 10.3. Passive-anisotropic (reluctance)-rotors for SMs: a.) multiple flux barrier rotor, b.) axially-laminated-anisotropic (ALA) rotor

10.3. PULSATING TORQUE

It is a known fact that an ideal SM, with sinusoidal m.m.f. and constant airgap — when fed with sinusoidal symmetric currents in the stator at frequency $\omega_1 = \omega_r$ (ω_r - rotor speed) — produces a constant torque.

In reality, pulsating torques may occur due to:

a. stator (and rotor) slot openings;

b. magnetic saturation caused by flux harmonics;

c. current waveforms;

d. PM field pulsations due to stator slot openings (cogging torque).

Items a to c cause the so-called electromagnetic pulsating torques while d causes the zero stator current (or cogging) torque.

Rotor pole (or PM) span correlation with stator slot openings, stator slot inclination or PM pole inclination, fractional q (slots per pole and phase) and, finally, special current waveform shaping through PEC control are all methods to reduce these, basically reluctance, parasitic torques [4] to less than 1% of rated torques. High performance smooth torque drives capable of operating below 20 rpm in sensorless control or under 1 rpm with position sensor control are thus obtained.

Note: Pulsating torque investigation requires, in most cases, FEM analysis - two-, quasi-two or three-dimensional [4].

10.4. THE PHASE COORDINATE MODEL

The phase coordinate model is based on phase equations using stator and rotor circuits. For q ≥ 2 (slots per pole per phase) - distributed windings - the inductance matrix contains sinusoidal terms. We will take the case of the salient pole rotor as the nonsalient pole rotor is a particular case of the former.

The phase inductance matrix [L] is (see Figures 10.1 and 10.2)

$$[L] = \begin{bmatrix} & a & b & c & f & d_r & q_r \\ a & L_{aa} & L_{ab} & L_{ac} & L_{af} & L_{ad_r} & L_{aq_r} \\ b & L_{ab} & L_{bb} & L_{bc} & L_{bf} & L_{bd_r} & L_{bq_r} \\ c & L_{ac} & L_{bc} & L_{cc} & L_{cf} & L_{cd_r} & L_{cq_r} \\ f & L_{af} & L_{bf} & L_{cf} & L_{ff} & L_{fd_r} & 0 \\ d_r & L_{ad_r} & L_{bd_r} & L_{cd_r} & L_{fd_r} & L_{d_r d_r} & 0 \\ q_r & L_{aq_r} & L_{bq_r} & L_{cq_r} & 0 & 0 & L_{q_r q_r} \end{bmatrix} \qquad (10.4)$$

For distributed windings (q ≥ 2) all the stator self, mutual, and stator-rotor inductances are rotor position, θ_{er}, dependent

$$L_{aa} = L_{sl} + L_0 + L_2 \cos 2\theta_{er}$$
$$L_{bb} = L_{sl} + L_0 + L_2 \cos(2\theta_{er} + 2\pi/3)$$
$$L_{cc} = L_{sl} + L_0 + L_2 \cos(2\theta_{er} - 2\pi/3) \qquad (10.5)$$
$$L_{bc} = -L_0/2 + L_2 \cos 2\theta_{er}$$
$$L_{ac} = -L_0/2 + L_2 \cos(2\theta_{er} + 2\pi/3)$$
$$L_{ab} = -L_0/2 + L_2 \cos(2\theta_{er} - 2\pi/3)$$

$$L_{ad_r} = L_{sd_r} \cos\theta_{er}, \ L_{bd_r} = L_{sd_r} \cos(\theta_{er} - 2\pi/3),$$
$$L_{cd_r} = L_{sd_r} \cos(\theta_{er} + 2\pi/3)$$
$$L_{af} = L_{sf} \cos\theta_{er}, \ L_{bf} = L_{sf} \cos(\theta_{er} - 2\pi/3),$$
$$L_{cf} = L_{sf} \cos(\theta_{er} + 2\pi/3)$$
$$L_{aq_r} = -L_{sq_r} \sin\theta_{er}, \ L_{bq_r} = -L_{sq_r} \sin(\theta_{er} - 2\pi/3),$$
$$L_{cq_r} = -L_{sq_r} \sin(\theta_{er} + 2\pi/3)$$

Rotor inductances are evidently independent of rotor position as the saliency is on the rotor itself

$$L_{ff} = L_{fl} + L_{fm}$$
$$L_{d_r d_r} = L_{d_r l} + L_{d_r m}, \ L_{q_r q_r} = L_{q_r l} + L_{q_r m} \qquad (10.6)$$

where $L_{fl}, L_{d_r l}, L_{q_r l}$ are leakage inductances while the others are related to main flux path.

On the other hand, for *concentrated* windings (q = 1), when nonsalient pole PM rotors are used, all stator inductances are independent of rotor position, the rotor cage is eliminated.

Only the motion-related inductances between the constant field-current PM equivalent circuit and the stator windings depend on rotor position.

$$[L]_{\substack{PM \\ q=1}} = \begin{bmatrix} L_s & L_{ab} & L_{ab} & L_{af}(\theta_{er}) \\ L_{ab} & L_s & L_{ab} & L_{bf}(\theta_{er}) \\ L_{ab} & L_{ab} & L_s & L_{cf}(\theta_{er}) \\ L_{af}(\theta_{er}) & L_{bf}(\theta_{er}) & L_{cf}(\theta_{er}) & 0 \end{bmatrix} \tag{10.7}$$

The stator self-inductance L_s and the stator-stator mutual inductances L_{ab} are all equal to each other but different from $-L_0 / 2$ (as they would become for nonsalient rotors ($L_2 = 0$)). To a first approximation, $L_{ab} = -L_0 / 3$ [1].

The voltage-current matrix equation in phase coordinates (stator coordinates for stator, rotor coordinates for rotor) are

$$[V] = [r] \cdot [i] + \frac{d[\lambda]}{dt} \tag{10.8}$$

with

$$[\lambda] = \left[L(\theta_{er}) \right] \cdot [i] \tag{10.9}$$

$$[i] = \left[i_a, i_b, i_c, i_{f_0} \right] \tag{10.10}$$

$$[r] = Diag[r_s, r_s, r_s, 0] \tag{10.11}$$

$$[\lambda] = [\lambda_a, \lambda_b, \lambda_c, 0] \tag{10.12}$$

Finally, the electromagnetic torque T_e may be calculated from the coenergy derivative with respect to rotor position

$$T_e = \frac{dW_{co}}{d(\theta_{er})} = \frac{d}{d(\theta_{er})} \int_0^{[i]} [\lambda] d[i]^T \tag{10.13}$$

After neglecting magnetic saturation, we multiply (10.8) by $[i]^T$

$$[i]^T \cdot [V] = [r] \cdot [i] \cdot [i]^T + \frac{d}{dt} \left(\frac{1}{2} [L] \cdot [i] \cdot [i]^T \right) + \\ + \frac{1}{2} \cdot [i]^T \cdot \frac{\partial}{\partial(\theta_{er})} \left[L(\theta_{er}) \right] \cdot [i] \cdot \frac{d(\theta_{er})}{dt} \tag{10.14}$$

The last term is the electromagnetic power P_{elm}

$$T_e = \frac{P_{elm} \cdot p}{\omega_r} = \frac{p}{2} \cdot [i]^T \cdot \frac{\partial}{\partial \theta_{er}} \left[L(\theta_{er}) \right] \cdot [i] \tag{10.15}$$

The motion equations are

$$\frac{J}{p} \frac{d\omega_r}{dt} = T_e - T_{load}; \quad \frac{d(\theta_{er})}{dt} = \frac{\omega_r}{p} \tag{10.16}$$

As for the induction motor, we ended up with a set of eight nonlinear differential equations with time varying coefficients (basically, inductances). As such, these equations are used for special cases for machines with some asymmetry or for unbalanced supply voltage operation.

Also, for the concentrated stator winding (q = 1) and the PM nonsalient cageless rotor, the phase variable model is the only one to be used as $L_{af}(\theta_{er})$, $L_{bf}(\theta_{er})$ and $L_{cf}(\theta_{er})$ are far from sinusoidal functions. In order to get rid of the inductance dependence on rotor position for $q \geq 2$, the space-phasor (d-q) model is used.

10.5. THE SPACE-PHASOR (d-q) MODEL

Proceeding as for the induction motor, we may define the stator current space-phasor, $\bar{i}_s$ in stator coordinates

$$\bar{i}_s^{\ s} = \frac{2}{3} \cdot \left(i_a + a \cdot i_b + a^2 \cdot i_c \right) \tag{10.17}$$

On the other hand, from (10.9) with (10.4), the phase a flux linkage λ_a is

$$\lambda_a = L_{aa} i_a + L_{ab} i_b + L_{ac} i_c + L_{af} i_f + L_{ad_r} i_{d_r} + L_{aq_r} i_{q_r} \tag{10.18}$$

Making use of the inductance definition of (10.5) we find

$$\lambda_a = L_{sl} \, \mathrm{Re}\left(\bar{i}_s\right) + \frac{3}{2} L_0 \, \mathrm{Re}\left(\bar{i}_s\right) + \frac{3}{2} L_2 \, \mathrm{Re}\left(\bar{i}_s^{\ *} e^{2j\theta_{er}}\right) +$$
$$L_{af} \, \mathrm{Re}\left(i_f^{\ r} e^{j\theta_{er}}\right) + L_{sd_r} \, \mathrm{Re}\left(i_{d_r}^{\ r} e^{j\theta_{er}}\right) - \mathrm{Re}\left(j L_{sq_r} i_{q_r}^{\ r} e^{j\theta_{er}}\right) \tag{10.19}$$

The stator flux space-phasor $\bar{\lambda}_s$ is

$$\bar{\lambda}_s^{\ s} = \frac{2}{3}\left(\lambda_a + a\lambda_b + a^2 \lambda_c \right) \tag{10.20}$$

Making use of (10.18) for λ_b, λ_c, (10.20) yields

$$\bar{\lambda}_s^{\ s} = L_{sl} \bar{i}_s^{\ s} + \frac{3}{2} L_0 \bar{i}_s^{\ s} + \frac{3}{2} L_2 \bar{i}_s^{\ s*} e^{2j\theta_{er}} +$$
$$L_{sf} i_f^{\ r} e^{j\theta_{er}} + L_{sd_r} i_{d_r}^{\ r} e^{j\theta_{er}} + L_{sq_r} j i_{q_r}^{\ r} e^{j\theta_{er}} \tag{10.21}$$

Multiplying (10.21) by $e^{-j\theta_{er}}$ we obtain

$$\bar{\lambda}_s^{\ s} e^{-j\theta_{er}} = \left(L_{sl} + \frac{3}{2} L_0 \right) \bar{i}_s^{\ s} e^{-j\theta_{er}} + \frac{3}{2} L_2 \left(\bar{i}_s^{\ s} e^{-j\theta_{er}} \right)^* +$$
$$+ L_{sf} i_f^{\ r} + L_{sd_r} i_{d_r}^{\ r} + j L_{sq_r} i_{q_r}^{\ r} \tag{10.22}$$

We should now note that

$$\overline{\lambda}_s = \overline{\lambda}_s{}^s e^{-j\theta_{er}}; \quad \overline{i}_s = \overline{i}_s{}^s e^{-j\theta_{er}} \qquad (10.23)$$

are space-phasors in rotor coordinates (aligned with rotor d axis).
   With (10.23), (10.22) becomes

$$\overline{\lambda}_s = \left(L_{sl} + \frac{3}{2} L_0 \right) \overline{i}_s + \frac{3}{2} L_2 \overline{i}_s{}^* + L_{sf} i_f{}^r + L_{sd_r} i_{d_r}{}^r + j L_{sq_r} i_{q_r}{}^r \qquad (10.24)$$

Next, the stator-phase equations in stator coordinates are

$$r_s i_a - V_a = -\frac{d\lambda_a}{dt}$$

$$r_s i_b - V_b = -\frac{d\lambda_b}{dt} \qquad (10.25)$$

$$r_s i_c - V_c = -\frac{d\lambda_c}{dt}$$

We may translate them in space-phasors as

$$r_s \overline{i}_s{}^s - \overline{V}_s{}^s = -\frac{d\overline{\lambda}_s{}^s}{dt} = -\frac{d}{dt} \left(\overline{\lambda}_s e^{j\theta_{er}} \right) = -e^{j\theta_{er}} \frac{d\overline{\lambda}_s}{dt} - j\omega_r \overline{\lambda}_s e^{j\theta_{er}} \qquad (10.26)$$

The final form of (10.26) is

$$r_s \overline{i}_s - \overline{V}_s = -\frac{d\overline{\lambda}_s}{dt} - j\omega_r \overline{\lambda}_s; \quad \omega_r = \frac{d\theta_{er}}{dt} \qquad (10.27)$$

and

$$V_0 = r_s i_0 + L_{sl} \frac{di_0}{dt}; \quad i_0 = \left(i_a + i_b + i_c \right)/3 \qquad (10.28)$$

Equation (10.27) is, in fact, identical to that for the induction machine. Only the flux expression (10.24) is different.

$$\overline{\lambda}_s = \lambda_d + j\lambda_q; \quad \lambda_d = L_{sl} i_d + \lambda_{dm}; \quad \lambda_{dm} = L_{dm} i_d{}^r + L_{sf} i_f{}^r + L_{sd} i_{d_r}{}^r \qquad (10.29)$$

$$\lambda_q = L_{sl} i_q + \lambda_{qm}; \quad \lambda_{qm} = L_{qm} i_d + L_{sq} i_{q_r}{}^r \qquad (10.30)$$

with

$$L_{dm} = \frac{3}{2} \left(L_0 + L_2 \right)$$

$$L_{qm} = \frac{3}{2} \left(L_0 - L_2 \right) \qquad (10.31)$$

L_{dm} and L_{qm} are called the d-q magnetizing inductances.
We may reduce the rotor to stator currents

$$\frac{i_f}{i_f^r} = \frac{L_{sf}}{L_{dm}} = K_f; \quad \frac{i_{d_r}}{i_{d_r}^r} = \frac{L_{sd}}{L_{dm}} = K_d; \quad \frac{i_{q_r}}{i_{q_r}^r} = \frac{L_{sq}}{L_{dm}} = K_q; \quad (10.32)$$

$$\lambda_{dm} = L_{dm} i_{dm}; \quad i_{dm} = i_d + i_f + i_{d_r}$$
$$\lambda_{qm} = L_{qm} i_{qm}; \quad i_{qm} = i_q + i_{q_r}; \quad i_m = \sqrt{i_{dm}^2 + i_{qm}^2} \quad (10.33)$$

Magnetic saturation may be determined by unique $\lambda_{dm}(i_m)$ and $\lambda_{qm}(i_m)$ functions to be either calculated or measured [5].

The stator equation (10.27) in d-q coordinates becomes

$$V_d = r_s i_d + \frac{d\lambda_d}{dt} - \omega_r \lambda_q \quad (10.34)$$

$$V_q = r_s i_q + \frac{d\lambda_q}{dt} + \omega_r \lambda_d \quad (10.35)$$

Now we should add the rotor equations in rotor d-q coordinates (as along axes d and q, the rotor windings are not at all symmetric)

$$V_f = r_f i_f + \frac{d\lambda_f}{dt}; \quad \lambda_f = L_{fl} i_f + \lambda_{dm} \quad (10.36)$$

$$0 = r_{d_r} i_{d_r} + \frac{d\lambda_{d_r}}{dt}; \quad \lambda_{d_r} = L_{d_r l} i_{d_r} + \lambda_{dm} \quad (10.37)$$

$$0 = r_{q_r} i_{q_r} + \frac{d\lambda_{q_r}}{dt}; \quad \lambda_{q_r} = L_{q_r l} i_{q_r} + \lambda_{qm} \quad (10.38)$$

The torque, T_e, is

$$T_e = \frac{3}{2} p \, Re(j\bar{\lambda}_s \bar{i}_s^*) = \frac{3}{2} p \left(\lambda_d i_q - \lambda_q i_d \right) \quad (10.39)$$

Finally, the d-q variables are related to the abc variables by the Park transformation

$$\bar{V}_s = V_d + jV_q = \frac{2}{3} \left(V_a + aV_b + a^2 V_c \right) e^{-j\theta_{er}}; \quad a = e^{j2\pi/3} \quad (10.40)$$

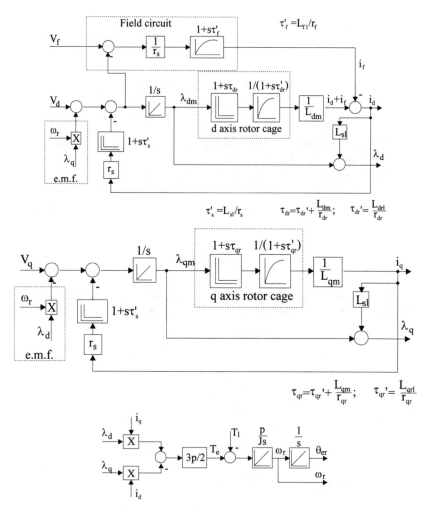

Figure 10.4. Signal flow diagram of SMs

Note also that all rotor variables are reduced to the stator

$$V_f = V_f^{\,r} / K_f; \quad r_f = r_f^{\,r} / K_f^{\,2}; \quad L_{fl} = L_{fl}^{\,r} / K_f^{\,2} \tag{10.41}$$

$$r_{dr} = r_{dr}^{\,r} / K_d^{\,2}; \quad L_{drl} = L_{drl}^{\,r} / K_d^{\,2} \tag{10.42}$$

$$r_{qr} = r_{qr}^{\,r} / K_q^{\,2}; \quad L_{qrl} = L_{qrl}^{\,r} / K_q^{\,2} \tag{10.43}$$

The motion equations (10.16) are to be added.

The d-q model is also of 8th order and basically nonlinear, but the coefficients are position (time) independent. The signal flow diagram of the d-q model is shown in Figure 10.4.

PM or reluctance rotor SMs used for variable speed lack the excitation winding and i_f = const. (zero for RSM).

They also lack the damper cage ($\tau_{dr} = \tau_{rd}' = \tau_{qr} = \tau_{qr}' = 0$) (Figure 10.5).

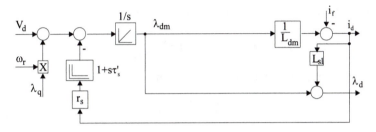

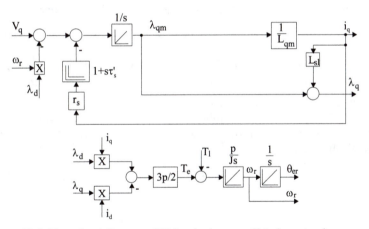

Figure 10.5. Flow signal diagram of PM and reluctance SMs (no rotor damper cage)

The signal flow diagrams may be arranged into equivalent circuits (Figure 10.6).

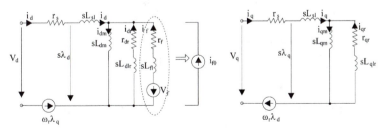

Figure 10.6. SM equivalent circuits

For the PM-SM (Figure 10.6), the field winding circuit is replaced by an ideal (constant) current source i_{f0}; $i_{f0} = 0$ for the RSM.

Finally, for the cageless rotor, $r_{dr} = r_{qr} = \infty$ ($i_{dr} = i_{qr} = 0$). For steady-state, simply s = 0 as we will demonstrate in the following section.

10.6. STEADY-STATE OPERATION

For steady-state the stator phase voltages are given by

$$V_{abc} = \sqrt{2}V\cos\left[\omega_1 t - (i-1)\frac{2\pi}{3}\right]; \quad i = 1,2,3 \qquad (10.44)$$

where ω_1 = const., and thus the rotor position θ_{er} is

$$\theta_{er} = \int\omega_r dt = \int\omega_1 dt = \omega_1 t + \theta_0 \qquad (10.45)$$

Note that $\omega_r = \omega_{r0} = \omega_1$ for steady-state.

Using (10.40) the d-q voltage components V_d, V_q in rotor coordinates are

$$\overline{V}_s = V_d + jV_q = V\sqrt{2}\cos\theta_0 - jV\sqrt{2}\sin\theta_0 \qquad (10.46)$$

As V_d, V_q are time independent so will be the currents in the stator i_d, i_q. Also d/dt = 0 (s = 0) and thus from (10.38), $i_{d_r} = i_{q_r} = 0$ (zero damper cage currents) and $i_f = i_{f0} = V_f / r_f$ = const.

The space-phasor diagram follows from ((10.34)-(10.35)) which may be written in space-phasor form as the stator is symmetric electrically and magnetically (Figure 10.7).

$$\overline{V}_s = r_s \overline{i}_s + j\omega_r \overline{\lambda}_s; \quad \omega_r = \omega_1 \qquad (10.47)$$

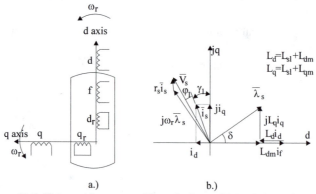

a.)   b.)

Figure 10.7. SM - steady-state - with excitation or PMs a.) d-q axis model, b.) space-phasor diagram

For steady-state all variables in the space-phasor (or d-q) model are d.c. quantities. Space angles translate into time-phase shift angles between stator quantities (like φ_1 - the power factor angle) in phase coordinates.

Using the d-q model for speed control is thus ideal, as steady-state means d.c.

10.7. TO VARY SPEED VARIABLE FREQUENCY IS MANDATORY

SMs keep a tight relationship between speed and frequency as the rotor m.m.f. (if any) is of d.c. (or PM) character

$$n = f_1 / p; \quad \omega_r = \omega_1; \quad \Omega_r = \omega_r / p \qquad (10.48)$$

Varying the number of pole pairs p is unusual so, in fact, to vary speed variable frequency is mandatory. How to vary frequency in relation to stator voltage V_s and field current i_f (or voltage V_{ex}) — if any — to obtain steady-state and (or) transient performance suitable for various applications is the key to SM drives control.

While this problem will be dealt with in detail in the following chapters, the main principles will be illustrated through three numerical examples.

Example 10.1. Unity power factor and constant stator flux.

Let us consider an excited synchronous motor with the data: $V_n = 660V$ (line voltage, rms, star connection), $I_n = 500/\sqrt{2}$ (rms), $r_s = 0.016\Omega$, $L_d = 2L_q = 0.0056H$, $L_{sl} = 0.1L_d$, $\omega_1 = 2\pi60rad/s$, p = 2 pole pairs.

Determine:

- The steady-state space-phasor diagram for unity power factor ($\varphi_1 = 0$)
- Calculate the values of i_f, i_d, i_q for unity power factor and rated phase current, I_n.
- Calculate the torque for $\varphi_1 = 0$.
- Derive and plot the torque/speed curve for rated and half-rated stator flux for variable frequency and constant stator voltage.

Solution:

a. The required space-phasor diagram comes from Figure 10.7b with $\varphi_1=0$.

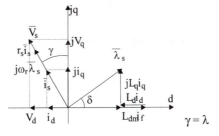

Figure 10.8. Space-phasor diagram for unity power factor

b. From Figure 10.8 we have

$$V_s = r_s i_s + \omega_r \lambda_s \tag{10.49}$$

$$\overline{\lambda}_s = \lambda_d + j\lambda_q = L_d i_d + L_{dm} i_f + jL_q i_q \tag{10.50}$$

For no-load $V_{s0} = \omega_r L_{dm} i_f; \ \left(i_d = i_q = 0\right); \ \omega_r = \omega_1 \tag{10.51}$

$$V_s = \frac{660\sqrt{2}}{\sqrt{3}} = 537.92V; \ i_s = i_n\sqrt{2} = \frac{500}{\sqrt{2}}\sqrt{2} = 500A \tag{10.52}$$

From (10.49) the rated stator flux λ_s is

$$\lambda_s = \frac{V_s - r_s i_s}{\omega_r} = \frac{537.92 - 0.016 \cdot 500}{2\pi 60} = 1.406 \ Wb \tag{10.53}$$

$$i_q = i_s \cos\delta \tag{10.54}$$

$$L_q i_q = \lambda_s \sin\delta \tag{10.55}$$

Thus $\lambda_s \tan\delta = L_q i_s \tag{10.56}$

$$\tan\delta = \frac{L_q i_s}{\lambda_s} = \frac{0.0028 \cdot 500}{1.406} \approx 1 \Rightarrow \delta = 45^0 \tag{10.57}$$

$$-i_d = i_q = \frac{i_s}{\sqrt{2}} = \frac{500}{\sqrt{2}} = 354.61 \ A \tag{10.58}$$

$$\lambda_d = \lambda_q = \frac{\lambda_s}{\sqrt{2}} = \frac{1.406}{\sqrt{2}} = 0.997 \ Wb \tag{10.59}$$

$$\lambda_d = L_d i_d + L_{dm} i_f \tag{10.60}$$

The field current i_f (reduced to the stator) is

$$i_f = \frac{\lambda_d - L_d i_d}{L_d - L_{sl}} = \frac{0.997 - 0.0056 \cdot (-354.61)}{0.0056(1 - 0.1)} = 591.82 \ A \tag{10.61}$$

The electromagnetic torque T_e is

$$T_e = \frac{3}{2}p\left(\lambda_d i_q - \lambda_q i_d\right) =$$

$$= \frac{3}{2}\cdot 2\cdot\left(0.997\cdot 354.61 - 0.997\cdot\left(-354.61\right)\right) = 2121 \text{ Nm}$$

(10.62)

Note that for $\cos\varphi_1 = 1$ the torque expression (10.39) may be written as

$$T_e = \frac{3}{2}p\lambda_s i_s$$

(10.63)

The stator flux and current space-phasors are rectangular, as in a d.c. motor. Consequently, for constant stator flux λ_s, the torque is proportional to stator current.

Varying the frequency $\omega_1 = \omega_r$, such that $\lambda_s = $ const and $\cos\varphi_1 = 1$ appears as an optimum way to speed control.

The torque/speed curve may be derived from (10.49) with (10.63) by eliminating the stator current i_s

$$i_s = \frac{2T_e}{3p\lambda_s}$$

(10.64)

$$V_s = \omega_r\lambda_s + \frac{2r_s T_e}{3p\lambda_s}$$

(10.65)

As $\omega_1 = \omega_r$, varying speed means, implicitly, that the frequency changes.

The torque/speed curve is a straight line as in a separately excited d.c. brush motor. The ideal no-load speed ($T_e = 0$) ω_{r0} is

$$\omega_{r0} = \frac{V_s}{\lambda_s}$$

(10.66)

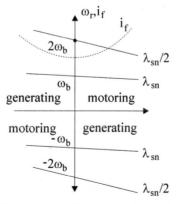

Figure 10.9. Speed/torque curves for constant flux and unity power factor

For rated flux and full voltage, V_{sn}, the base speed ω_b is obtained. Above the base speed, ω_{r0} may be modified only through flux weakening as in a d.c. brush motor. Halving the flux doubles the ideal no-load speed ω_{r0}. The torque speed curves are illustrated on Figure 10.9.

Four-quadrant operation is implicit as negative frequency $\omega_1 = \omega_r$ means negative order (a $\rightarrow$ c $\rightarrow$ b instead of a $\rightarrow$ b $\rightarrow$ c) of stator phases fed from a PEC.

Note that in order to maintain unity power factor, the field current has to be increased with torque (power) — Figure 10.9.

A tight coordination of stator and rotor currents control is required to maintain the above conditions, but once this is done — through vector control — ideal (linear) torque/speed curves are obtained.

Low speed, large power, SMs for cement mills or the like, fed from cycloconverters, operate, in general, under such conditions (Chapter 14).

Example 10.2. Leading power factor and constant stator flux.
Consider an excited SM with the same data as in example 10.1.
Determine:
- The steady-state-space-phasor diagram for a leading power factor angle $\varphi_1 = -12°$;
- Calculate the values of i_f, i_d, i_q for rated current and $\varphi_1 = -12°$.
- Calculate the stator flux λ_s, its components λ_d and λ_q and the corresponding electromagnetic torque T_e.
- Calculate the value of field current for half the rated torque and same (base) speed $\omega_b = 2\pi 60$ rad/s.
- Plot the torque/speed curve for rated and half-rated stator flux and $\varphi_1 = -12°$.
Solution:
The space-phasor diagram of Figure 10.7b is redrawn with $\varphi_1 = -12°$ (leading) as in Figure 10.10.

From Figure 10.10 we may write approximately

$$V_s \approx r_s i_s \cos\varphi_1 + \omega_r \lambda_s \qquad (10.67)$$

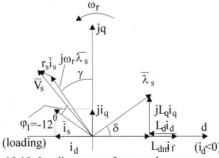

Figure 10.10. Leading power factor and constant stator flux

The torque T_e is

$$T_e = \frac{3}{2} p \, \text{Re}\left(j \overline{\lambda}_s \overline{i_s}^* \right) = \frac{3}{2} p \lambda_s i_s \cos \delta \approx \frac{3}{2} p \lambda_s i_s \cos \varphi_1 \qquad (10.68)$$

$$\lambda_s \approx \frac{V_s - r_s i_s \cos \varphi_1}{\omega_r} =$$
$$\frac{537.92 - 500 \cdot 0.016 \cdot \cos(-12°)}{2\pi 60} = 1.4068 \text{ Wb} \qquad (10.69)$$

The corresponding torque T_e is

$$T_e = \frac{3}{2} \cdot 2 \cdot 1.4068 \cdot 500 \cdot 0.978 = 2063.83 \text{ Nm} \qquad (10.70)$$

$$i_q \approx i_s \cos\left(\delta + |\varphi_1|\right) \qquad (10.71)$$

$$L_q i_q = \lambda_s \sin \delta \qquad (10.72)$$

Consequently,

$$\frac{\sin \delta}{\cos\left(\delta + |\varphi_1|\right)} = \frac{L_q i_s}{\lambda_s} = \frac{0.0028 \cdot 500}{1.4068} = 0.99 \qquad (10.73)$$

$\delta = 39^0$ (it was 45^0 for $\cos\varphi_1 = 1$ in example 10.1)

Now
$$i_q = 500 \cdot \cos(39° + 12°) = 413.66 \text{ A} \qquad (10.74)$$

$$i_d = -500 \cdot \sin(39° + 12°) = -388.57 \text{ A} \qquad (10.75)$$

$$\lambda_q = L_q i_q = 0.0028 \cdot 314.66 = 0.881 \text{ Wb} \qquad (10.76)$$

$$\lambda_d = \lambda_s \cos \delta = 1.4068 \cdot \cos 39° = 1.09388 \text{ Wb} \qquad (10.77)$$

Finally, from (10.58) the field current is obtained

$$i_f = \frac{\lambda_d - L_d i_d}{L_{dm}} = \frac{1.09388 - 0.0056 \cdot (-388.57)}{0.0056 \cdot 0.9} = 648.78 \text{ A} \qquad (10.78)$$

This value has to be compared with $i_f = 591.82$A for $\cos\varphi_1 = 1$ in (10.61).

The torque/speed curve may be obtained from (10.67)-(10.68) under the form

$$V_s = \omega_r \lambda_s + \frac{2r_s T_e}{3p\lambda_s} \tag{10.79}$$

Thus the torque curve is quite similar to that of (10.65) for $\cos\varphi_1 = 1$.

The main difference is that for the same stator current, the same stator voltage and speed, we obtain slightly less torque and we need a higher field current to provide for leading power factor angle $\varphi_1 < 0$. This value of $\varphi_1 = -12 < 0$ might be preserved for various loads if the field current is continuously adjusted. This type of coordinated frequency-stator flux-field current change method is currently used in high speed high power excited-rotor SM drives with current-source inverters. It will be detailed in the chapter dedicated to large power drives (Chapter 14).

Example 10.3. PM-SM, operation for constant i_d.

A nonsalient pole rotor PM-SM has the data: $V_n = 180V$ (rms, line voltage, star connection of phases), $L_d = 0.4L_q = 0.05H$ $r_s = 1\Omega$, no-load line voltage (rms) at $n_n = 1500rpm$ ($f_1 = 60Hz$), $V_{on} = 180V$ (rms).

Determine:
- The PM flux linkage λ_{PM};
- The torque/speed curve for $i_d = 0$, $i_d = \pm 5A$ and the ideal no-load speed in the three cases, stator flux and torque for $i_q = 10A$;
- Draw d-q equivalent circuits for steady-state and introduce the iron losses in the model.

Solution:

The no-load voltage V_{on} "translated" into space-vector terms, V_{s0}

$$V_{s0} = \frac{V_{on}\sqrt{2}}{\sqrt{3}} = \frac{180\sqrt{2}}{\sqrt{3}} = 146.7 \text{ V} \tag{10.80}$$

According to (10.50)

$$(\lambda_d)_{i_d=0} = L_{dm}i_{f0} = \lambda_{PM} \tag{10.81}$$

$$(\lambda_q)_{i_q=0} = 0 \tag{10.82}$$

$$V_{s0} = \omega_r L_{dm}i_{f0} = \omega_r \lambda_{PM} \tag{10.83}$$

$$\lambda_{PM} = \frac{V_{s0}}{\omega_r} = \frac{146.70}{2\pi 60} = 0.389 \text{ Wb} \tag{10.84}$$

As $$\lambda_d = L_d i_d + \lambda_{PM} \qquad \lambda_q = L_q i_q \tag{10.85}$$

the electromagnetic torque T_e is

$$T_e = \frac{3}{2}p\left(\lambda_d i_q - \lambda_q i_d\right) = \frac{3}{2}p\left[\lambda_{PM} + \left(L_d - L_q\right)i_d\right]i_q \qquad (10.86)$$

The d-q voltage components ((10.34)-(10.35)), with d/dt = 0, are

$$\begin{aligned} V_d &= r_s i_d - \omega_r \lambda_q \\ V_q &= r_s i_q + \omega_r \lambda_d \end{aligned} \qquad (10.87)$$

Now we replace i_q from (10.86) into (10.87). Thus we have

$$V_d^2 + V_q^2 = V_s^2 = r_s^2\left(i_d^2 + i_q^2\right) + 2\omega_r r_s\left(\lambda_d i_q - \lambda_q i_d\right) + \\ + \omega_r^2\left[\left(L_d i_d + \lambda_{PM}\right)^2 + \left(L_q i_q\right)^2\right] \qquad (10.88)$$

or

$$V_s^2 = r_s^2 i_d^2 + \left(r_s^2 + \omega_r^2 L_q^2\right) \cdot \frac{4}{9}\frac{T_e^2}{p^2}\frac{1}{\left[\lambda_{PM} + \left(L_d - L_q\right)i_d\right]^2} + \\ + \frac{4}{3}\frac{\omega_r}{p}r_s T_e + \omega_r^2\left(L_d i_d + \lambda_{PM}\right)^2 \qquad (10.89)$$

The ideal no-load ($T_e = 0$) speed ω_{r0} is

$$\omega_{r0} = \frac{\sqrt{V_s^2 - r_s^2 i_d^2}}{L_d i_d + \lambda_{PM}} \qquad (10.90)$$

For $i_d = 0$

$$\left(\omega_{r0}\right)_{i_d=0} = \frac{V_{sn}}{\lambda_{PM}} = \frac{180\sqrt{2}}{0.389\sqrt{3}} = 377.13 \, \text{rad/s} \qquad (10.91)$$

Also, $$\left(\omega_{r0}\right)_{i_d=-5A} = \frac{\sqrt{146.30^2 - (1.5)^2}}{\left(0.05 \cdot (-5) + 0.389\right)} \approx 984 \, \text{rad/s} \qquad (10.92)$$

$$\left(\omega_{r0}\right)_{i_d=5A} = \frac{\sqrt{146.30^2 - (1.5)^2}}{\left(0.05 \cdot 5 + 0.389\right)} \approx 229 \, \text{rad/s} \qquad (10.93)$$

Note that $i_d < 0$ means a demagnetizing effect. However, the torque expression (10.86) shows that as $L_d < L_q$ only $i_d < 0$ produces a positive torque contribution.

A positive torque contribution for $i_d < 0$ is accompanied by a reduction in the stator flux as

$$\left(\lambda_s\right)_{\substack{i_q=0 \\ i_d=-5A}} = \lambda_d = L_d i_d + \lambda_{PM} = -0.05\cdot 5 + 0.389 = 0.149\ \text{Wb} \quad (10.94)$$

$$\left(\lambda_s\right)_{\substack{i_q=0 \\ i_d=5A}} = 0.05\cdot 5 + 0.389 = 0.639\ \text{Wb} \quad (10.95)$$

The torque for i_d = 0, -5A, +5A, i_q = 10A is

$$\left(T_e\right)_{\substack{i_d=0 \\ i_q=10A}} = \frac{3}{2}p\lambda_{PM}i_q = \frac{3}{2}\cdot 2\cdot 0.389\cdot 10 = 11.67\ \text{Nm} \quad (10.96)$$

$$\left(T_e\right)_{\substack{i_d=-5A \\ i_q=10A}} = \frac{3}{2}\cdot 2\cdot\left(0.389 + (0.05 - 0.125)\cdot(-5)\right)\cdot 10 = 22.92\ \text{Nm}\,(10.97)$$

$$\left(T_e\right)_{\substack{i_d=5A \\ i_q=10A}} = \frac{3}{2}\cdot 2\cdot\left(0.389 + (0.05 - 0.125)\cdot 5\right)\cdot 10 = 0.42\ \text{Nm} \quad (10.98)$$

Positive i_d values are to be avoided as they increase the flux level and decrease the torque considerably.

The d-q equivalent circuits for steady-state (s = 0) reflect, in fact, Equations (10.87) - Figure 10.11.

Though the core losses are related to the main flux components we may assume them to be produced by the stator flux.

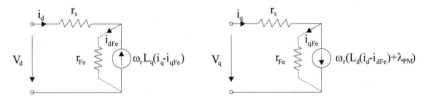

Figure 10.11. Steady-state d-q equivalent circuits of PM-SM with core losses

included

The core losses occurring in the stator are considered produced in a resistance r_{Fe}, to be determined through measurements.

Now, with core losses included, Equations (10.87) are no longer strictly valid but they may be modified simply based on the equivalent circuit. Introduced in parallel with e.m.f.s, r_{Fe} should not vary dramatically with frequency.

Example 10.4. The reluctance synchronous motor (RSM) steady-state
A high performance RSM has the data: L_d = 10 L_q = 0.1 H, p = 2, r_s = 1Ω, f_1 = 60 Hz.
Determine:
a. The space-phasor diagram;

b. The maximum torque per given stator current and for given stator flux;

c. Introduce a PM along axis q to reduce the q axis flux. Draw the new space-phasor diagram and discuss it. Calculate the torque for i_d = 3A, i_q = 15A without and with PM flux λ_{PM} = -$L_q i_q$.

d. Calculate the stator flux to produce the respective torques.

Solution:

The space-phasor diagram is obtained simply from the general one (Figure 10.7) with i_f = 0 and $L_d \gg L_q$ (Figure 10.12).

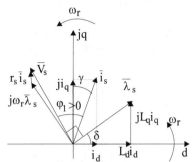

Figure 10.12. RSM - space-phasor diagram

The torque expression (10.86) with i_f = 0 and $L_d \gg L_q$ is

$$T_e = \frac{3}{2} p \left(L_d - L_q \right) i_d i_q \qquad (10.99)$$

The voltage equations (10.87) become

$$V_d = r_s i_d - \omega_r L_q i_q$$
$$V_q = r_s i_q + \omega_r L_d i_d \qquad (10.100)$$

The stator flux $\overline{\lambda}_s$ and stator current $\overline{i}_s$ are

$$\overline{\lambda}_s = \lambda_d + j\lambda_q = L_d i_d + jL_q i_q \qquad (10.101)$$

$$\overline{i}_s = i_d + ji_q \qquad (10.102)$$

The maximum torque for given stator current, i_s

$$i_s = \sqrt{i_d^2 + i_q^2} \qquad (10.103)$$

may be obtained from (10.99) and (10.103)

$$\frac{\partial T_e}{\partial i_d} = 0 \qquad (10.104)$$

After eliminating i_q we get

$$i_{di} = i_{qi} = i_s / \sqrt{2}; \quad T_{ei} = \frac{3}{2} p (L_d - L_q) \frac{i_s^2}{2} \qquad (10.105)$$

Consequently, $\gamma = 45°$ in Figure 10.12.

If $i_s = i_n$ (rated current) $i_{di} = i_{sn} / \sqrt{2}$ the machine is driven into heavy saturation as i_{di} is too high in comparison with the rated no-load phase current i_0, with the exception of very low power motors when $i_0 / i_n > 0.707$. Though it leads to maximum torque per ampere it may be used for torques lower than 50-60% of rated torque. This may be the case for low speed operation in some variable speed drive applications.

On the other hand, for given stator flux (10.101), the maximum torque is obtained for

$$i_{dk} L_d = i_{qk} L_q = \lambda_s / \sqrt{2} \qquad (10.106)$$

The maximum torque is

$$T_{ek} = \frac{3}{2} p (L_d - L_q) \frac{\lambda_s^2}{2 L_d L_q} \qquad (10.107)$$

This time the d-q flux angle is $\delta = 45°$ in Figure 10.12.

As it provides maximum torque per given flux, this is the limit condition at high speeds when the flux is limited (flux weakening zone).

Introducing the PM in axis q

$$\lambda_q = L_q i_q - \lambda_{PMq}; \quad \lambda_d = L_d i_d \qquad (10.108)$$

In (10.108) $\lambda_{PMq} \geq L_q i_{q\,max}$ to avoid PM demagnetization. The torque expression (10.86) becomes

$$i_{dk} = i_{di} = i_0 / \sqrt{2}; \quad T_e = \frac{3}{2} p \left[(L_d - L_q) i_q + \lambda_{PM} \right] i_d \qquad (10.109)$$

This time, to change the torque sign (for regenerative braking) the sign of i_d (rather than i_q) has to be changed in variable speed drives. The space-phasor diagram of PM-RSM is shown in Figure 10.13.

Note that for motoring $\delta < 0$ (the flux vector is lagging (not leading) the d axis) and the power factor has been increased, together with torque.

Only low remnant flux density (moderate cost) PMs are required as $L_q i_{sn}$ is still a small flux ($L_d / L_q \gg 1$).

Let us now calculate the torque and stator flux for the two situations

$$(T_e)_{\lambda_{PMq}=0} = \frac{3}{2} \cdot p (L_d - L_q) i_d i_q = \frac{3}{2} \cdot 2 \cdot 0.1 \cdot \left(1 - \frac{1}{10} \right) 3 \cdot 15 = 12.15 \, \text{Nm} \qquad (10.110)$$

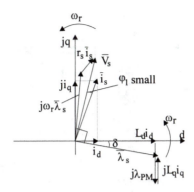

Figure 10.13. PM-SRM - space-phasor diagram

$$\lambda_s = \sqrt{L_d^2 i_d^2 + L_q^2 i_q^2} = \sqrt{0.1^2 \cdot 3^2 + \left(\frac{0.1}{10} \cdot 15\right)^2} = 0.335 \, \text{Wb} \quad (10.111)$$

$$\left(T_e\right)_{\lambda_{PM_q} \neq 0} = \frac{3}{2} \cdot 2 \cdot \left(0.1 \cdot \left(1 - \frac{1}{10}\right) \cdot 15 + \frac{0.1}{10} \cdot 15\right) \cdot 3 = 13.15 \, \text{Nm} \quad (10.112)$$

$$\lambda_s = L_d i_d = 0.1 \cdot 3 = 0.3 \, \text{Wb} \quad (10.113)$$

So, the PMs in axis q produce almost a 10% torque increase for a 10% decrease in flux level.

The improvements are more spectacular in lower saliency machines (L_d / L_q = 5-8), which are easier to build.

10.8. COGGING TORQUE AND TOOTH-WOUND PMSMs

PMSMs develop a pulsating torque at zero stator current due to the PM magnetic coenergy variation with rotor position in the presence of stator slot openings.

This zero current torque which has a zero average value per 360° is called cogging torque.

Cogging torque stems from the PM airgap flux density variation due to stator slot openings and depends also on other factors such as:
- PM shape and placement
- The number of stator slots Ns and rotor poles 2p

Figure 10.14 shows the PM airgap flux density obtained from Finite Element Analysis (FEA) at the center of airgap for a 27 stator slot 6 rotor PM poles PMSM [6].

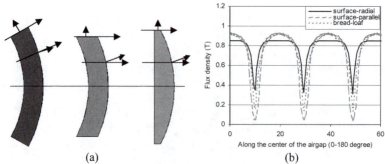

(a)   (b)

Figure 10.14. PM shapes (a) and their airgap flux density (b) [6]

The cogging torque may be calculated, within FEA by the coenergy method or by the Maxwell stress tensor method.

There is also a relationship between the e.m.f. harmonics content and the cogging torque as evident in Table 10.1 for the same Ns = 27 slot, 2p = 6 pole PMSM and surface radial (SR), surface parallel (SP) and bread-loaf (BL) shape PMs (Figure 10.14).

Table 10.1 Cogging torque and e.m.f. harmonics [6]

PM shape	Tcoggpp (mNm)	Line to line emf at 1000 rpm [RMS]		
		1st (V)	5th (V)	7th (V)
SR	47.88	2.857	0.05604	0.012256
SP	15.63	2.8681	0.00648	0.008132
BL	5.562	2.8691	0.03459	0.004897

While the emf fundamental and consequently the electromagnetic torque for given sinusoidal current would be the same for all three shapes, the cogging torque is notably smaller for the bread-loaf PMs. The e.m.f. waveform and cogging torque versus position curves depend also on the number of slots per pole per phase q. For q=1 (N_s=24, $2p_1$=8 poles) the e.m.f., cogging torque and electromagnetic torque for sinusoidal current are all shown in Figure 10.15.

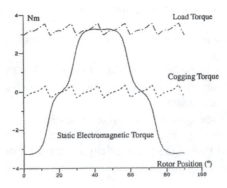

Figure 10.15. E.m.f., cogging torque, total torque (for N_s=24, $2p_1$=8 poles) versus rotor position for sinusoidal current [7]

It is evident that the number of cogging torque fundamental periods N is the lowest common multiplier of N_s and 2p:

$$N = LCM(N_s, 2p) \qquad (10.114)$$

The larger N, the smaller the peak value of cogging torque. In the case on Figure 10.15, N =N_s = 24.

There are applications such as automotive power steering where the total torque pulsations should be well less than 1% of peak electromagnetic torque of the motor.

Torque pulsations include cogging torque but also refer to e.m.f. and current time harmonics; magnetic saturation of the magnetic circuit plays an important role [6].

Among the many ways to reduce the cogging torque we mention here:

- A higher $N = LCM(N_s, 2p)$
- PM shape and span optimization (Figure 10.16)
- Rotor or stator skewing (the electromagnetic torque is also reduced)

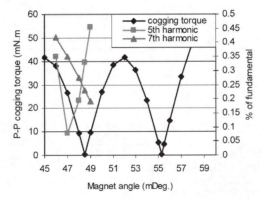

Figure 10.16. Cogging torque and line e.m.f. harmonics content with PM angle

for surface PM rotor poles (N_s, 2p=6)

There are two PM span angles for which the cogging torque is zero; the e.m.f. harmonics are not zero for those situations, but also small.

The slot-wound stator windings have been introduced to reduce stator cogging losses by reducing the end-connection length and to reduce the copper torque by increasing the value of $N = LCM(N_s, 2p)$. They may also be called factionary windings as q (slot/pole/phase) < 1.

They may be built in one layer or two layers (Figure 10.18).

Table 10.2 Winding factor / single layer winding [8]

Q	p 2	4	6	8	10	12	14	16
3	*	*	*	*	*	*	*	*
6	*	**0,866**	*	0,866	0,500	*	*	*
9	*	0,736	**0,667**	**0,960**	0,960	0,667	0,218	0,177
12	*	*	*	0,866	**0,966**	*	**0,966**	0,866
15	*	*	0,247	0,383	0,866	0,808	0,957	0,957
18	*	*	*	0,473	0,676	**0,866**	0,844	**0,960**
21	*	*	*	0,248	0,397	0,622	0,866	0,793
24	*	*	*	*	0,430	*	0,561	0,866

Table 10.3 Winding factor/double layer winding [8]

Table 10.4 Dimension and proprieties of the 2 layer winding 6 slot/4 pole PMSM

Parameter	value	unit
Topology	Inner rotor IPMSM	
Number of phases	3	-
Number of stator slots	6	-
Number of rotor poles	4	-
Geometry		
Stator outer diameter	56	mm
Stator inner diameter	28	mm
Airgap (minimal)	0.5	mm
Stack length	45	mm
Magnet width	12	mm
Magnet height	3.5	mm
Winding		
Nb. slots/pole/phase	0.5	-
Nb. winding layer	2	-
Nb. turns per phase	20	-
Materials		
Core material	M800-50A	
Magnet type	NdFeB (1.2 T)	

The e.m.f. waveform and cogging torque of a 6/4 slot/pole combination PMSM is shown in Figure 10.17. The e.m.f. is almost sinusoidal. The complete geometrical data are given in Table 10.4 [9].

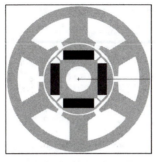

(a) Cross-section of IPMSM

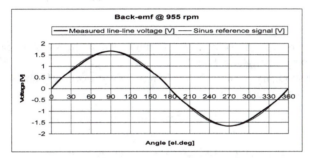

(b) Shape of the back e.m.f. at 955 rpm

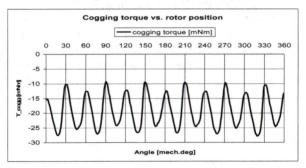

(c) Cogging torque vs. rotor position angle

Figure 10.17. E.m.f. of a 2 layer winding 6 slot/ 4 poles PMSM

Despite the fact that the PM airgap flux density has a rather rectangular distribution in the airgap the e.m.f. may be rather sinusoidal by geometrical optimization, either with surface PM or with interior PMs on the rotor [9].

It is also feasible to produce a trapezoidal e.m.f. waveform, in general, for unilayer windings with nonuniform slots [9, 10].

Figure 10.18. Windings for a $2p_1 = 10$ pole machine: a) conventional: $N_s = 30$, q = 0.4, b) one layer tooth wound: $N_s = 12$, q = 0.4, c) two layer tooth-wound, $N_s = 12$, q = 0.4

To secure a large enough e.m.f. fundamental the winding factor K_{w1} should be large enough (> 0.866).

For fractionary q and N_s / 2pm = z/n slots/pole/phase, the space harmonics that occurs with these windings are:

$$v = \pm \frac{1}{n}(2mg + 2); g = 0, \pm 1, \pm 2, \pm 3 \text{ - for n an even number;}$$

$$v = \pm \frac{1}{n}(2mg + 1); g = 0, \pm 1, \pm 2, \pm 3 \text{ - for n an odd number.}$$

Integer and fractionary (sub)harmonics may be present.

The magnetomotive force (m.m.f.) space harmonics produce additional components in the linkage inductance of the slot-wound PMSM while subharmonics may also produce torque pulsations of low frequency.

Double layer such windings are known to having lower subharmonics.

The fundamental winding – factor for some main stator slot N_s, pole number $2p_1$ combinations for single layer and double layer slot-wound PMSMs are shown in Tables 10.2, 10.3.

Consequently slot-wound PMSM, not only enjoy larger efficiency (because of shorter coil end-turns) but also may be controlled for either sinusoidal current or for rectangular current.

They are already applied in industry, from hard disks to servodrives, automotive wheel steering or low speed high torque motor/generators.

Note: Other nonoverlapping coil winding PM brushless motor configurations have been proposed recently but have not reached the markets yet.

They are:
- Transverse flux PM brushless machines (TFM) [11, 12, 13] (with ring shape coil for a large number of poles)
- Claw pole stator composite magnetic core PM brushless machines [14] to cut the manufacturing costs in small motors.

- Flux reversal PM brushless machines (FRM) [15] with standard laminations and stator PMs (in general) to cut manufacturing costs in medium – large torque low speed applications.
- Axial - airgap PMSM, have been introduced to increase torque / volume for various applications [16].

10.9. THE SINGLE-PHASE PMSM

For low power (say below 500 W) home and automotive actuator applications, the single phase PMSM is sometimes used in order to cut the costs of the power electronics converter for variable speed, by reducing the number of power switches.

A typical cylindrical type single phase PMSM is shown in Figure 10.19.

Figure 10.19. Single phase PMSM with bifilar stator winding.

The stator magnetic core is made of silicon laminations (0.5 mm or less in thickness).

There are two semicoils for the positive polarity of current and other two semicoils for the negative polarity of current. The bifilar winding is thus obtained. Though it leads to an increase of copper losses it allows for a low power switch count controller (Chapter 11).

The 2 slot / 2 pole configuration leads to a 2 period cogging torque per revolution.

The machine will not start as the initial position sees the PMs aligned to the stator poles where the electromagnetic (interaction) torque is zero.

To produce a nonzero electromagnetic torque position a few solutions may be applied: a parking magnet pair is placed off-settled between the main poles (Figure 10.11) or a notch is placed asymmetrically on the stator pole towards airgap; alternatively, a stepped airgap under the stator pole solution may be applied.

In all cases only motion in one direction is provided. But once the motor has started in one direction it may be braked and then moved eventually into the opposite direction based on a Hall proximity sensor signal (Chapter 10), if the PEC allows for a.c. current in the phase coils.

10.10. STEADY STATE PERFORMANCE OF SINGLE PHASE PMSM

At least for sinusoidal e.m.f., the single phase PM brushess motor is, in fact, a synchronous machine. So all the equations already developed in this chapter apply here, but for a single stator electric circuit.

For steady state the phasor form of stator voltage equation is straightforward:

$$\underline{V}_s = R_s \underline{I}_s + \underline{E}_s + j\omega L_s \underline{I}_s ; \underline{E}_s = j\omega_r \lambda_{PM} \qquad (10.115)$$

With a sinusoidal e.m.f. E_s:

$$E_s(t) = E_{s1} \cos(\omega_r t) \qquad (10.116)$$

and V_s, I_s, stator voltage and current, L_s – stator inductance (independent of rotor position for surface PM rotors), λ_{PM} – the PM flux linkage, R_s – in the stator resistance.

For steady state:

$$I_s(t) = I_{s1} \cos(\omega_r t - \gamma) \qquad (10.117)$$

The electromagnetic (interaction) torque is:

$$T_e = p\frac{E_s(t)I(t)}{\omega_r} = \frac{p_1}{\omega_r}\frac{E_{s1}I_{s1}(t)}{2}(\cos\gamma + \cos(2\omega_r t - \gamma)) \qquad (10.118)$$

As for the single phase IM or for the a.c. brush motor the electromagnetic torque pulsates at double stator current frequency (speed).

Now the cogging torque, with $N_s = 2$ slots, $2p = 2$ poles, has $N = LCM$ $(2, 2) = 2$ periods:

$$T_{cogg} \approx T_{cog\,max} \cos(2\omega_r t - \gamma_{cog}) \qquad (10.119)$$

It may be feasible that with such a cogging torque variation to secure $\gamma_{cog} = \gamma$ and, for a certain load torque:

$$T_{cog\,max} = -\frac{pE_{s1}I_{s1}}{2\omega_r} \qquad (10.120)$$

This way the torque pulsations are cancelled (the large cogging torque (10.120) may be used for rotor parking at a proper angle to secure safe starting).

At lower load, pulsations occur into the total torque as their complete elimination is performed only for the rated load.

The phasor diagram that corresponds to equation (10.115) is shown in Figure 10.20.

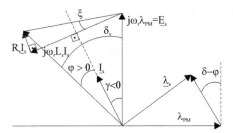

Figure 10.20. The steady state phasor diagram of the single phase PMSM

The core losses p_{iron} may be expressed as proportional to speed and stator flux squared:

$$P_{iron} \approx \frac{\omega_r^2 \lambda_s^2}{P_{iron}} \qquad (10.121)$$

R_{iron} may be measured or calculated at the design stage.

In general the copper losses are larger than the iron losses and may be calculated afterwards and added to the efficiency formula.

To calculate the steady state performance for given voltage power angle δ_v, stator voltage V_{s1} and ω_r ($E_s = \omega_r \lambda_{PM}$ – given also) we obtain:

$$I_s = \frac{\sqrt{V_s^2 + E_s^2 - 2V_s E_s \cos \delta_v}}{Z} \qquad (10.122)$$

$$Z = \sqrt{R_s^2 + j\omega_r^2 L_s^2}\,; \tan \varphi = \frac{R_s}{\omega_r L_s}\,; \cos(\varphi + \xi) = \frac{E}{I \cdot Z}\sin \delta_v \qquad (10.123)$$

$$\lambda_s^2 = \lambda_{PM}^2 + L_s^2 i_s^2 - 2\lambda_{PM} L_s i_s \cos(\delta_v - \varphi) \qquad (10.124)$$

The mechanical power P_{mec} is:

$$P_{mec} \approx V_s I_s \cos \varphi - I_s^2 R_s - \frac{\omega_r^2 \lambda_s^2}{R_{core}} - p_{mec} \qquad (10.125)$$

p_{mec}= mechanical losses.
Efficiency η is:

$$\eta = \frac{P_{mec}}{V_s I_s \cos \varphi} \qquad (10.126)$$

The electromagnetic torque Te versus power angle for a 150 W single phase PMSM is shown in Figure 10.21 in P.U. for zero and nonzero stator resistance R_s.

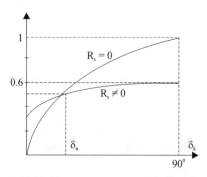

Figure 10.21. Torque versus power (voltage) angle

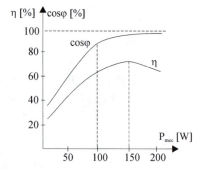

Figure 10.22. Efficiency and power factor versus output power P_{mec} (W)

It is evident that the stator resistance should not be neglected because otherwise the peak torque (at $\delta_v = 90°$) is grossly overestimated.

In V/f scalar control variable speed drives the torque should not surpass the real peak torque, but better be notably smaller (by 30 – 40%) to secure rather stable operation to moderate load torque perturbation. Typical efficiency and power factor curves for a 150 W motor [17] show mildly good performance (Figure 10.22).

Example 10.5:
A single phase PMSM is fed at Vsn = 120 V (RMS) at the frequency f_1= 60 Hz and has two rotor poles 2p = 2. At the corresponding speed, the no load voltage E_s, when the motor is driven by a drive is: E_s=0.95V_{sn}. The stator resistance R_s= 5 Ω and the stator inductance L_s= 0.05 H. Calculate:
a) The speed at f_1= 60 Hz;
b) The stator current for δ_v= 0,45°, 90°.
c) The power factor versus δ_v.
d) The average torque for the voltage power angle δ_v=0,45°, 90°, with core and mechanical losses neglected.

Solution:
a) The speed n is:

$$n = f_1/p = 60/1 = 60 \text{ rps} = 3600 \text{ rpm}$$

b) The stator current from (10.122) is:

$$I_s = \frac{\sqrt{V_s^2 + E_s^2 - 2V_s E_s \cos \delta_v}}{Z} =$$

$$= \sqrt{120^2 + (120 \cdot 0.95)^2 - 2 \cdot 120 \cdot 120 \cdot 0.95 \cdot \cos \delta_v} =$$

$$= \begin{cases} 18/19.5 = 0.92\text{A}, & \delta_v = 0 \\ \dfrac{86.583}{19.5} = 4.44\text{A}, & \delta_v = 45° \\ \dfrac{157.49}{19.5} = 8.07, & \delta_v = 90° \end{cases}$$

with : $Z = \sqrt{5^2 + (2\pi 60 \cdot 0.05)^2} = 19.5\Omega$

c) The power factor angle φ comes from (10.123) with:

$$\xi = \tan^{-1} \frac{R_s}{\omega_r L_s} = \tan^{-1} \frac{5}{377 \cdot 0.05} = 14.38°$$

$$\cos(\varphi + \xi) = \frac{E_s}{Z \cdot I_s} \sin \delta_v$$

For $\delta_v = 0$: $(\varphi + \xi) = 90°$, $\varphi = 90 \quad -14.38 = 75.32°$;
$\delta_v = 45°$: $\cos(\varphi + \xi) = 0.833, (\varphi + \xi) = 33.6°$, $\varphi = 33.6 -14.38 = 19.20°$;
$\delta_v = 90°$: $\cos(\varphi + \xi) = 0.648, (\varphi + \xi) = 49.59°$, $\varphi = 49.59 - 14.38 = 35.21°$;

d) The electromagnetic torque T_e is:

$$T_e = \frac{pE_s I_s \cos(\delta_v - \varphi)}{\omega_r} =$$

$$\begin{cases} \delta_v = 0, T_e = \dfrac{1}{377} 102 \cdot 0.92 \cdot \cos(0 - 75.62) = 0.0618\text{Nm} \\ \delta_v = 45°, T_e = \dfrac{1}{377} 102 \cdot 4.44 \cdot \cos(45 - 19.20) = 1.0815\text{Nm} \\ \delta_v = 90°, T_e = \dfrac{1}{377} 102 \cdot 8.07 \cdot \cos(90 - 35.21) = 1.2589\text{Nm} \end{cases}$$

From $\delta_v = 45°$ to $\delta_v = 90°$ the torque increases by less than 25%, in part due to the stator resistance presence (the increase would be 30% with zero resistance).

10.11. SUMMARY

- Synchronous motors (SMs) are, in general, three-phase a.c. motors with d.c. excited or PM excited or variable reluctance rotors.
- Because the stator m.m.f. travels along the periphery with a speed $n_1 = f_1 / p$ (f_1 - stator frequency, p - pole pairs), the rotor will rotate at the same speed as only two m.m.f.s at relative standstill to produce nonzero average torque.
- Changing the speed means changing frequency.
- Excited rotor SMs require copper-rings and brushes to be connected to a d.c. controlled power supply. Alternatively, a rotary transformer with diodes on the rotor (secondary) side may provide for brushless energization of the excitation (field) windings.
- From all variable frequency-fed SMs only excited rotor SMs, fed from current-source PECs (as shown in Chapter 14) have a squirrel cage on the rotor, to reduce the machine commutation inductance.
- Unity or leading power factor operation at variable speed and torque is possible only with excited-rotor SMs based on continuous field current control, to nullify the input reactive power to the machine. For this case, if the stator flux is also constant, the speed/torque curve becomes linear, as for a d.c. brush motor with separate excitation. Flux weakening is performed above base speed.
- PM-rotor SMs may be operated at variable speeds by their control in d-q orthogonal axes coordinates, through adequate coordination of i_d-i_q relationship with torque and speed requirements. Negative (demagnetizing) i_d is required for positive reluctance torque as $L_d < L_q$.
- Multiple flux barrier rotors with conventional or axial laminations, eventually with PMs in axis q (of highest reluctance, $L_d > L_q$), have been shown to produce power factors and efficiencies similar to those of induction machines in the same stator.
- The d-q (space-phasor) model of SMs seems the most adequate vehicle for the design and analysis of variable speed drives with SMs.
- To reduce copper losses and frame size tooth wound PMSMs have been introduced for various applications: from hard disks, through industrial servodrives to active power steering.
- Single phase PMSM are recommended for low power, low system cost home appliance or automotive electric actuation technologies.
- The next chapter will deal only with PM and reluctance SM drives — low and medium powers (up to hundreds of kW), fed from PWM-IGBT converters. Large power industrial drives with excited rotor SMs will be treated in Chapter 14 as they require special PECs.

10.12. PROBLEMS

10.1. A PM-SM with interior PMs ($L_d < L_q$) is fed with a step voltage along the q axis, ΔV_q, while V_d = constant and the speed is constant. Using (10.34)-(10.35) and Figure 10.5, determine the Laplace form of the current Δi_d, Δi_q and torque responses.

10.2. A PM-SM with surface PMs and $q \geq 2$ has the data: $L_d = L_q = 0.01H$, r_s = 0.5Ω, p = 2 (pole pairs), rated current (rms, phase current), no-load voltage (at n_0 = 1800rpm), V_0 = 80V (rms, phase voltage).
Calculate:
 a. The PM flux linkage in the d-q model λ_{PM};
 b. The torque at rated current with $i_d = 0$ and the corresponding rated terminal voltage for n = 1800 rpm;
 c. The ideal no-load speed for $i_q = 0$ and i_d corresponding to rated current;
 d. The torque at rated current and voltage for n' = 3600 rpm.

10.3. An RSM has the data: rated voltage V_n = 120V (rms, phase voltage), ω_{1n} = 120rad/s, i_{sn} = 20A (rms, phase current) and the inductances l_d, l_q in relative units: l_d = 3 p.u., l_q = 0.3 p.u., r_s = 0.5 p.u., p = 2 pole pairs.
Calculate:
 a. L_d, L_q, r_s in Ω if $1(P.U.) = L\left(\dfrac{V_n}{I_n\omega_{1n}}\right)^{-1}$;
 b. For $i_d / i_q = \sqrt{L_q / L_d}$ and rated current and voltage determine the base speed ω_b, torque, input power and power factor angle φ_1.
 c. Add a PM in axis q (acting against i_q) such that the no-load voltage, at base speed ω_b, e_0 = 0.15 and calculate the ideal no-load speed for $i_q = i_{sn}$.

10.13. SELECTED REFERENCES

1. **T.J.E. Miller**, Brushless permanent magnet and reluctance motor drives, OUP, 1989
2. **S.A. Nasar, I. Boldea, L. Unnewehr**, Permanent magnet, reluctance and self-synchronous motors, CRC Press, Florida, 1993.
3. **I. Boldea**, Reluctance synchronous machines and drives, OUP, 1996.
4. **R.P. Deodhar, D.A. Staton, T.M. Jahns, T.J.E. Miller**, Prediction of cogging torque using the flux - m.m.f. diagram technique, IEEE Trans., vol.IA-32, no.3, 1996, pp.569-576.
5. **I. Boldea and S.A. Nasar**, Unified treatment of core losses and saturation in orthogonal axis model of electric machines, Proc.IEE, vol.134, Part.B, 1987, pp.355-363.
6. **M.S. Islam, S. Mir, T. Sebastian, S. Underwood**, "Design consideration of sinusoidal excited permanent magnet machines for low torque ripple applications", Record of IEEE – IAS – 2004, Seattle, USA.
7. **J. Cross, P. Viarouge**, "Synthesis of high performance PM motors with concentrated windings", IEEE Trans. Vol. EC-17, no. 2, 2002, pp. 248-253.
8. **F. Magnussen, C. Sadarangani**, "Winding factors and Joule losses of permanent magnet machines with concentrated windigs", Record of IEEE – IEMDC – 2003, vol. 1, pp. 333-339.

9.   **D. Iles – Klumpner, I. Boldea,** "Optimization design of an interior permanent magnet synchronous motor for an automotive active steering system", Record of OPTIM 2004, vol. 2, pp. 129-134.

10. **D. Ishak, Z.Q. Zhu, D. Howe,** "Permanent magnet brushless machines with unequal tooth widths and similar slot and pole numbers", Record of IEEE – IAS – 2004, Seattle, USA.

11. **H. Weh, H. May,** "Achievable force densities for permanent magnet excited machines in new configurations", Record of IEM – 1986, vol. 3, pp. 1107-1111.

12. **E. Henneberger, I.A. Viorel,** Variable reluctance electrical machines, Shaker Verlag, Aachen, 2001.

13. **I. Luo, S. Huang, S. Chen, T. Lipo,** "Design and experiments of a novel axial circumpherential current permanent magnet (AFCC) machine with radial airgap", Record of IEEE – IAS – 2001.

14. **I. Cross, P. Viarouge,** "New structure of polyphase claw pole machines", IEEE Trans. Vol. IA – 40, no. 1, 2004, pp. 113-120.

15. **I. Boldea, I. Zhang, S.A. Nasar,** "Theoretical characterization of flux reversal machine in low speed servodrives – the pole – PM configuration", IEEE Trans. Vol. IA 38, no. 6, 2002, pp. 1549-1557.

16. **M. Cerchio, G. Griva, F. Profumo, A. Tenconi,** "'Plastic' axial flux machines: design and prototyping of a multidisk PM synchronous motor for aircraft applications", Record of ICEMS – 2004, Je Ju Island, Korea.

17. **V. Ostovic,** "Performance comparison of U core and round stator single phase PM motors for pump applications", IEEE Trans. Vol. IA – 38, no. 2, 2002, pp. 476-482.

Chapter 11

PM AND RELUCTANCE SYNCHRONOUS MOTOR DRIVES

11.1. INTRODUCTION

PM and reluctance synchronous motors are associated with PWM voltage-source inverters in variable speed drives.

While PM-SM drives now enjoy worldwide markets, reluctance synchronous motor (RSM) drives still have a small share of the market in the low power region, though recently high saliency (and performance) RSMs with q axis PMs have been successfully introduced for wide constant power speed range applications (such as spindle drives).

High performance applications are, in general, provided with PM-SM drives. RSM drives may be used for general applications as the costs of the latter are lower than those of the former and, in general, close to the costs of similar IM drives.

In what follows, we will deal first with PM-SM drives and after that, with RSM drives. Constant power operation will be discussed towards the end of the chapter for both types of synchronous motors. Control of single-phase PMSM ends this chapter.

11.2. PM-SM DRIVES: CLASSIFICATIONS

Basically, we may distinguish three ways to classify PM-SM drives: with respect to current waveform, voltage-frequency correlation and motion sensor presence.

From the point of view of current waveform we distinguish:

- rectangular current control — Figure 11.1a — the so called brushless d.c. motor drive - q = 1 slot per pole per phase, surface PM rotor;
- sinusoidal current control — Figure 11.1b — the so-called brushless a.c. drive - q ≥ 2 slots per pole per phase.

From the point of view of motion sensor presence there are:

- drives with motion sensors;
- drives without motion sensors (sensorless).

Finally, sinusoidal current (brushless a.c.) drives may have:

- scalar (V / f) control — a damper cage on the rotor is required;
- vector control (current or current and voltage);
- direct torque and flux control (DTFC).

The stator current waveforms — rectangular or sinusoidal (Figure 11.1) — have to be synchronized with the rotor position.

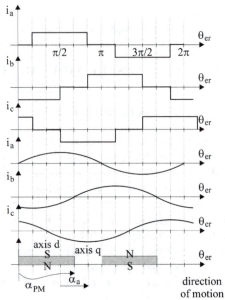

Figure 11.1. Rectangular or sinusoidal current control; α_a - advance angle

This may be done by turning on and off the SCRs in the PWM inverter in relation to the rotor position θ_{er}.

In general, to compensate for the e.m.f. increase with speed and for the machine inductance delay in current variation, the phase commutation is advanced by an angle α_a which should increase with speed. Current waveforming is obtained through current controllers.

Rectangular current control is preferred when the PM e.m.f. is nonsinusoidal (trapezoidal), q = 1, concentrated coil stator windings — to reduce torque pulsations and take advantage of a simpler position (proximity) sensor or estimator.

Scalar control (V/f) is related to sinusoidal current control without motion sensors (sensorless) (Figure 11.2).

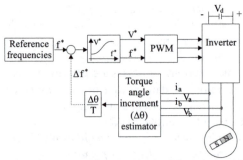

Figure 11.2. V/f (scalar) control for PM-SM (and for RSM) with torque angle increment compensation

The torque angle increment $\Delta\theta$ is estimated and the reference frequency is increased by Δf^* to compensate for the torque variation and keep the motor currents in synchronism with the rotor during transients. A rotor cage is useful to damp the oscillations as ramping the frequency is limited.

Low dynamics applications may take advantage of such simplified solutions. For faster dynamics applications vector control is used (Figure 11.3).

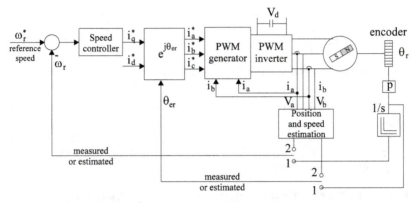

Figure 11.3. Basic vector control of PM-SM (or for RSM):

1 - with motion sensor; 2 - sensorless

The rotor position and speed are either measured or estimated (for sensorless drives) and used as speed feedback (ω_r) and vector rotator (θ_{er}) to generate reference phase currents. Closed-loop or open-loop PWM is used to "construct" the current (or voltage) waveforms locked into synchronism with the rotor. The PM-SM is controlled along the d-q model in rotor coordinates, which corresponds to d.c. for steady-state (Chapter 10).

Correlating i_d^* with i_q^* is a matter of optimization, according to some criterion. Vector control is considerably more complicated in comparison with V/f control, but superior dynamic performance is obtained (quicker torque control, in essence).

To simplify the motor control, the direct torque and flux control (DTFC for IMs) has been extended to PM-SM (and to RSMs) as torque vector control (TVC) in [10], Figure 11.4.

The direct stator flux and torque control leads to a table of voltage switching (voltage vector sequence). Vector rotation has been dropped but flux and torque observers are required. While speed is observed, rotor position estimation is not required in sensorless driving.

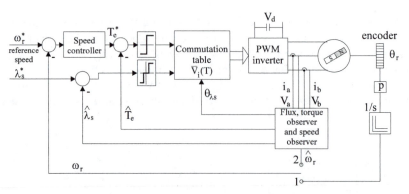

Figure 11.4. Direct torque and flux control (DTFC) of PM-SMs (and RSMs)

Again, fast flux and torque control may be obtained even in sensorless driving.

Rectangular current control and sinusoidal current control (through vector control or DTFC) are going to be detailed in what follows for motion sensor and sensorless driving.

11.3. RECTANGULAR CURRENT CONTROL (BRUSHLESS D.C. MOTOR DRIVES)

Rectangular current control is applied to nonsinusoidal (trapezoidal) PM-e.m.f. waveform, typical for concentrated coil stator windings (three slots per pole: q = 1). To reduce torque pulsations, rectangular current is needed. With rectangular current, however, the reluctance torque production is not very efficient and thus nonsalient pole (surface PM) rotors are preferred.

The motor phase inductance L_a is [1]

$$L_a = L_{sl} + L_{ag} \tag{11.1}$$

$$L_{ag} \approx \frac{\mu_0 W_1^2 L\tau}{2pg_e}; \quad g_e \approx K_c g + h_{PM} / \mu_{rec} \tag{11.2}$$

where: W_1 - turns per phase, L - stack length, τ - pole pitch, g - airgap, K_c - Carter coefficient, h_{PM} - PM radial thickness, p - pole pairs, L_{sl} - leakage inductance. The mutual inductance between phases, L_{ab}, is

$$L_{ab} \approx -L_{ag} / 3 \tag{11.3}$$

Thus the cycling inductance L_s (per phase) for two-phase conduction is

$$L_s = L_a - L_{ab} = L_{sl} + \frac{4}{3} L_{ag} \tag{11.4}$$

11.3.1. Ideal brushless d.c. motor waveforms

In principle the surface PM extends over an angle α_{PM} less than 180° (which represent the pole pitch, Figure 11.1). The two limits of α_{PM} are $2\pi/3$ and π. Let us suppose that the PM produces a rectangular airgap flux distribution over $\alpha_{PM} = \pi$ (180°) (Figure 11.5a). The stator phase m.m.f. is supposed to be rectangular, a case corresponding to q = 1 (three slots per pole). Consequently, the PM flux linkage in the stator winding $\lambda_{PM}(\theta_{er})$ varies linearly with rotor position (Figure 11.5b). Finally the phase e.m.f. E_a is rectangular with respect to rotor position (Figure 11.5c).

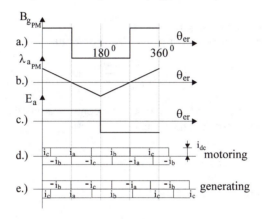

Figure 11.5. Ideal waveforms for BLDC (brushless d.c. motor):

a.) PM airgap flux density; b.) PM flux per phase a; c.) e.m.f. in phase a;

d.) ideal currents for motoring; e.) ideal currents for generating

For zero advance angle $\alpha_a = 0$ (Figure 11.1) — the phase currents are in phase with the e.m.f. (i_a, E_a) for motoring (Figure 11.5d).

At any instant, due to assumed instantaneous current commutation, only two phases are in conduction. Consequently, only two thirds of the PMs are utilized. So the back core flux is 50% larger than necessary. On the other hand, if the currents extend over 180° (with three phases conducting at any time) for 120° wide PMs, it would mean 50% more than necessary copper losses for a given torque.

For 120° wide magnets, delta connection is used; while 180° wide magnets require star connection. To reduce the fringing (leakage) flux between neighboring magnets their span is 150°-160°.

Returning to the linear phase flux linkage variation with rotor position (Figure 11.5b) we have:

$$\lambda_{aPM}(\theta_{er}) = \left(1 - \frac{2}{\pi}\theta_{er}\right)\lambda_{PM} \tag{11.5}$$

The maximum flux linkage per phase λ_{PM} is:

$$\lambda_{PM} = W_1 B_{gPM} \tau L \tag{11.6}$$

The phase e.m.f., E_a, is:

$$E_a = -\frac{d\lambda_{aPM}(\theta_{er})}{d\theta_{er}} \cdot \frac{d\theta_{er}}{dt} = \frac{2}{\pi}\lambda_{PM}\omega_r \tag{11.7}$$

where ω_r is the electrical angular speed.

As two phases conduct at any time the ideal torque T_e is constant.

$$T_e = 2E_a i_{dc}\frac{p}{\omega_r} = \frac{4}{\pi}\lambda_{PM} p i_{dc} \tag{11.8}$$

Between any two current instantaneous commutations the phase current is constant ($i_d = i_{dc}$) and thus the voltage equation is:

$$V_d = 2r_s i_{dc} + 2E_a = 2r_s i_{dc} + \frac{4}{\pi}\lambda_{PM}\omega_r \tag{11.9}$$

From (11.9) and (11.8) we obtain:

$$\omega_r = \omega_{r0}\left(1 - \frac{T_e}{T_{esc}}\right) \tag{11.10}$$

with:

$$\omega_{r0} = \frac{\pi}{4}\frac{V_d}{\lambda_{PM}}; \quad T_{esc} = \frac{4}{\pi}\lambda_{PM} p i_{sc}; \quad i_{sc} = \frac{V_d}{2r_s} \tag{11.11}$$

The ideal speed/torque curve is linear (Figure 11.6) as for a d.c. brush PM motor.

The speed may be reduced and reversed by reducing the level and changing the polarity of d.c. voltage through supplying each motor phase by proper commutation in the PWM inverter (Figure 11.6). The current level is reduced by chopping.

The value of maximum flux linkage in a phase (λ_{PM}) may be reduced by advancing the current in the phase by the angle $\alpha_a \neq 0$. On the other hand, for an advancing angle $\alpha_a = \pi$, the electromagnetic power becomes negative and thus regenerative braking mode is obtained (Figure 11.5e).

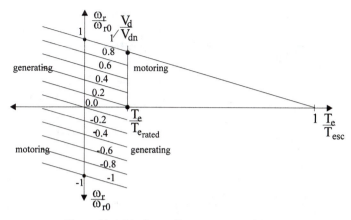

Figure 11.6. Ideal speed/torque curves of BLDC

11.3.2. The rectangular current control system

In general, a rectangular current control system contains the BLDC motor, the PWM inverter, the speed and current controllers and the position (speed) sensors (or estimators, for sensorless control) and the current sensors (Figure 11.7).

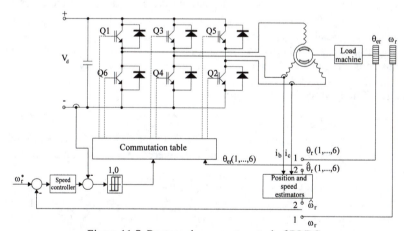

Figure 11.7. Rectangular current control of BLDC

The currents sequence, produced through inverter adequate control, with 120° current waveforms as in Figure 11.8 also shows the position of the 6 elements of the proximity sensors with respect to the axis of phase a for a zero advance angle $\alpha_a = 0$.

The location of proximity sensors P (a+, a–, b+, b–, c+, c–) is situated 90° (electrical) behind the pertinent phase with respect to the direction of motion.

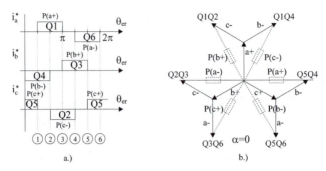

Figure 11.8. a.) Current sequencing; b.) phase connection

With two phases conducting, the stator active m.m.f. is on from 60° to 120° with respect to the rotor position. The ideal voltage vector (Figure 11.8) also jumps 60° for any phase commutation in the inverter. Each phase is on 120° out of 180° for the 120° conducting strategy.

To reverse the speed, the addresses (IGBTs) of the proximity sensor elements action are shifted by 180° (P(a+) → P(a−); P(b+) → P(b−); P(c+) → P(c−)). The proximity sensor has been located for zero advance angle to provide similar performance for direct and reverse motion. However, through electronic means, the advance angle may be increased as speed increases to reduce the peak PM flux in the stator phase and thus produce more torque, for limited voltage, at high speeds.

Using the same hardware we may also provide for 180° conduction conditions, at high speeds, when all three phases conduct at any time.

11.3.3. The hysteresis current controller

The d.c. current control requires only one current sensor, placed in the d.c. link, to regulate the current level. The distribution of the current through the six groups of two phases at a time is triggered by the proximity position sensor.

For 180° conduction, with three phases working at a time, a single current sensor in the d.c. link may not fully "represent" the three phase currents, especially if the dead time in the inverter is not properly compensated.

Quite a few current controllers may be applied for the scope. We prefer here the hysteresis controller as it allows a quick understanding of motor-inverter interactions [2]. Once the current in a phase is initiated (as triggered by the proximity sensor) it increases until it reaches the adopted maximum value i_{max}. In that moment, the phase is turned off until the current decreases to i_{min}. The duration of on and off times, t_{on} and t_{off}, is determined based on the hysteresis band: ($i_{max}-i_{min}$) (Figure 11.9).

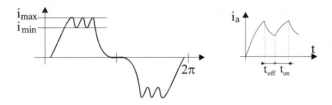

Figure 11.9. Current chopping

Let us consider the motor equations for the on (t_{on}) and off (t_{off}) intervals with Q1 and Q4 and D1 D4 in conduction (Figure 11.10), that is a + b - conduction.

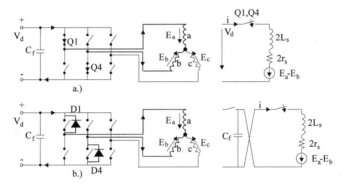

Figure 11.10. Conduction of phases a and b: a.) on-time; b.) off-time;

During the on-time, Figure 11.10a, the a+b- equation is

$$V_d = 2r_s i + 2L_s \frac{di}{dt} + E_a - E_b \qquad (11.12)$$

$$E_a - E_b \approx 2E_a = E_0 = \text{const.} \qquad (11.13)$$

Note that, if the turn-on is advanced by α_a, during part of the on-time $E_a - E_b = 0$ and thus a faster current increase is possible.

The solution of (11.12) is

$$i(t) = \frac{V_d - E_0}{2r_s}\left(1 - e^{-t\frac{r_s}{L_s}}\right) + i_{min} e^{-t\frac{r_s}{L_s}} \qquad (11.14)$$

To allow for current rising $V_d > E_0$. Above a certain speed $V_d < E_0$ and thus current chopping is not feasable any more. The current waveform contains, in this case, a single on-off pulse triggered by the proximity sensor (estimator).

During the off-time (diodes D1 and D4 conducting, in Figure 11.10b) the voltage equation is

$$0 = 2r_s i + 2L_s \frac{di}{dt} + (E_a - E_b) + \frac{1}{C_f} \int i \, dt + V_{c0} \qquad (11.15)$$

V_{c0} is the capacitor voltage at the end of the on - time (11.15), or at the beginning of the off-time.

The solution of (11.15) with t' = t–t_{on} is

$$i(t) = -\frac{V_{c0} + E_0}{2\omega L_s} e^{-\alpha_1 t'} \sin \omega t' - i_{max} \frac{\omega_0}{\omega} e^{-\alpha_1 t'} \sin(\omega t' - \phi) \qquad (11.16)$$

with $$\omega_0 = \sqrt{\frac{1}{2C_f L_s}}; \quad \alpha_1 = \frac{r_s}{2L_s}; \quad \omega = \sqrt{\omega_0^2 - \alpha_1^2}; \quad \phi = \tan^{-1} \frac{\omega}{\alpha_1} \qquad (11.17)$$

The torque $T_e(t)$ expression is

$$T_e = \frac{(E_a - E_b)i(t)}{\frac{\omega_r}{p}} \qquad (11.18)$$

So if the e.m.f. is constant in time the electromagnetic torque reproduces the current pulsations between i_{min} and i_{max}.

Example 11.1.

A BLDC motor is fed through a PWM inverter from a 300V d.c. source (V_d = 300V). Rectangular current control is performed at an electrical speed $\omega_r = 2\pi 10$rad/s. The no-load line voltage at ω_r is $E_0 = 48V$ = const. The cyclic inductance L_s = 0.5mH, r_s = 0.1Ω and the stator winding has q = 1 slot per pole per phase and two poles (2p = 2).

The filter capacitor C_f = 10mF, the current chopping frequency f_c = 1.25kHz and t_{on} / t_{off} = 5/3.

Determine:

a) the minimum and maximum values of current (i_{min}, i_{max}) during current chopping;

b) calculate the torque expression and plot it.

Solution:

a) To find i_{min} and i_{max} we have to use (11.14) for i(t) = i_{max} and t = t_{on} and (11.16) for i(t) = i_{min} and t' = t_{off}.

$$i_{max} = \frac{(300 - 48)}{2 \cdot 0.1} \left(1 - e^{-0.5 \cdot 10^{-3} \cdot 0.1/(0.5 \cdot 10^{-3})} \right) + i_{min} e^{-0.5 \cdot 10^{-3} \cdot 0.1/(0.5 \cdot 10^{-3})} \qquad (11.19)$$

Note that: t_{on} = 0.5ms and t_{off} = 0.3ms   (11.20)

Also from (11.17),

$$\omega_0^2 = \frac{1}{2 \cdot 0.5 \cdot 10^{-3} \cdot 10^{-2}} = 10^5; \quad \alpha_1 = \frac{0.1}{2 \cdot 0.5 \cdot 10^{-3}} \qquad (11.21)$$

$$\omega = \sqrt{\omega_0^2 - \alpha_1^2} = \sqrt{10^5 - 100^2} = 300 \text{rad/s} \qquad (11.22)$$

$$\phi = \tan^{-1}(300/100) = 71.56° \qquad (11.23)$$

$$i_{min} = -\frac{(300+48)}{2 \cdot 300 \cdot 0.5 \cdot 10^{-3}} \cdot e^{-100 \cdot 0.3 \cdot 10^{-3}} -$$
$$i_{max} \cdot \frac{316.22}{300} e^{-100 \cdot 0.3 \cdot 10^{-3}} \sin(300 \cdot 0.3 \cdot 10^{-3} - 71.56 \cdot \pi/180) \qquad (11.24)$$

From (11.19) and (11.24) we may calculate I_{max} and I_{min}

$$I_{max} = 205.32A \qquad I_{min} = 54.29A \qquad (11.25)$$

The average current is: $I_0 = (I_{max} + I_{min})/2 \approx 130$ A.

b) The average torque T_{av} is

$$T_{av} = \frac{2E_0 I_0}{\dfrac{\omega_r}{p}} = \frac{2 \cdot 48 \cdot 130}{2 \cdot \pi \cdot \dfrac{10}{1}} = 198.72 \text{ Nm} \qquad (11.26)$$

The instantaneous torque (11.18) includes current pulsations (as from (11.14) and (11.16), Figure 11.11).

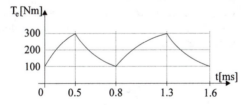

Figure 11.11. Torque pulsations due to current chopping

It should be noted that the chopping frequency is low for the chosen speed ($E_0 \ll V_d$) and thus the current and torque pulsations are large.

Increasing the chopping frequency will reduce these pulsations. To keep the current error band ($I_{max}-I_{min}$) within reasonable limits, the chopping frequency should vary with speed (higher at lower speeds and lower at medium speeds). Though the high frequency torque pulsations due to current chopping are not followed by the motor speed, due to the much larger mechanical time constant, they produce flux density pulsations and, thus, notable additional core and copper losses (only the average current I_0 is, in fact, useful).

11.3.4. Practical performance

So far the phase commutation transients — current overlapping — have been neglected. They, however, introduce notable torque pulsations at $6\omega_r$ frequency (Figure 11.12), much lower than those due to current chopping. To account for them, complete simulation or testing is required [3].

There are also some spikes in the conducting phase when the other two phases commute (points A and B on Figure 11.12).

Also not seen from Figures (11.11-11.12) is the cogging torque produced at zero current by the slot-openings in the presence of rotor PMs. Special measures are required to reduce the cogging torque to less than 2-5% of rated torque for high performance drives.

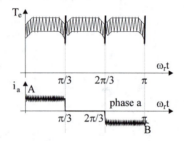

Figure 11.12. Torque pulsations due to phase commutation

While at low speeds current chopping is feasible, at high speeds one current pulse remains (Figure 11.13). The current controller gets saturated and the required current is not reached.

As the advance angle is zero ($\alpha_a = 0$), there is a delay in "installing" the current and thus, as the e.m.f. is "in phase" with the reference current, a further reduction in torque occurs.

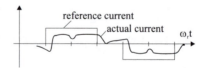

Figure 11.13. Current waveform at high speeds

11.3.5. Extending the torque/speed domain

Extending the torque/speed domain may be obtained (for a given drive) by advancing the phase commutation time by an angle α_a dependent on speed. This phase advancing allows fast current rise before the "occurence" of the e.m.f. (assuming a PM span angle $\alpha_{PM} < 150°\text{-}160°$). An approximate way to estimate the advance angle required α_a, for 120° conduction, may be based on linear current rise [4] to the value I

$$(\alpha_a)_{120°} = \omega_r \frac{L_s I}{V_{dc}}; \quad \omega_r = 2pn\pi \qquad (11.27)$$

where n is the rotor speed in rps.

Torque at even higher speeds may be obtained by switching from 120° to 180° current conduction (three phases working at any time). The current waveform changes, especially with advancing angle (Figure 11.14).

This time the e.m.f. is considered trapezoidal, that is close to reality.

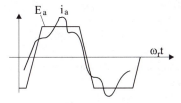

Figure 11.14. 180° conducting with advancing angle at high speed

The advancing angle α_a may be, for high speeds, calculated assuming sinusoidal e.m.f. [4] and current variation

$$(\alpha_a)_{180°} = \tan^{-1}\left(\frac{\omega_r L_s}{r_s}\right) \qquad (11.28)$$

It has been demonstrated [3,4] that 120° conduction is profitable at low to base speeds while 180° conduction with advancing angle is profitable for high speeds (Figure 11.15).

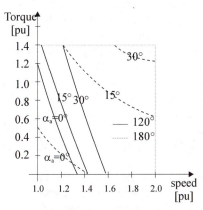

Figure 11.15. Torque/speed curves for various advancing angle α_a.

A smooth transition between 120° and 180° conduction is required to fully exploit the torque/speed capabilities of brushless d.c. motor drives.

Example 11.2.
Digital simulation: a brushless d.c. motor drive

We will present here the simulation results on a permanent magnet brushless DC motor drive (BLDC). The motor equations (see (10.6) in Chapter 10) are

$$L_s \cdot s \begin{vmatrix} i_a \\ i_b \\ i_c \end{vmatrix} = \begin{vmatrix} 1 & 0 & 0 \\ 0 & 1 & 0 \\ 0 & 0 & 1 \end{vmatrix} \cdot \begin{vmatrix} V_{an} \\ V_{bn} \\ V_{cn} \end{vmatrix} - \begin{vmatrix} R_s & 0 & 0 \\ 0 & R_s & 0 \\ 0 & 0 & R_s \end{vmatrix} \cdot \begin{vmatrix} i_a \\ i_b \\ i_c \end{vmatrix} - \begin{vmatrix} e_a(\theta_{er}) \\ e_b(\theta_{er}) \\ e_c(\theta_{er}) \end{vmatrix} \qquad (11.29)$$

with
$$L_s = L_a + M \qquad (11.30)$$

L_a, M self and mutual inductances. Introducing the null point of d.c. link (0), the phase voltages are

$$V_a = V_{a0} - V_{n0}$$
$$V_b = V_{b0} - V_{n0} \qquad (11.31)$$
$$V_c = V_{c0} - V_{n0}$$

For star connection of phases

$$V_{n0} = \left[\left(V_{a0} + V_{b0} + V_{c0} \right) - \left(e_a + e_b + e_c \right) \right] / 3 \qquad (11.32)$$

Here n represents the star connection point of stator windings. V_{a0}, V_{b0}, V_{c0} could easily be related to the d.c. link voltage and inverter switching state.

The simulation of this drive was implemented in MATLAB/SIMULINK. The motor model was integrated in a block (PM_SM).

The changing of motor parameters for different simulations is as simple as possible. After clicking on this block, a dialog box appears and you can change them by modifying their default values.

The drive system consists of a PI speed controller (Ki = 20, Ti = 0.05s), a reference current calculation block, hysteresis controller and motor blocks. The study examines system behavior for starting, load perturbation and speed reversal. The motor operates at desired speed and the phase currents are regulated within a hysteresis band around the reference currents (as functions of rotor position).

The integration step (50μs) can be modified from the Simulink's *Simulation / Parameters*.

To find out the structure of each block presented above, unmask it (*Options/Unmask*). Each masked block contains a short help screen describing that block (inputs / outputs / parameters).

The block structure of the electric drive system is presented in Figure 11.16 and the motor block diagram in Figure 11.17.

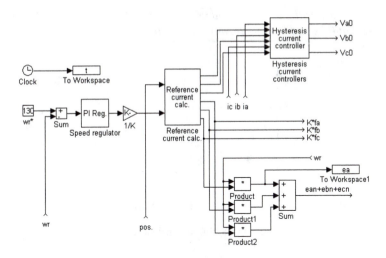

Figure 11.16. The BLDC rectangular current-drive controller

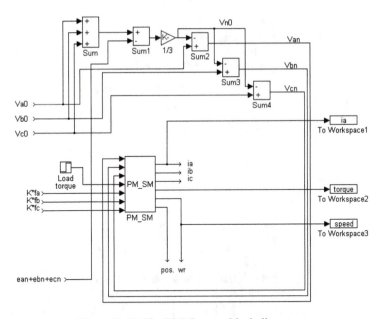

Figure 11.17. The BLDC motor block diagram

The drive and motor used for this simulation have the following parameters: V_{dc} = 220V, 2p = 2, R_s = 1Ω, L_s = 0.02H, M = −0.006667H, J = 0.005kgm², K = 0.763, hb (hysteresis band) = 0.2.

The following figures represent the speed (Figure 11.18), torque (Figure 11.19) and current (Figure 11.20) responses, and e.m.f. waveform (Figure

11.21), for a *starting process, loading (6Nm) at 0.2s, unloading at 0.4s, speed reversal at 0.5s and loading (6Nm) again at 0.8s.*

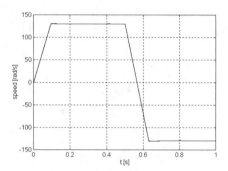

Figure 11.18. Speed transient response

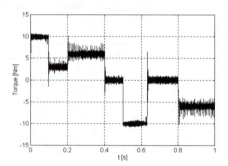

Figure 11.19. Torque response

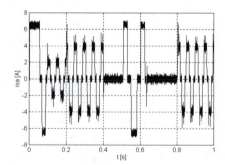

Figure 11.20. Current waveform (i_a).

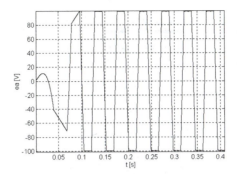

Figure 11.21. Induced voltage waveform (e_a).

Details of part of the program are given in Listing 1 for easy reference:

Listing 1:

```
function [sys, x0]=bldc0(t,x,u,flag,K)
if flag==3,
%Inputs: position, speedcontroller output/K (reference current amplitude);
sq3=1.732050808;
pos=u(1);
ir=u(2);
sq=0.866025;
per=fix(pos/2/pi)*360;
%Sector selection from 0 degrees (0-60;60-120...);
    if (sin(pos)<sq3/2)&(cos(pos)>0.5)&(sin(pos)>=0), iar=ir;ibr=-ir;icr=0;fa=1;fb=-
1;fc=((per-pos*180/pi)/30+1);end;
    if (sin(pos)>sq3/2)&(abs(cos(pos))<0.5),    iar=ir;ibr=0;icr=-ir;fa=1;fb=((pos*180/pi-60-
per)/30-1);fc=-1;end;
    if (sin(pos)<sq3/2)&(cos(pos)<-0.5)&(sin(pos)>=0), iar=0;ibr=ir;icr=-ir;fa=((per+120-
pos*180/pi)/30+1);fb=1;fc=-1;end;
    if (sin(pos)>-sq3/2)&(cos(pos)<-0.5)&(sin(pos)<0), iar=-ir;ibr=ir;icr=0;fa=-
1;fb=1;fc=((pos*180/pi-180-per)/30-1);end;
    if (sin(pos)<-sq3/2)&(abs(cos(pos))<=0.5),  iar=-ir;ibr=0;icr=ir;fa=-1;fb=((per+240-
pos*180/pi)/30+1);fc=1;end;
    if (sin(pos)>-sq3/2)&(cos(pos)>0.5)&(sin(pos)<0), iar=0;ibr=-ir;icr=ir;fa=((pos*180/pi-
300-per)/30-1);fb=-1;fc=1;end;
    ean=K*fa;
    ebn=K*fb;
    ecn=K*fc;
    sys=[iar ibr icr ean ebn ecn];
elseif flag==0,
    x0=[];
    sys=[0 0 6 2 0 1]';
else sys=[];
end
```

11.4. VECTOR (SINUSOIDAL) CONTROL

The vector (sinusoidal) control is applied both for surface PM and interior PM rotors and distributed stator windings (q ≥ 2).

According to Chapter 10, the torque T_e expression is

$$T_e = \frac{3}{2} p \left[\lambda_{PM} i_q + \left(L_d - L_q \right) i_d i_q \right]$$ (11.33)

The space vector equation (Chapter 10) is

$$\overline{V}_s = r_s \overline{i}_s + \frac{d\overline{\lambda}_s}{dt} + j\omega_r \overline{\lambda}_s$$ (11.34)

$$\overline{V}_s = V_d + jV_q; \quad \overline{i}_s = i_d + ji_q; \quad \overline{\lambda}_s = \lambda_d + j\lambda_q$$ (11.35)

$$\lambda_d = \lambda_{PM} + L_d i_d; \quad \lambda_q = L_q i_q$$ (11.36)

In general, $L_d \leq L_q$ and thus the second (reactive) torque component is positive only if $i_d \leq 0$. For $i_d \leq 0$ the PM flux is diminished by the i_d m.m.f. (Figure 11.22), but the PM always prevails to avoid complete PM demagnetization.

For vector control, in general, the rotor does not have a damper cage and thus there is no rotor circuit, and, as a direct consequence, no current decoupling network is necessary, in contrast to the IM case.

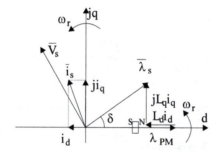

Figure 11.22. Space vector diagram of PM-SM with negative i_d ($i_d < 0$)

In vector control the PM-SM is controlled in d-q coordinates and then the reference d-q currents (or voltages) are transformed into stator coordinates through the vector rotator (inverse Park transformation) to be then realized by PWM in the inverter.

11.4.1. Optimum i_d-i_q relationships

Since the current decoupling network is missing and the drive command variable is the reference torque T_e^*, as required by a speed (speed and position) controller, we should now choose i_d^* and i_q^* from the torque equation (11.33). Evidently we need one more equation. This additional information may be obtained through an optimization criterion such as: maximum torque per current, maximum torque per flux, maximum efficiency, etc.

As at low speeds the drive is current limited and above base speed ω_b, it is flux limited, we may use these two criteria combined for a high performance drive. The maximum torque / ampere criterion is applied by using (11.34) and

$$i_s^* = \sqrt{i_d^{*2} + i_q^{*2}} \tag{11.37}$$

$$T_e^* = \frac{3}{2}p\left[\lambda_{PM} + (L_d - L_q)i_d^*\right]\sqrt{i_s^{*2} - i_d^{*2}} \tag{11.38}$$

to find

$$i_{di}^2 - i_{di}\frac{\lambda_{PM}}{L_d - L_q} - i_{qi}^2 = 0 \tag{11.39}$$

For $L_d = L_q$, starting with (11.33), it follows that

$$(i_{di})_{L_d=L_q} = 0 \tag{11.40}$$

On the other hand, for maximum torque per flux

$$\lambda_s^* = \sqrt{(\lambda_{PM} + L_d i_d)^2 + (L_q i_q)^2} \tag{11.41}$$

Proceeding as above we obtain a new relationship between i_{d_λ} and i_{q_λ}

$$(\lambda_{PM} + L_d i_{d_\lambda})^2(2L_d - L_q) - L_d L_q i_{d_\lambda}(\lambda_{PM} + L_d i_{d_\lambda}) + (L_d - L_q)\lambda_s^{*2} = 0 \tag{11.42}$$

where
$$\lambda_s^* \approx \frac{V_s^*}{\omega_r} \tag{11.43}$$

for $L_d = L_q$ (11.41) becomes (starting all over with (11.33))

$$(i_{d_\lambda})_{L_d=L_q} = -\frac{\lambda_{PM}}{L_d} \tag{11.44}$$

Equation (11.44) signifies complete cancellation of PM flux. The PM is still not completely demagnetized as part of the $L_d i_d$ is leakage flux which does not flow through the PM. Also note that constant current i_s means a circle in the i_d-i_q plane and constant stator flux λ_s means an ellipse.

We may now represent (11.38) and (11.40) and (11.41) and (11.42) as in Figure 11.23.

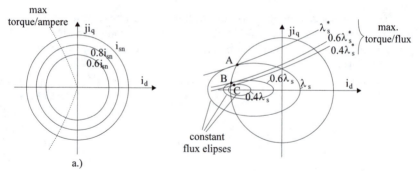

a.)

Figure 11.23. i_d-i_q a.) for given current i_s^* b.) for given flux λ_s^*

So, for each value of reference torque T_e^*, according to one of the two optimization criteria, unique values of i_d^* and i_q^* are obtained, provided the current limit i_s^* and the flux limit λ_s^* are not surpassed.

Note that the flux limit is related to speed ω_r (through 11.43). So, in fact, the torque T_e^* is limited with respect to stator current and flux (speed) (Figure 11.23a).

$$i_d^* = f_{d_{i,\lambda}}\left(T_e^*\right) \tag{11.45}$$

$$i_q^* = f_{q_{i,\lambda}}\left(T_e^*\right) \tag{11.46}$$

T_e^* limit is dependent on speed (Figure 11.24b).

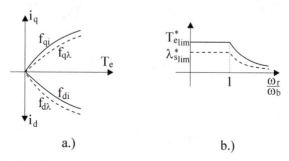

a.)   b.)

Figure 11.24. i_d-i_q optimum relationship and torque and flux limits with speed

Example 11.3. A PM-SM with $V_{sn} = 120\sqrt{2}$ V, $i_{sn} = 20$A, $l_d = 0.6$ p.u., no-load voltage $e_0 = 0.7$p.u., $l_q = 1.2$p.u., $n_n = 15$rps, $p = 1$ (pole pairs), $r_s = 0$, is driven at a current of 10A up to base speed n_b, according to the criterion of maximum torque per current.

Calculate the torque T_{ei} up to base speed and the base speed ω_b.
Solution:
First, we have to calculate l_d, l_q, e_0 in absolute units L_d, L_q, E_0

$$L_d = l_d \cdot \frac{V_{sn}}{i_{sn} \cdot \omega_n} = \frac{0.6 \cdot 120\sqrt{2}}{2\pi 15 \cdot 20} \approx 0.0539 \text{ H} \qquad (11.47)$$

$$L_q = l_q \cdot \frac{V_{sn}}{i_{sn} \cdot \omega_n} = \frac{1.2 \cdot 120\sqrt{2}}{2\pi 15 \cdot 20} \approx 0.1077 \text{ H} \qquad (11.48)$$

$$E_0 = e_0 \cdot V_{sn} = 0.7 \cdot 120\sqrt{2} = 118.44 \text{ V} \qquad (11.49)$$

The PM flux λ_{PM} is

$$\lambda_{PM} = \frac{E_0}{\omega_b} = \frac{118.44}{2\pi 15} = 1.257 \text{ Wb} \qquad (11.50)$$

Making use of (11.39) with (11.50)

$$2i_{di}{}^2 + i_{di} \cdot \frac{1.257}{0.0539 - 0.1077} - 10^2 = 0 \qquad (11.51)$$

$$i_{di} = -3.645 \text{ A} \qquad (11.52)$$

$$i_{qi} = \sqrt{100 - 3.645^2} = 9.312 \text{ A} \qquad (11.53)$$

and the torque

$$T_{ei} = \frac{3}{2} p \left[\lambda_{PM} + \left(L_d - L_q \right) i_d \right] \cdot i_q = \qquad (11.54)$$
$$= \frac{3}{2} \cdot 1 \cdot \left[1.257 + \left(0.0539 - 0.1077 \right) \left(-3.645 \right) \right] \cdot 9.312 = 20.3 \text{ Nm}$$

Now we have to find the maximum speed for which this torque may be produced

$$\lambda_{di} = \lambda_{PM} + L_d i_d = 1.257 + 0.0539 \cdot (-3.645) = 1.06 \text{ Wb}$$
$$\lambda_{qi} = L_q i_q = 0.1077 \cdot 9.312 = 1.00 \text{ Wb} \qquad (11.55)$$

$$\lambda_{si} = \sqrt{\lambda_{di}{}^2 + \lambda_{qi}{}^2} = \sqrt{1.06^2 + 1.00^2} = 1.457 \qquad (11.56)$$

Consequently, for rated voltage (maximum inverter voltage) the base speed ω_b is

$$\omega_b = \frac{V_{sn}}{\lambda_{si}} = \frac{120\sqrt{2}}{1.457} = 116.108 \text{ rad/s} \qquad (11.57)$$

$$n_b = \frac{\omega_b}{2\pi p} = \frac{116.108}{2\pi 1} = 18.454 \text{ rps} \approx 1110 \text{ rpm} \qquad (11.58)$$

In a similar way we may proceed above base speed using the maximum torque / flux criterion.

Note: The value of T_{ei}, if provided for up to base speed, does not leave any extra voltage for acceleration (at base speed).

This is only a small item that sheds light on the intricacies of a high performance drive design.

11.4.2. The indirect vector current control

Indirect vector current control means, in fact, making use of precalculated $f_{d_{i,\lambda}}$ and $f_{q_{i,\lambda}}$ ((11.45) - (11.46)), to produce the reference d -q currents i_d^*, i_q^*. Then, with Park transformation, the reference phase current controllers are used to produce PWM in the inverter. A rotor position sensor is required for position and speed feedback (11.25).

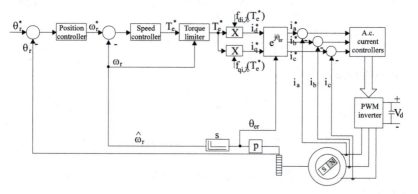

Figure 11.25. Indirect vector current control of PM-SM

As for the IMs, the a.c. controllers may be replaced by d.c. current controllers (in rotor coordinates) to improve the performance at high speeds especially. Though, in principle, direct vector control is possible, it is hardly practical unless the drive is sensorless.

11.4.3. Indirect voltage and current vector control

As expected, the vector current control does not account for the e.m.f. effect of slowing down the current transients with increasing speed. This problem may be solved by using the stator voltage equation for voltage decoupling

$$V_d^* = (r_s + sL_d)i_d^* - \omega_r L_q i_q^* \qquad (11.59)$$

$$V_q^* = (r_s + sL_q)i_q^* + \omega_r(\lambda_{PM} + L_d i_d^*) \qquad (11.60)$$

The d.c. current controllers may replace the first terms in (11.59)-(11.60) and thus only the motion-induced voltages are feedforwarded. Then PWM is performed open-loop, to produce the phase voltages at the motor terminals through the PWM inverter (Figure 11.26).

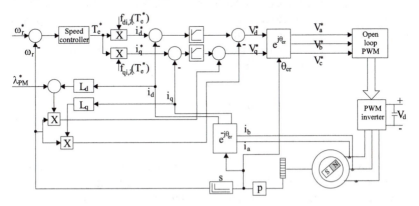

Figure 11.26. Indirect voltage and current vector control of PM-SM

with d-q (d.c.) current controllers

As expected, the open-loop (voltage) PWM has to observe the voltage limit

$$(V_d^*)^2 + (V_q^*)^2 \le V_{sn}^2 \qquad (11.61)$$

At low speeds the current controllers prevail, while at high speeds the voltage decoupler takes over. Results obtained with such a method [5], based only on f_{di}, f_{qi} (maximum torque per current criterion) proved a remarkable enlargement of the torque/speed envelope (Figure 11.27).

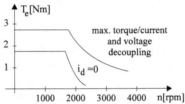

Figure 11.27. Torque/speed envelope for $i_d = 0$ and for maximum torque/current and voltage decoupling

Note: The e.m.f. compensation in indirect current vector control may also be solved by desaturating the i_q controller based on d axis controller saturation for a given interval of time [6]. Similar results with those in Figure 11.27 have been thus obtained with a notably smaller on line computation effort.

11.4.4. Fast response PM-SM drives: surface PM rotor motors with predictive control

Interior PM-SM drives have been, so far, treated for vector control. As L_q is rather high in such motors, at least i_q response is rather slow in comparison with IM drives. However, for surface PM rotor PM-SMs, $L_d = L_q = L_s$ is small and thus fast current (and torque) response may be obtained.

The current increment Δi_s is

$$\Delta \overline{i}_s(t) \approx \frac{\overline{V}_s - \overline{E}}{L_s} \Delta t$$

$$\Delta \overline{i}_s = \overline{i}_s^{\;*} - \overline{i}_s$$

(11.62)

So the current controller error, $\Delta \overline{i}_s$, with the e.m.f. space vector $\overline{E}$ estimated, is used to calculate the required (predicted) stator voltage vector which, then, may be open-loop pulse width modulated to control the drive.

This classical method [7] has produced spectacular results with position sensors (encoders) — Figure 11.28.

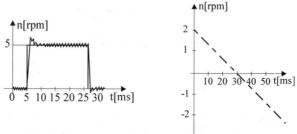

Figure 11.28. Speed responses at low speed

The low speed/time linearity [7] is obtained also by using special measures to reduce all torque pulsations to less than 1.5% of rated torque.

Example 11.4.
Digital simulation: Indirect vector a.c. current control
For the PM-SM motor in example 11.3, making use of the indirect vector current control system of Figure 11.25, perform digital simulations.

The simulation of this drive was implemented in MATLAB/SIMULINK. The motor model was integrated in a block (PM_SM) (Figures 11.29, 11.30, 11.31).

The changing of motor parameters for different simulations is as simple as possible. By clicking on this block, a dialog box appears and you can change the parameters by modifying their default values.

The reference values for the current components (i_d* and i_q*) have been obtained from the torque/current criterion (11.39). You can modify the speed regulator (*PI regulator*) parameters (amplification and integration time constant) to study the behavior of the system. A *Commutation Frequency* block has been included to satisfy the sampling criteria (to control the commutation frequency). The integration step can be modified from the Simulink's *Simulation/Parameters*.

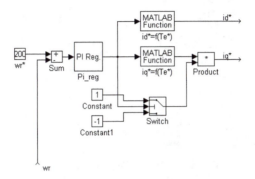

Figure 11.29. i_d^*, i_q^* referencers

To find out the structure of each block presented above, unmask them (*Options/Unmask*). The *abc_to_alfabeta* and *alfabeta_to_dq* blocks produce the coordinate transform.

The block structure of the electric drive system is presented in Figures (11.29-11.31).

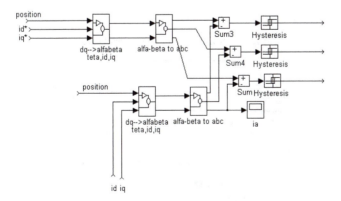

Figure 11.30. A.c. current controllers

The motor used for this simulation has the following parameters: P_n = 900W, U_n = 220V, 2p = 4, n = 1700rpm; I_n = 3A, λ_{PM} = 0.272Wb, R_s = 4.3Ω, L_d = 0.027H, L_q = 0.067H, J = 0.00179kgm².

The following figures represent the speed (Figure 11.32), torque (Figure 11.33) and current waveform (Figure 11.34), for a *starting process and load torque (2Nm) applied at 0.4s.*

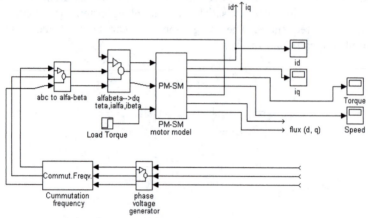

Figure 11.31. PM-SM block diagram

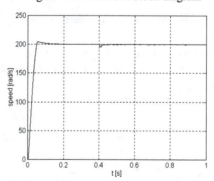

Figure 11.32. Speed transient response

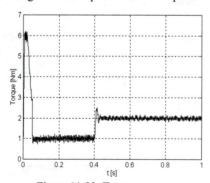

Figure 11.33. Torque response

The d-q reference current relationships to torque for maximum torque per current are given in Figure 11.35.

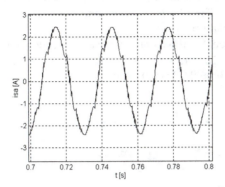

Figure 11.34. Current waveform (i_a) under steady-state

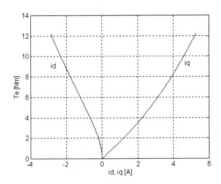

Figure 11.35. The d-q currents versus torque

Example 11.5.

Digital simulation: indirect vector d.c. current control of PM-SM.

The simulation of this drive (Figure 11.36) was implemented in MATLAB/SIMULINK simulation program. The motor model was integrated in the PM_SM block (Figure 11.37).

Changing of motor parameters for different simulations is as simple as possible. By clicking on the block, a dialog box appears and you can change the parameters by modifying their default values.

The reference values for the current components (i_d* and i_q*) have been obtained from the torque/current criterion (11.39). This part of simulation has been developed as a Matlab routine, Listing2 and Figure 11.38.

You can modify the speed regulator (*PI regulator*) parameters (amplification and integration time constant) to study the behavior of the system. A *Commutation Frequency* block was introduced to satisfy the sampling criteria (to control the commutation frequency). The integration step can be modified from the Simulink's *Simulation/Parameters*.

To find out the structure of each block presented above, unmask them (*Options/Unmask*). The *abc_to_alfabeta* and *alfabeta_to_dq* blocks produce the coordinate transform.

The motor used for this simulation has the following parameters: $P_n = 900W$, $U_n = 220V$, $2p = 4$, $n = 1700rpm$; $I_n = 3A$, $\lambda_{PM} = 0.272Wb$, $R_s = 4.3\Omega$, $L_d = 0.027H$, $L_q = 0.067H$, $J = 0.000179kgm^2$.

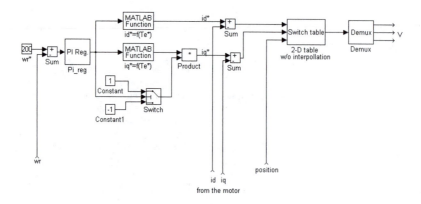

Figure 11.36. The indirect vector d.c. current controller

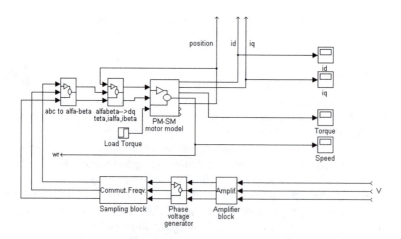

Figure 11.37. PM-SM model

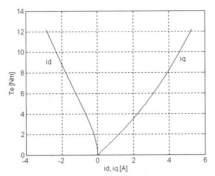

Figure 11.38. i_d- i_q reference currents versus torque for maximum torque / current criterion

The following figures represent the speed (Figure 11.39), torque (Figure 11.40) and d-q current responses (Figures 11.41, 11.42), for a *starting process and load torque (2Nm) applied at 0.4s.*

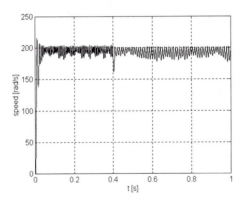

Figure 11.39. Speed transient response

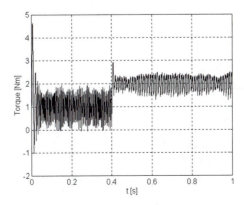

Figure 11.40. Torque response

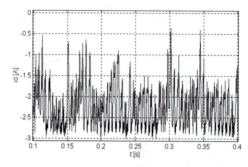

Figure 11.41.Current waveform (i_d) under steady-state

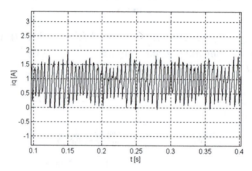

Figure 11.42.Current waveform (i_q) under steady-state

Listing2:

```
% This program creates the table for id*=f(Te*) and iq*=f(Te*).
% Variable definition
  clear;
  lPM=0.272;
  Ld=0.027;
  Lq=0.067;
  p1=4;
  Isn=6;
k=1;
for is=0:0.1:Isn
  p=[2 lPM/(Ld-Lq) -is*is];
  R=roots(p);
  if (R(1)<R(2)), id=R(1);
  else
   id=R(2);
  end
  iq1=sqrt(is*is-id*id);
  iq2=-iq1;
  Te1=1.5*p1*(lPM+(Ld-Lq)*id)*iq1;
  Te2=-Te1;
  Vid(k)=id;
  Viq1(k)=iq1;
  Viq2(k)=iq2;
  VTe1(k)=Te1;
```

```
    VTe2(k)=Te2;
    i=is;
    k=k+1;
end
M1=[VTe1;Vid]';
M21=[VTe1;Viq1]';
M22=[VTe2;Viq2]';
plot(Vid,VTe1,'-',Viq1,VTe1,'-g')
xlabel('id,iq')
ylabel('Te')
title('Variation of Id & Iq with Te')
```

As seen from Figure 11.36 the d-q current error drives a switch table to produce the desired (adequate) voltage vector in the inverter.

11.5. DIRECT TORQUE AND FLUX CONTROL (DTFC) OF PM-SMs

DTFC, as presented for the IM, implies direct torque and stator flux control through direct triggering of one (or a combination of) voltage vector(s) in the PWM voltage source inverter that supplies the PM-SM.

A general configuration of a DTFC system for PM-SMs is shown in Figure 11.43.

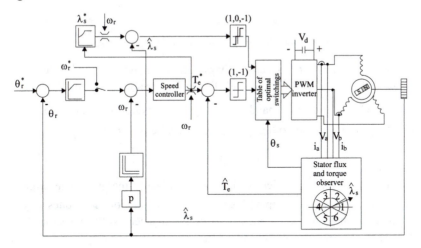

Figure 11.43. DTFC for PM-SMs

The table of optimal switchings (TOS) is obtained based on the principle that the stator flux space vector variation goes in the direction of the applied voltage space vector

$$\overline{\lambda}_s \approx \overline{\lambda}_{s0} + \overline{V}(i) \cdot T \qquad (11.63)$$

Also, to increase torque, we have to advance the flux vector in the direction of motion. Flux increasing means an angle above 90° between $\overline{V}(i)$ and $\overline{\lambda}_{s0}$. Consequently, the adequate voltage vector is also chosen based on the 60° wide sector (θ_i) where the initial flux vector is located (Figure 11.44).

The complete table of optimal switching (TOS) is the same as for the IM (Chapter 9, Table 9.2.). The TOS is "visited" with a certain frequency and thus constant switching frequency is required.

Applying all voltage vectors for a given (constant) time interval $T = 1 / f_c$ may not be the most adequate solution for low speeds where short nonzero voltage vectors and long zero vectors are required.

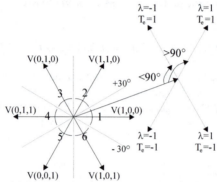

Figure 11.44. Voltage vector selection for the first sector

The T_{on} time of nonzero voltage vectors may thus be made speed dependent, provided $T_{on} < T$ and $T_{on} > T_{min}$ (minimum conduction time allowable). On the other hand, if the "exact" position of stator flux $\hat{\theta}_s$ is estimated, the required voltage vector $\overline{V}_s{}^*$ may be predicted

$$\overline{V}_s{}^* = r_s \overline{i}_s{}^* + \Delta\overline{\lambda}_s / T \tag{11.64}$$

This time the space vector flux error (not only the amplitude error) is required. Once $\overline{V}_s{}^*$ is computed, space vector modulation techniques may be applied to calculate the timings t_1, t_2 of the adjacent voltage vectors V(i), V(i+1) and of the zero voltage vector $V_0(V_7)$

$$\overline{V}_s{}^* = \overline{V}(i) \cdot t_1 + \overline{V}(i+1) \cdot t_2 + V_0(0) \cdot t_0 \tag{11.65}$$

$$t_0 + t_1 + t_2 = T \tag{11.66}$$

Smoother torque and flux control is obtained this way for given commutation frequency. The flux/torque relationship (Figure 11.43) is derived directly from the maximum torque/current and maximum torque/flux criteria.

Also the maximum reference flux and torque are limited depending on speed to keep the current and voltage within available limits.

The key component of DTFC is represented by the stator flux and torque observer (Figure 11.45) which represents the "price" for its control simplicity and robustness.

11.5.1. The stator flux and torque observer

It is, in general, assumed that the stator currents i_a, i_b and voltages V_a, V_b (or the d.c. link voltage V_d) are measured. Also, an encoder provides both rotor position θ_r and speed feedback in all high performance drives [8]. The observers considered for IMs may be applied here also. Among the real, practical ones we mention the voltage-current, full order observers, Kalman filters and MRAS closed-loop configurations.

The voltage-current closed-loop observer is presented here in some detail.

The voltage model in stator coordinates is

$$\frac{d\overline{\lambda}_{sv}{}^s}{dt} = \overline{V}_s - r_s\overline{i}_s \tag{11.67}$$

while the current model, in rotor coordinates, is;

$$\overline{\lambda}_{dq} = \overline{\lambda}_{si}{}^r = L_d i_d + \lambda_{PM} + jL_q i_q; \quad \overline{i}_{dq} = \overline{i}_s{}^r = i_d + ji_q \tag{11.68}$$

The current in rotor coordinates $\overline{i}_s{}^r$ is

$$\overline{i}_s{}^r = \overline{i}_s{}^s e^{-j\theta_{er}} \tag{11.69}$$

The stator flux $\overline{\lambda}_{si}{}^r$ is transformed back to stator coordinates as

$$\overline{\lambda}_{si}{}^s = \overline{\lambda}_{si}{}^r e^{j\theta_{er}} \tag{11.70}$$

It is evident from the above that the voltage model is plagued by the stator resistance error and the integration drift at low frequencies.

The current model is influenced by parameter detuning (due to magnetic saturation) and by the position error, but it works from zero speed (frequency). A combination of the two with a PI compensator is designed such that the current model is predominant at low speeds, while the voltage model takes over at high speeds (Figure 11.45).

The PI compensator provides for this discrimination. The frequency band for the transition is provided by an adequate choice of K_i and τ_i based on the given observer poles ω_1 and ω_2 [9], real and negative

$$K_i = -(\omega_1 + \omega_2); \quad \tau_i = \frac{|K_i|}{\omega_1\omega_2} \tag{11.71}$$

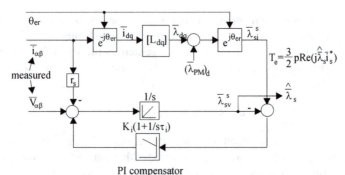

PI compensator

Figure 11.45. Voltage and current stator flux observer

Typical values of ω_1 and ω_2 are $\omega_1 = -(3 - 10)$ rad/s, $\omega_2 = -(3 - 10)|\omega_1|$.

Low speed and torque response for such a DTFC drive with sliding mode speed controller and $L_d = 4.1$mH, $L_q = 8.2$mH, $\lambda_{PM} = 0.2$Wb, $r_s = 0.6\Omega$, J = 0.005Kgm^2, p = 4pole pairs, $f_c = 10$KHz, $V_d = 200$V, $T_{emax} = 12$Nm are given in Figure 11.46 [9]. The flux and torque hysteresis bands are 0.5% and 1%, respectively.

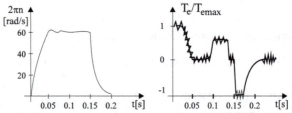

Figure 11.46. Low speed and torque transients with DTFC of PM-SM

Fast torque control is demonstrated, as expected. The absence of speed overshoot is due to the use of a sliding mode speed controller.

The torque is calculated easily from flux and current (Figure 11.45). Providing for low and high speed good performance, DTFC is a strong contender for vector voltage and current control as it is simpler and more robust.

11.6. SENSORLESS CONTROL OF PM-SMs

Sensorless control is targeted so far for speed controlled drives where a less than 100 to 1 speed control range is required. It reduces hardware costs and improves mechanical reliability. High performance drives — for more than 100 to 1 speed range and for precision positioning — still require motion sensor feedback. Unfortunately, vector control requires, even for speed control only, rotor position information.

As for IM sensorless, control may be open-loop or closed-loop. Open loop (V / f) runs into problems of stability and is adequate for low dynamics applications that require a maximum of 10 to 1 speed control range with good precision. Such drives need a damper cage on the rotor for improved stability (Figure 11.2).

Feedforward torque compensation may be added to slightly improve the dynamic response. On the other hand, closed-loop sensorless PM-SM or BLDC motor drives require rotor and position observers to produce a 100 to 1 speed control range with fast dynamics. Closed-loop sensorless control is approached in what follows.

BLDC motor drives (with surface PM rotors, trapezoidal e.m.f. and rectangular current control) require special proximity (position) estimators. A classification of such estimators includes zero crossing of open phase e.m.f., the back e.m.f. integration approach, third harmonic method, and conduction time of diodes in the PWM inverter.

A collection of papers on this subject is available in [10], Section 2. Only 10 / 1 speed control range is ususally claimed with BLDC motor drives as, especially, speed estimation from position estimation is coarse at low speeds. On the other hand, vector controlled PM -SM use, for position and speed estimation, voltage and current or full order observers, Kalman filters or MRAS.

As DTFC requires a flux and torque observer, for sensorless control only speed estimation is additionally required. Note that DTFC does not require explicit rotor position for speed control while for vector control it does (for vector rotators). However, implicit rotor position is required for speed calculation.

An example of a speed calculator is obtained with respect to Figure 11.47.

$$\hat{\omega}_r = \frac{d\hat{\theta}_{\lambda s}}{dt} - \frac{d\hat{\delta}}{dt}$$

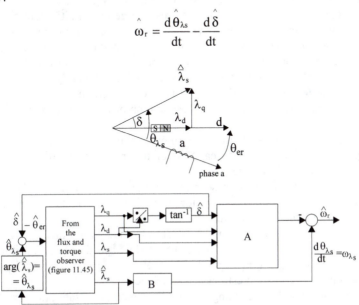

Figure 11.47. Rotor speed estimator for DTFC control of PM-SMs

where blocks A and B contain

$$
\begin{aligned}
\text{A:} \quad & \frac{d\hat{\delta}}{dt} = \frac{\lambda_d(k-1)\lambda_q(k) - \lambda_q(k-1)\lambda_d(k)}{\lambda_s^2 T} \\
\text{B:} \quad & \frac{d\theta_{\lambda_s}}{dt} = \frac{\lambda_{s\alpha}(k-1)\lambda_{s\beta}(k) - \lambda_{s\beta}(k-1)\lambda_{s\alpha}(k)}{\lambda_s^2 T}
\end{aligned}
\tag{11.72}
$$

As the d-q flux angle δ varies only in the first and fourth quadrants ($\lambda_d \geq 0$, to avoid PM demagnetization), its estimation is straightforward. The derivatives required to estimate rotor speed $\hat{\omega}_r$ are done based on $\sin\alpha$ and $\cos\alpha$ derivative properties in discrete form with T as the sampling time (Figure 11.47).

Note that the speed estimator does not use inertia and, consequently, the motion equation has not been used for the scope. The latter may be used either to identify the load perturbation torque or somehow to correct the speed estimated from electrical equations.

11.6.1. Initial rotor position detection

Most position and speed observers are not capable of detecting the initial rotor position. Special starting methods or initial position identification, without moving the rotor to a known position, are required for safe, stable starting. Initial rotor identification may be performed by the drive itself through sending short voltage vector signals V_1, V_3, V_5 for a few microseconds until the phase currents reach a certain limit. From the time required for the current rising in the three cases, in case of interior PM rotors, the rather exact initial rotor position may be obtained based on the sinusoidal inductance-position dependence. On the other hand, for surface-PM rotor, additionally, negative m.m.f. is required for each position (V_2, V_4, V_6) to find out in which 60°-wide sector the rotor is placed based on the slight magnetic saturation (for positive m.m.f.) and desaturation of PMs (for negative m.m.f.) [11].

Alternatively, it is to find the rotor position sector (within 60°) by comparing the phase voltage decay through the free wheeling diodes after the turn off of a nonzero voltage vector.

DTFC sensorless tends to selfstart from any position after a short lived oscillation as a certain voltage vector is applied until the flux surpasses the reference value for the first time.

For electric drives with nonhesitant start from any rotor position and very low speed (below 10 rpm) high performance (low speed error and fast torque response), signal injection methods have been introduced for rotor position estimation. Initial position estimation is included.

They are based on tracking the machine natural or saturation produced saliency.

In advanced solutions special saliency tracking observers with online magnet polarity discrimination are applied.

Rotating carrier-signal injection or a.c. carrier-signal injection at a given frequency, larger than the rated fundamental frequency of the drive, are used for the scope, while the current response to these signals is checked for the components with rotor position information [21].

Precision of a few mechanical degrees has been obtained this way for the initial rotor position estimation.

Above a certain low speed the injection voltage signal has to be eliminated and a different, fundamental voltage based, rotor position estimation is required. Switching from one position observer to the other should be smooth and nonhesitant [22, 23, 24].

Making use of special PWM voltage sequences to analyze the current transient response and thus find the initial rotor position and the rotor position at low speeds is another way to handle this problem.

Sensorless PM-SM drives with 100 to 1 speed control range, speed precision of 1% of rated speed and fast torque response (milliseconds) are now close to reaching markets and thus much progress on the subject is expected in the near future.

11.7. RELUCTANCE SYNCHRONOUS MOTOR (RSM) DRIVES

RSMs are characterized by the absence of rotor currents and, in rotor configurations with multiple flux barriers (or anisotropic axially laminated (ALA), Figure 11.48a, b), high L_d / L_q rates provide for the torque density (Nm/kg), power factor and efficiency comparable or slightly superior to induction motors. For $L_d / L_q > 8$, the RSM is fully competitive with the IM. The absence of rotor currents, especially when an encoder is available for rotor position feedback, makes the RSM drive control notably simpler than for IMs.

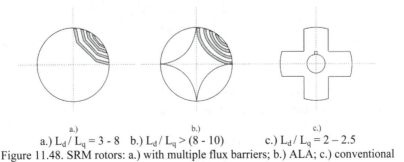

a.)   b.)   c.)
a.) $L_d / L_q = 3 - 8$ b.) $L_d / L_q > (8 - 10)$ c.) $L_d / L_q = 2 - 2.5$
Figure 11.48. SRM rotors: a.) with multiple flux barriers; b.) ALA; c.) conventional

As ideally the core losses in the rotor are zero (in reality some harmonics flux loss currents exist), the RSM may work safely at low speeds and high torque, provided the stator is properly cooled. On the other hand, the slightly fragile rotor limits the peripheral rotor speeds to, probably, 50-60m/s. So SRM might be considered from 100 W up to 1 MW.

Naturally, for conventionally laminated anisotropic rotors (Figure 11.48c), $L_d / L_q < 3$, low power factor (about 0.45-0.50) at low power at higher peripheral speeds are acceptable.

11.7.1. RSM vector control principles

The RSM may be considered as a nonlinear system with two entries; the stator flux $\bar{\lambda}_s$ and the torque T_e (Chapter 10, Equations 10.116-10.117)

$$\bar{\lambda}_s = L_d i_d + j\left(L_q i_q - \lambda_{PMq}\right)$$

$$T_e = \frac{3}{2} p\left[\lambda_{PM_q} + \left(L_d - L_q\right) i_q\right] i_d \qquad (11.73)$$

A PM in axis q (of lower inductance) has been included.

Note that the PM flux is opposite to and higher than any $L_q i_q$ contribution: $\lambda_{PMq} \geq L_q i_q$.

Also, i_q should not change sign if positive reluctance torque in the presence of PMs is desired. To switch from positive to negative torque, the sign of i_d is changed. Conversely, if $\lambda_{PMq} = 0$, in general, to change the torque sign, the polarity of i_q is changed. As $L_d > L_q$, i_d transients are slower than i_q transients, so faster torque response is obtained for i_q control.

As there are no rotor currents, no current decoupling network is required. Instead, it is paramount to find i_d, i_q relationships to torque. This is done based on optimization performance criteria, as already explained for PM-SMs earlier in this chapter.

Practical criteria would be maximum torque/current at low speeds, maximum torque/flux at high speeds and maximum power factor or efficiency.

Three basic criteria are discussed separately for the "pure" RSM ($\lambda_{PMq} = 0$).

For maximum torque / current, using (11.73) we obtain

$$i_{di} = i_{qi} = i_s / \sqrt{2}$$

$$T_{ei} = \frac{3}{2} p\left(L_d - L_q\right) i_{di}^{\,2} \qquad (11.74)$$

As $i_{di} < 0.5 i_n \cdot \sqrt{2}$ (i_n - rated phase current, rms) to avoid heavy magnetic saturation, about half of the rated torque may be expected when using this criterion.

The maximum power factor criterion ($r_s = 0$) is based on the definition

$$\tan\varphi_1 = \frac{L_d i_d^{\,2} / 2 + L_q i_q^{\,2} / 2}{\left(L_d - L_q\right) i_d i_q} \qquad (11.75)$$

Finally, we find (for $d\tan\varphi_1 / d\left(i_d / i_q\right) = 0$)

$$\left(\frac{i_d}{i_q}\right)_{\varphi_{1min}} = \sqrt{\frac{L_q}{L_d}} \tag{11.76}$$

$$T_{e\varphi_1} = \frac{3}{2}p\left(L_d - L_q\right)i_{di}^2\sqrt{\frac{L_d}{L_q}} \tag{1.77}$$

$$\left(\cos\varphi_1\right)_{max} = \frac{1 - L_q/L_d}{1 + L_q/L_d} \tag{11.78}$$

For maximum torque per flux

$$\frac{i_{d\lambda}}{i_{q\lambda}} = \frac{L_q}{L_d}; \quad \lambda_{d\lambda} = \lambda_{q\lambda} = \lambda_s/\sqrt{2} \tag{11.79}$$

Again, for $i_{d\lambda} = i_{di}$

$$T_{e\lambda} = \frac{3}{2}p\left(L_d - L_q\right)i_{di}^2\frac{L_d}{L_q} \tag{11.80}$$

or $$T_{e\lambda} = \frac{3}{2}p\left(L_d - L_q\right)\frac{\lambda_s^2}{2L_dL_q} \tag{11.81}$$

The highest torque may be obtained with the criterion of maximum torque/flux, but for the lowest power factor (slightly lower than 0.707). The base speed ω_b, defined for maximum torque available, is thus:

$$\omega_b = \frac{V_{sn}}{L_d i_{di}\sqrt{2}} \tag{11.82}$$

For a more detailed analysis of steady-state and transient torque capabilities of RSM, see [12].

In essence, for servo drives without constant power (flux weakening) zone, and fast speed response, the criterion of maximum power factor may be used for steady-state. For fast transients, the maximum torque / flux approach is adequate. As expected, the PWM inverter has to be rated for the rather high currents required for fast speed transients.

Trajectories of i_d-i_q currents and λ_d-λ_q fluxes for the three main optimal control strategies are shown in Figure 11.49.

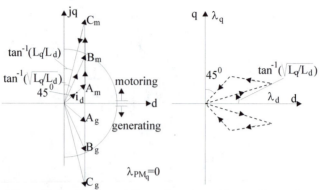

Figure 11.49. Current and flux trajectories with speed increasing

It has been shown that the influence of magnetic saturation (of L_d mainly) is small for maximum power factor control [13].

On the contrary, for maximum torque/flux control, the influence of magnetic saturation on performance is notable. For small torques and below base speed, the control may follow the d-q current trajectory $0A_mB_m$ while above base speed, for constant power control, the trajectory $0B_mC_m$ is to be followed. Consequently, the i_d, i_q angle γ

$$\gamma = \tan^{-1} \frac{i_d}{i_q} \qquad (11.83)$$

should be made variable with torque (and speed).

The same conditions may be met by working with flux linkage d-q angles (Figure 11.49b). In general, for a wide ($\omega_{max} / \omega_b > 2$) constant power zone, PMs in axis q are required [14]. For low speeds vector current control is more adequate while above base speed, vector voltage control is better.

A combination of the two is, perhaps, the practical solution for wide speed range control. Vector current control [15, 16], eventually combined with voltage decoupling in d-q orientation or in stator flux orientation [17], has been proposed. They are very similar to the corresponding methods developed for PM-SMs. In what follows, we will only illustrate the vector control of RSM through a vector d-q current control system with i_d constant up to base speed ω_b and speed dependent above ω_b [18]. We will also treat direct torque and flux control DTFC [19] of RSM.

11.7.2. Indirect vector current control of RSM

The indirect vector current control scheme (Figure 11.50) is similar to that for the PM-SMs.

Either speed or position control may be performed. The reference magnetization current i_d^* is constant up to base speed ω_b and decreases inversely with speed above ω_b.

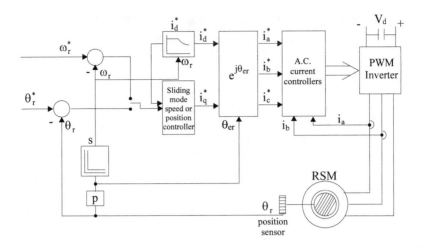

Figure 11.50. Indirect vector current control of RSM

Let us introduce some details on the sliding mode (SM) speed controller simulation and test results. The SM speed controller output produces the reference torque current i_q^* and the simplest control law is

$$\begin{cases} i_q^* = +i_{qk} \text{ for } s_\omega > \omega_h \text{ or } \left(|s_\omega| < \omega_h \text{ but } \dot{s}_\omega < 0\right) \\ i_q^* = -i_{qk} \text{ for } s_\omega \leq -\omega_h \text{ or } \left(|s_\omega| < \omega_h \text{ but } \dot{s}_\omega > 0\right) \end{cases} \qquad (11.84)$$

where ω_h is the hysteresis band of the controller and s_ω is the SM functional

$$s_\omega = K_{p\omega}\left(\omega_r^* - \omega_r\right) - K_{d\omega}\frac{d\omega_r}{dt} \qquad (11.85)$$

Besides meeting the conditions of touching the straight line $s_\omega = 0$ (in the state space plane) we need to know $K_{p\omega}$ and $K_{d\omega}$. To do so, we admit that i_q varies linearly in time and the targeted speed ω_r^* is reached after s_ω changes sign at least once (Figure 11.51).

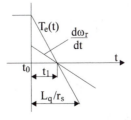

Figure 11.51. Ideal transients (constant i_d, linear variation of i_q)

The ratio $K_\omega = K_{d\omega} / K_{p\omega}$ is to be determined first.

If, for $t = 0$, $s_\omega = 0$, for no-load, (11.85) yields

$$\omega_r^* - \omega_r(t_0) = K_\omega \frac{d\omega_r}{dt} = K_\omega \frac{p}{J} T_e(t_0) \qquad (11.86)$$

where T_e is the torque, J - inertia and p - pole pairs.

If i_q varies linearly, the torque T_e does the same thing (i_d = const) and, from t_0 to t_0+t_1 (Figure 11.51), we have

$$\frac{d\omega_r}{dt} = \frac{p}{J} T_e(t_0) \cdot \left[1 - \frac{t - t_0}{t_1} \right] \qquad (11.87)$$

From (11.87), through integration, we have

$$\omega_r^* - \omega_r(t_0) = \frac{p}{2J} t_1 T_e(t_0) \qquad (11.88)$$

Comparing with (11.86) yields

$$K_\omega = t_1 / 2 \qquad (11.89)$$

A good approximation could be $t_1 = L_q / 2r_s$ and, thus,

$$K_\omega = L_q / 4r_s \qquad (11.90)$$

By limiting the steady-state error we may find the second relation, to calculate $K_{p\omega}$ and $K_{d\omega}$ and thus the SM controller is fully designed.

In a similar way, but for linear speed deceleration during braking, an SM position controller may be designed. Its constants would be $K_{p\theta}$ and $K_{d\theta}$ [18].

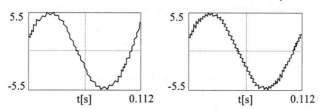

Figure 11.52. Steady-state phase current: a.) digital simulation; b.) test results

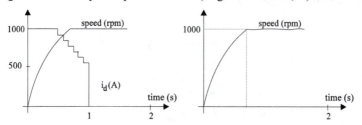

Figure 11.53. Step speed responses: a.) digital simulation; b.) test results

For an RSM with data $2p = 6$, $L_q = 24$mH, $L_d = 88$mH for $i_d^* = 3$A, $r_s = 0.8\Omega$, $J = 0.0157$kgm^2, $V_0 = 80$V, $f_c = 5$kHz, for $i_{dm}^* = 3$A and $i_{qm}^* = 5$A, $K_{p\omega} = 5$, $K_\omega = 7.5 \times 10^{-3}$s, $K_{d\omega} = 0.01$s. $\omega_h = 5$rad/s, $J_{max} = 5$J, $K_{p\theta} = 5$, $K_{d\theta} = 0.75$, $\theta_h = 3\pi/50$ rad, the digital simulation and test results are shown for comparison in Figures (11.52-11.53).

Note that the acceleration decreases above 500rpm as step-wise i_d decrease with speed is performed (Figure 11.53a).

As the position SM controller is designed for 5J it is no wonder that the responses for J and 5J do not differ notably (Figure 11.54). This kind of robustness, with the elimination of the overshoot and with fast response, are intrinsic merits of SM controllers.

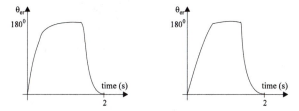

Figure 11.54. Positioning responses (calculated):

a.) rated inertia J; b.) 5J

Note: A complete simulation application is not given here as it may be treated as a particular case of the previous examples for PM-SMs.

11.7.3. Direct torque and flux control (DTFC) of RSM

The DTFC of RSM is similar to that of PM-SM. The basic configuration is shown on Figure 11.55.

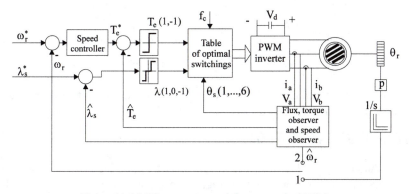

Figure 11.55. Direct torque and flux control of RSM:

1. with motion sensor; 2. sensorless

The flux and torque observers are paramount to DTFC. A speed observer (estimator) is required, in addition, for sensorless control (Figure 11.55). The rationale and table of optimal switchings developed for PM-SMs remain valid here (Section 11.5). Only the flux and torque observer bear some pecularities in the sense that the PM flux is dropped from axis d (Figure 11.45) and eventually subtracted from axis q (if a PM along axis q is added). Also $L_d > L_q$.

Digital simulation results [14] with DTFC are now presented for a 1.5kW, 2-pole motor with L_d = 140.77mH, L_q = 7.366mH r_s = 0.955Ω, J = 2.5x10⁻³kgm², switching frequency f_c = 15kHz in Figure 11.57. The time constant is τ_s in the speed SM controller functional s_ω

$$s_\omega = \left(\omega_r{}^* - \omega_r\right) - \tau_s \frac{d\omega_r}{dt} \qquad (11.91)$$

τ_s = 1.2ms

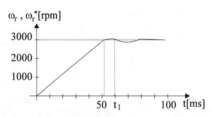

Figure 11.56. Speed step response with step load 3Nm at t_1 = 60ms.

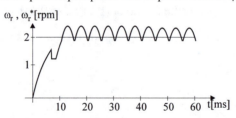

Figure 11.57. Low speed step response

Fast speed response without overshoot is evident both for low and high speeds (Figure 11.56-11.57).

11.7.4. Sensorless control of RSM

Although sensorless control of RSM may be open-loop (V / f or scalar) eventually, with some load torque angle compensation (Figure 11.2), closed-loop sensorless vector control is favored due to superior dynamic performance and wider speed control range (100 to 1).

For vector sensorless control, both rotor position θ_r and speed ω_r have to be estimated (or observed). Various methods to do this have been proposed (for a collection of papers on the subject see [10], Section 3).

They are based on the combined voltage and current observer, current slope, inductance measurements, third flux harmonics, extension of the zero crossing of phase current, etc.

We mention here only the voltage and current model for flux and torque observer (Figure 11.45 - with $\lambda_{PM} = 0$ and $L_d > L_q$). We have to add to it the rotor speed estimator (Figure 11.47) and a position feedback (Figure 11.58).

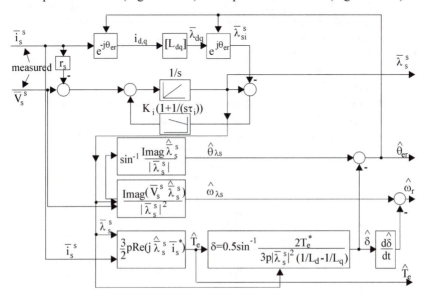

Figure 11.58. Position, speed, flux and torque observer for RSM (wide speed range)

For DTFC the rather involved observer ($\hat{\theta}_{er}$) in Figure 11.58 is required to determine the rotor position used here only in the current model part of the flux observer for the rotor speed $\hat{\omega}_r$ and, finally, for flux and torque computation. Also in Figure 11.58

$$\frac{d\delta}{dt} \approx \left(\cos\delta(k-1)\sin\delta(k) - \sin\delta(k-1)\cos\delta(k)\right)/T \qquad (11.92)$$

If $L_d \gg L_q$ the whole observer is only loosely dependent on machine parameters. Still, corrections for stator resistance variation due to temperature and L_d and L_q variation with stator currents i_d and i_q might be added for higher immunity to parameter variations.

As expected, the above observer may be used also for most vector control sensorless schemes.

11.7.5. DTFC–SVM sensorless control of RSM — a generic implementation

DTFC-SVM stands for direct torque and flux control with space vector modulation in the PWM converter that supplies the RSM. A similar method is applicable to PMSMs.

A generic sensorless RSM drive block diagram [25] is presented in Fig. 11.59.

Components description:

- RSM and VSI (voltage source inverter — VLT 3008-Danfoss) with current and DC link voltage measurements — necessary for flux, torque and speed estimators (together with the known inverter switching states);
- Stator flux and torque estimator;
- Kalman filter based speed estimator;
- PI speed controller;
- DTC SW-TABLE and SVM switching and voltage selection table;
- Hysteresis controllers on stator flux and torque errors;
- Reference speed and stator flux as input.

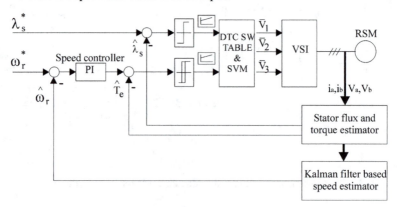

Figure 11.59. Sensorless RSM drive

The RSM model in d-q coordinates is:

$$\begin{cases} V_d = R_s i_d + \dfrac{d\lambda_d}{dt} - \omega_r \lambda_q \\[2mm] V_q = R_s i_q + \dfrac{d\lambda_q}{dt} + \omega_r \lambda_d \end{cases} \qquad (11.93)$$

$$T_e = \frac{3}{2} p (L_d - L_q) i_d i_q \qquad (11.94)$$

where ω_r is rotor electrical angular speed (rad/s), λ_d is d axis stator flux ($L_d i_d$), λ_q is q axis stator flux ($L_q i_q$), i_d, i_q, are d, q axis stator currents, p equals pole pair number.

$$\begin{cases} \lambda_d = L_{s\sigma} i_d + L_{md} i_d \\ V_q = L_{s\sigma} i_q + L_{mq} i_q \end{cases} \tag{11.95}$$

$L_{s\sigma}$, L_{md}, L_{mq} are the leakage and d-q axis magnetizing inductances.

For a good estimation, accurate current and voltage measurements are necessary. In this case the available variables are DC link voltage and two phase currents (star connection winding). The phase voltages are calculated from DC link measurement and the inverter switching states.

a) Estimators:

Flux estimator

The flux estimator is based on the combined voltage - current model of stator flux in RSM. The block scheme of this estimator is presented in Figure 11.60.

The voltage model is supposed to be used for higher speeds and the current model for low speed region. A PI regulator makes the smooth transition between these 2 models, at medium speed, on the current error. For the current model we need the rotor position angle θ_e, based on the stator flux position.

The equations describing these flux estimators are:

$$\hat{\underline{\lambda}}_s^s = \int \left(\underline{V}_s - R_s \cdot \underline{i}_s + U_{comp} \right) dt \tag{11.96}$$

$$U_{comp} = \left(K_p + K_i / s \right) \left(\hat{\underline{i}}_s^s - \underline{i}_s \right) \tag{11.97}$$

The coefficients K_P and K_I may be calculated such that, at zero frequency, the current model stands alone, while at high frequency the voltage model prevails:

$$K_p = \omega_1 + \omega_2, \ K_i = \omega_1 \cdot \omega_2 \tag{11.98}$$

Values such as $\omega_1 = 2\text{-}3\,\text{rad/s}$ and $\omega_2 = 20\text{-}30\,\text{rad/s}$ are practical for a smooth transition between the two models; $K_p = 50$ and $K_i = 30$ were used in experiments.

The torque estimator

The torque estimation uses the estimated stator flux and the measured currents, all in stator coordinates.

$$\hat{T}_e = \frac{3}{2} p(\lambda_{s\alpha} i_{s\beta} - \lambda_{s\beta} i_{s\alpha}) \tag{11.99}$$

The speed estimator

The speed estimator is based on a Kalman filter approach with rotor position as input. This way a position derivative calculation is eliminated.

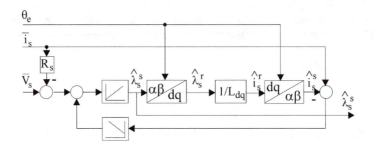

Figure 11.60. The stator flux estimator

The corresponding equations are:

$$\begin{cases} \varepsilon_k = \sin\theta_e \cdot \cos\theta_k - \sin\theta_k \cdot \cos\theta_e \\ \theta_k = \theta_k + T_e \cdot \hat{\omega}_r + K_1 \cdot \varepsilon_k \\ \text{if } \theta_k > \pi \Rightarrow \theta_k = -\pi \\ \text{if } \theta_k < -\pi \Rightarrow \theta_k = \pi \\ \hat{\omega}_r = \hat{\omega}_r + w_k + K_2 \cdot \varepsilon_k \\ w_k = w_{k-1} + K_3 \cdot \varepsilon_k \end{cases} \tag{11.100}$$

where K_1, K_2, K_3 are Kalman filter parameters; $\hat{\omega}$ is the estimated rotor speed; θ_e -estimated rotor position.

The rotor position necessary for the speed estimator has been calculated from the stator flux position.

b) DTFC and SVM

The speed controller is of digital PI type:

$$T_e^* = \left(K_{ps} + 1/sT_{is}\right) \cdot \varepsilon_\omega \tag{11.101}$$

where $\varepsilon_\omega = \omega_r^* - \hat{\omega}_r$. The output of speed controller is the reference torque. PI parameters are variable depending on speed error in 3 steps: $K_{ps} = 10; 3; 1$ and $K_{is} = 0.01; 0.1; 0.5$ (for $abs(\varepsilon_\omega) \geq 1$; $abs(\varepsilon_\omega) < 1$ and $abs(\varepsilon_\omega) \leq 0.5$). The PI controller has the output limit at ± 15Nm.

Now, that we have the reference and estimated stator flux and torque values, their difference is known as $\varepsilon_{\lambda s}$ and ε_{Te}.

$$\varepsilon_{\lambda s} = \lambda_s^* - \hat{\lambda}_s \qquad (11.102)$$

$$\varepsilon_{Te} = T_e^* - \hat{T}_e \qquad (11.103)$$

A bi-positional hysteresis comparator for flux error and a three-positional one for torque error have been implemented. The two comparators produce the λ_s and T_e command laws for torque and flux paths:

$$
\begin{aligned}
\lambda_s &= +1 \ \text{ for } \varepsilon_{\lambda s} > 0 \\
\lambda_s &= -1 \ \text{ for } \varepsilon_{\lambda s} < 0 \\
Te &= +1 \ \text{ for } \varepsilon_{Te} > h_m \\
Te &= \ 0 \ \text{ for } |\varepsilon_{Te}| < h_m \\
Te &= -1 \ \text{ for } \varepsilon_{Te} < -h_m
\end{aligned}
\qquad (11.104)
$$

The band h_m is a relative torque value of a few percent.

The discrete values of flux and torque command of $+1$ means flux (torque) increase, -1 means flux (torque) decrease, while zero means zero voltage vector applied to inverter.

To detect the right voltage vector in the inverter, we need also the position of the stator flux vector in six 60-degree extension sectors, starting with the axis of phase a in the stator.

In each switching period only one voltage vector is applied to the motor and a new vector is calculated for the following period.

In the present implementation a combined DTFC-SVM strategy has been chosen in order to improve the steady state operation of the drive to reduce the torque-current ripples caused by DTFC.

DTFC-SVM is based on the fact that a SVM unit modulates each voltage vector selected from the DTFC SW TABLE before applying it to the inverter.

The SVM principle is based on switching between two adjacent active vectors and a zero vector during a switching period (Figure 11.61). An arbitrary voltage vector $\underline{V}$, defined by its length V and angle α, can be produced by adding two adjacent active vectors $\underline{V}_a(V_a, \alpha_a)$ and $\underline{V}_b(V_b, \alpha_b)$ and, if necessary, a zero vector V_0 or V_7. The duty cycles D_a and D_b for each active vector are calculated as a solution of the complex equations:

$$V(\cos\alpha + j\sin\alpha) = D_a V_a (\cos\alpha_a + j\sin\alpha_a) + D_b V_b (\cos\alpha_b + j\sin\alpha_b)$$

(11.105)

$$D_a = M_{index} \sin\alpha$$ (11.106)

$$D_b = \frac{\sqrt{3}}{2} M_{index} \cos\alpha - \frac{1}{2}D_a$$ (11.107)

where the modulation index M_{index} is:

$$M_{index} = \frac{\sqrt{3}V}{V_{dc}}$$ (11.108)

Because $\underline{V}_a$ and $\underline{V}_b$ are adjacent voltage vectors, $\alpha_b = \alpha_a + \pi/3$.
The duty cycle for the zero vectors is the remaining time inside the switching period:

$$D_0 = 1 - D_a - D_b$$ (11.109)

The vector sequence and the timing during one switching period are:

$$
\begin{array}{ccccccc}
V_0 & \underline{V}_a & \underline{V}_b & V_7 & \underline{V}_b & \underline{V}_a & V_0 \\
1/4D_0 & 1/2D_a & 1/2D_b & 1/2D_0 & 1/2D_b & 1/2D_a & 1/4D_0
\end{array}
$$ (11.110)

The sequence guarantees that each transistor inside the inverter switches once and only once during the SVM switching period. A strict control of the inverter's switching frequency can be achieved this way.

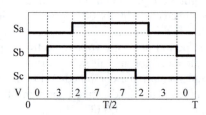

Figure 11.61. Switching pattern example

The duty cycle for each arm can be easily calculated. For SVM control, the reference value of voltage vector comes from its components as PI controller outputs on the flux error and the torque error:

$$\underline{V}^* = V^*(\cos\alpha + j\sin\alpha)$$ (11.111)

$$V^* = \sqrt{u_d^{*2} + u_q^{*2}} \qquad (11.112)$$

$$\alpha = \operatorname{arctg} \frac{u_q^*}{u_d^*} \qquad (11.113)$$

If the torque and/or flux error is large, the DTC strategy is applied with a single voltage vector through a switching period. In case of a small torque and/or flux error, the SVM strategy is enabled, based on the two PI controllers discussed before.

The DTFC strategy is suitable for transients because of its fast response; SVM control is suitable for steady state and small transients for its low torque ripple. The switching frequency is increased in the latter case, but the drive noise is reduced also.

c) Experimental setup and test results

The voltage source inverter used for the tests was a Danfoss VLT 3008 7KVA one, working at 9 KHz switching frequency.

The RSM has the following parameters: P_N = 2.2kW, V_N = 380/220V (Y/Δ), I_N = 4.98/8.5 A, f_N = 50 Hz, R_s = 3 Ω, L_d = 0.32 H, L_q = 0.1 H.

In what follows, test results are presented with sensorless control of RSM at different speeds and during transients (speed step response and speed reversing) using DTFC and/or SVM control strategies.

Simulation studies were made before tests.

The experimental setup of the RSM sensorless drive with DTFC-SVM is presented in Figure 11.62.

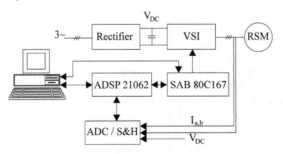

Figure 11.62. The experimental setup

The control system contains a 32-floating point DSP and a 16-bit microcontroller. The digital control system is supervised by a PC. The calculations for state estimation and control strategy are performed by ADSP-

21062 DSP. The inverter commands signals (space vector modulation) are produced by SAB 80C167 microcontroller.

The estimation and control algorithm is implemented using ANSI-C language. All differential equations are transformed to the digital form using the bilinear approximation:

$$s = \frac{2}{T_e}\left(\frac{z-1}{z+1}\right) \tag{11.114}$$

with the sampling time $T_e = 111\mu s$ for a 9 kHz switching frequency. All the calculations are done by DSP.

For current and voltage measurements, one 12-bit, 3.4 µs analog to digital converter (ADC) and a 8-channel simultaneous sampling and hold circuit is used. The sampling time is the same as the switching time. To avoid problems related to frequency aliasing, analog filters with 300 µs time constant are used for the current measurements path. The current and voltage measurement is done at the beginning of each switching period. The technique reduces the measurement noise because of the symmetry of the switching pattern. Each current is measured with a Hall sensor current transformer.

For speed comparisons, the speed was measured using an incremental encoder with a resolution of 1024 increments/revolution. The measured speed is filtered with an analog filter with 300 µs time constant.

To eliminate the nonlinear effects produced by the inverter, the dead time compensation was introduced. The inverter's dead time is 2.0 µs.

Steady state performance of RSM drive with DTFC and DTFC-SVM control strategy at 15 rpm is presented in Figures 11.63 and 11.64. The represented waveforms are phase current, estimated torque, and measured and estimated speed.

The reduction in torque and current ripple by SVM is evident.

Comparative results have been presented in Figures 11.65 and 11.66 for DTFC and DTFC-SVM control during transients (reference speed step from 450 rpm to 30 rpm). A slightly faster torque response has been obtained with a DTFC control strategy.

Test results have been compared in Figure 11.67 with the simulation ones in Figure 11.68 during speed reversal at 150 rpm, using DTFC control strategy.

With the combined DTFC-SVM strategy, low torque ripple operation has been obtained with RSM. Further improvement of the rotor position estimation is necessary in order to obtain very fast response with this machine. While the maximum switching frequency with DTFC control strategy was 9 kHz, with DTFC-SVM, the real commutation frequency is much higher and the torque pulsations and current harmonics are notably reduced.

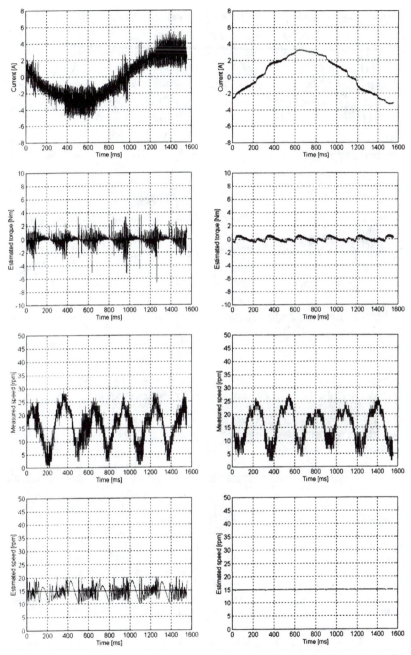

Figure 11.63. DTFC control of
RSM;15 rpm steady state
no load operation

Figure 11.64. DTFC-SVM control of
RSM; 15 rpm steady state
no load operation

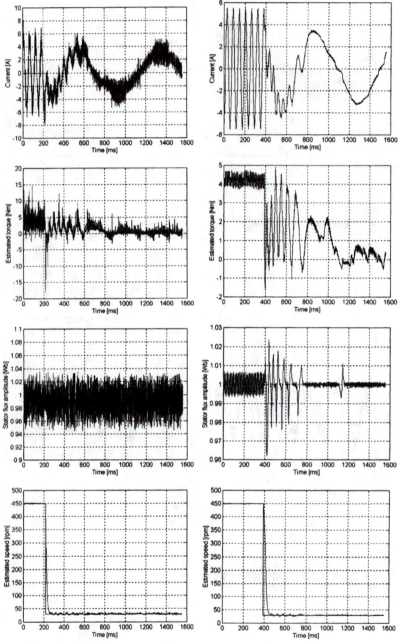

Figure 11.65. Speed, torque, stator flux and current transients during no load deceleration from 450 to 30 rpm with DTFC

Figure 11.66. Speed, torque, stator flux and current transients during no load deceleration from 450 to 30 rpm with DTFC-SVM

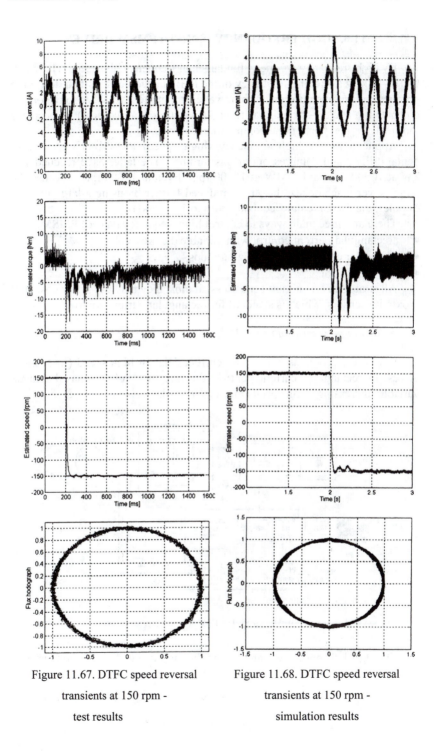

Figure 11.67. DTFC speed reversal
transients at 150 rpm -
test results

Figure 11.68. DTFC speed reversal
transients at 150 rpm -
simulation results

11.8. HIGH FREQUENCY (SPEED) PMSM DRIVES

By high frequency we mean fundamental frequency above 500 Hz, when the 10-15 kHz switching frequency of the IGBTs becomes less than desired to produce a close to sinusoidal current waveform in the PMSM. Also the online computation cycle — which has to be kept within 5° (electrical) — becomes too small for complex online computation controls.

Applications for high frequency (high speed) PMSM drives refer to dental drillers and compressors or gas turbines. The latter have reached 150 kW at 70000 rpm and 3 MW at 15000 rpm. Even at 15000 rpm, because the large power machine size has to be reduced further by using a large number of poles, the fundamental frequency goes to $1 - 1.5$ kHz.

There are three main ways to control high frequency (speed) PMSM:
- Rectangular current control with variable d.c. link voltage;
- V/f scalar control with stabilizing loops;
- Vector control.

All strategies are motion-sensorless, at most they may use 3 low cost Hall proximity sensors, which is standard for rectangular current control.

a) Rectangular current control

Due to the problems with rising current delay at high speed (frequency) drives, in rectangular current control, the d.c. link voltage may be made variable (increasing) with speed (Fig. 11.69) [26].

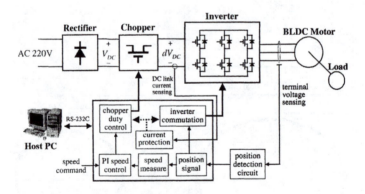

Figure 11.69. Variable d.c. link voltage PWM converter

The d.c. link output voltage V_0 of the step-down d.c.-d.c. converter is simply:

$$V_0 = dV_{dc} \qquad (11.116)$$

where V_{dc} is the input d.c. voltage and d is the duty ratio.

The value of d may be made dependent on speed error $\omega_r^* - \omega_r$:

$$d = \left(\omega_r^* - \omega_r\right) \cdot \left(K_p + \frac{K_i}{s}\right) \qquad (11.117)$$

This is a PI controller.

The speed ω_r has to be calculated (estimated). If Hall sensors are used, ω_r may be approximately calculated from the discrete information on rotor position, after proper filtration.

In discrete form (11.117) becomes:

$$d(k) = d(k-1) + K_p\left[\epsilon(k) - \epsilon(k-1)\right] + K_i T_s \epsilon(k-1)$$
$$0 \leq d(k) \leq 1; \quad \epsilon = \omega_r^* - \omega_r \qquad (11.118)$$

where T_s is the sampling time.

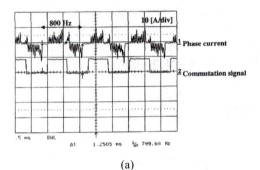

(a)

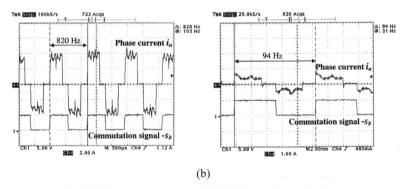

(b)

Figure 11.70. PWM converter control of PMSM: a) without d.c.-d.c. converter;

b) with d.c.-d.c. converter

It follows that the speed is regulated through the d.c. link voltage V_0. Consequently, only the commutation of phases every electrical 60° is provided through the PWM converter.

As a result, the commutation losses in the latter are reduced drastically, at the price of the losses in the d.c.-d.c. converter.

Ref. [26] shows the beneficial effect of the d.c.-d.c. converter with current control in the d.c. link upon the current pulse waveform of current control system (Fig. 11.70).

The sharp rise and fall of rectangular current at 45000 rpm is a clear indication of very good performance.

b) V/f control with stabilizing loops

As the fundamental frequency increases, the online available computation time decreases. A simplified control scheme is required. Fortunately the transient torque response quickness is not so demanding in high speed drives.

V/f open loop control is rather simple but it needs a stabilizing loop. It could be a current stabilizing loop or an active power variation damping that would modify the reference frequency f^* in a feedforwarding manner [27].

Alternatively, it is possible to assembly a dual stabilizing loop:
- One on reference voltage amplitude to control the stator flux amplitude.
- One on reference voltage phase angle to close loop control the current to PM flux linkage vector power angle γ (Fig. 11.71).

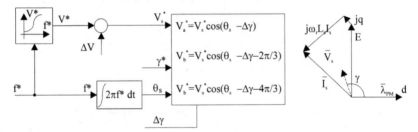

Figure 11.71. V/f control with flux and torque angle stabilizing loops.

$$\Delta V = \left(\lambda_s^* - \lambda_s\right)\cdot\left(K_{p\lambda} + \frac{K_{i\lambda}}{s}\right)$$

$$\Delta\gamma = \left(\gamma^* - \gamma\right)\cdot\left(K_{p\gamma} + \frac{K_{i\gamma}}{s}\right) \tag{11.119}$$

The maximum torque per current for surface - PM rotor ($L_d = L_q$) occurs for $\gamma^* = \pi/2$ (e.m.f. in phase with current).

The PM produced flux linkage λ_{PM} is straightforward:

$$\overline{\lambda}_{PM} = -L_s\overline{I}_s + \overline{\lambda}_s \tag{11.120}$$

But the stator flux linkage $\overline{\lambda}_s$ is:

$$\overline{\lambda}_s = \int\left(\overline{V}_s - R_s \overline{I}_s\right) \cdot dt \qquad (11.121)$$

Equations (11.120) - (11.121) represent the PM flux and stator flux estimators. As the speed is large the voltage model suffices while the integral off-set has to be eliminated.

Typical digital simulations with such a scheme at 80000 rpm with step load are shown in Fig. 11.72.

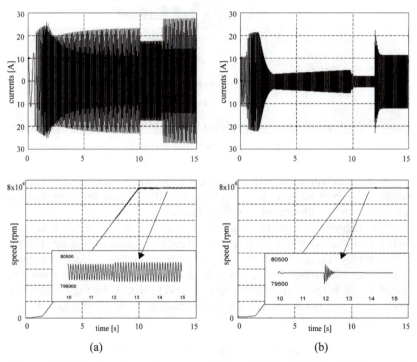

Figure 11.72. Steady state performance of V/f control of a PMSM at 80000 rpm: a) without stabilizing loop, b) with stabilizing loop.

The introduction of the power angle control stabilizes the drive while the flux loop control just improves the energy conversion ratio.

c) Vector control

If speed and rotor position angle θ_{er} may be observed by using simplified observers, even vector control may be applied up to 70000-100000 rpm.

In essence a combined voltage — current vector control (e.m.f. compensation) is used in [28]. Sample results are shown in Fig. 11.73.

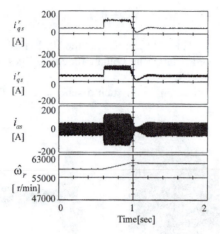

Figure 11.73. Acceleration characteristics of the PMSM with sensorless vector control scheme (from 58000 to 60000 rpm)

A problem with high frequency (speed) PMSM drives is the fact that the synchronous inductance L_s tends to be too small and thus the current ripple tends to be large with limited switching frequency.

The addition of external inductors in series with all phases leads to a reduction of current ripple [28]. Also it is feasible to increase the slot leakage inductance by filling only the bottom part of the semiclosed slots of the stator. Finally, for the surface-PM rotors at high speeds a resin coating should be placed above a 0.5-0.7 mm thick copper shield to reduce the rotor iron losses produced by the stator space and time harmonics. Such a copper shield also attenuates the current ripple.

11.9. SINGLE-PHASE PMSM CONTROL

Low power electric drives are applied mainly for home appliances and on automobiles. Air purifiers, air conditioning heater blowers, room temperatures sensors, vacuum pumps, electric remote control mirrors, automatic speed adjustment pumps, retractable head lamps, radiator cooling fans, oil cooling fans, throttle controls, sun roofs, variable shock absorbers, auto door locks, fuel pumps, rear wipers, windscreen wipers are all typical automobile small power electric drive applications.

PM brushless motors are a solution for these low power drives. As three phase PMSM drive control has already been treated earlier in this chapter, here only a typical single phase PMSM drive control is introduced.

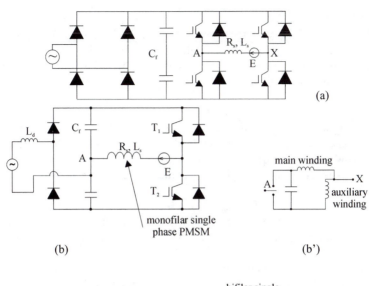

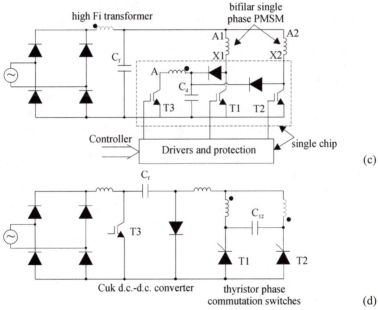

Figure 11.74. Representative power electronics converters for single-phase PMSM :
a.) with diode rectifier and 4 switch PWM inverter, single phase; b.) with half diode
bridge, voltage doubler and 2 switch PWM single phase converter (two phase
capacitor PMSM connection- (b')); c.) with diode rectifier, bifilar single phase
winding and two active switches PWM converter with C_{dump} and high frequency
transformer for sinusoidal a.c. source current, d.) with diode rectifier, Cuk chopper
and thyristor phase commutator

A lower SRC count power electronics control, typical for a single phase PMSM, is maybe the way for a lower global cost drive systems below 50 – 100 W. The radial or axial airgap single phase PMSM with safe-starting rotor parking provision (Chapter 10) with a small number of stator coils is typical for such a low power drive.

Typical converter topologies:
The typical power converter for such a drive is composed of a diode rectifier and a four active switch single phase PWM inverter (Figure 11.74a).

A lower count (two) switch PWM converter with splitted d.c. capacitor with voltage doubler and half bridge diode rectifier is shown in Figure 11.74b.

The voltage doubler (through L_d) is required to preserve the d.c. voltage of the typical 4 switch PWM converter (Figure 11.74a). The same converter may be used for a bidirectional motion drive with capacitor two phase PMSM (Figure 11.74b').

To simplify the drivers sources (by eliminating the voltage shifting) for the bifilar single phase PMSM, two active switches T_1 and T_2 in the lower leg are connected in series with the two parts of the a.c. winding (one of them serves one polarity) – Figure 11.74c. In addition, a third switch T3 with a C_d – dump capacitor and a high frequency transformer is used to channel the freewheeling energy when T_1 and T_2 are off, such that to restore the input a.c. current sinusoidality. C_d is designed at twice the rated d.c. voltage to provide fast phase turn off.

Another option of lower cost converter (Figure 11.74d) makes use of two thyristors (T_1, T_2) that provide phase commutation, while a single fast switch (T_3) d.c.-d.c. converter may be used also to restore the sinusoidality of the a.c. source input current according to today's standards (IEC 1000-3-2 for class D equipment, for example).

State space model:
Before addressing the problem of open or close loop speed control through power electronics, the state-space model of the single phase PMSM is required:

$$i_s R_s - V_s = -L_s \frac{di_s}{dt} - \omega_r \lambda_{PM}(\theta_{er})$$

$$\frac{J}{p} \frac{d\omega_r}{dt} = T_e + T_{cogging} - T_{load} \qquad (11.122)$$

$$T_e = p\lambda_{PM}(\theta_{er})i_s(t); \frac{d\theta_{er}}{dt} = \omega_r$$

The cogging torque $T_{cogging}(\theta_{er})$, which is the zero current torque, has to be specified, but also it has to provide for a parking position from which safe starting can be provided.

A FEM investigation of a two pole single phase PMSM with gradually variable airgap under the stator poles and surface PM rotor poles is shown in Fig. 11.75.

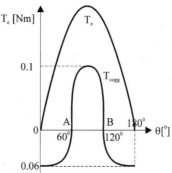

Figure 11.75. Cogging torque T_{cogg} versus position and interaction torque T_e (for constant current) with continuously increasing airgap under the stator poles.

It is evident that if the rotor parks in positions A, B there is enough interaction torque (T_e) and time to move the rotor in a single direction (from left to right) with a positive voltage pulse.

Control strategies:
One or two low cost Hall sensors may be placed on the stator to detect the rotor proximity. One of them would be used to sense the zero crossing e.m.f. E (or of maximum PM flux linkage) in the stator.

After a starting sequence, flux $\lambda_{PM}(\theta_{er})$ may be estimated:

$$\lambda_{PM}(\theta_{er}) = \int (V_s - R_s i_s) \cdot dt - L_s i_s \qquad (11.123)$$

This way the rotor position $\hat{\theta}_{er}$ and then the speed may be estimated. The Hall sensors may be used to operate corrections to the rotor position estimator.

A typical control system for sinusoidal or trapezoidal current in phase with the e.m.f. is shown in Figure 11.76.

Based on the PM flux estimation the $\cos\theta_{er}$, $\cos3\theta_{er}$ terms are found, but the stator current should be phase shifted by 90°, and thus contain terms in $\sin\theta_{er}$, $\sin3\theta_{er}$ which may be simply obtained. The d.c.-d.c. converter uses a single current close-loop controller, to produce the required current amplitude and waveform. The thyristors provide the phase commutation at zero current crossing as each of them serves half the winding.

To simplify the control it is possible to build a single phase I/f open loop speed control by carefully ramping the speed to maintain synchronism. A stabilizing loop may be used to damp speed oscillations brought by torque perturbations.

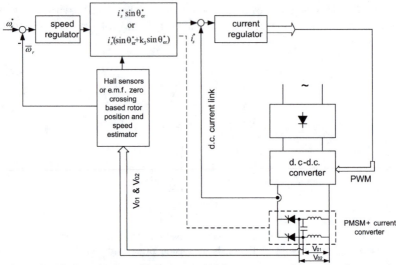

Figure 11.76. Generic sinusoidal or trapezoidal vector control (e.m.f. and current in phase) of a bifilar winding single phase PMSM.

A case study:

For a sinusoidal current control system with the rotor position available from an encoder, digital performance simulation on a single phase PMSM with the data:

$V_n = 200V$;

$f_n = 50$ Hz;

$P_n = 200W$;

$n_n = 3000$ rpm;

$2p = 2$;

$\eta_n = 0.8$;

$\cos\varphi_n = 0.8$;

$E_n = 0.8\, V_n$;

$J = 10^{-3}$ Kgm^2;

$T_{cogg} \approx T_{cog\,max} \cos(2\theta_{er} + \gamma)$

$\dfrac{T_{cog\,max}}{T_{en}} \approx 0.3$

has been run for the full count converter (Fig. 11.71a) and for the 2 thyristors plus one fast switch converter (Fig. 11.71d).

The computer programs written in Matlab/Simulink comprise the motor, the converter with sensors and the control system (it is detailed on the CD attached to the book as single PMSM and, respectively, single-phase PMSM-thyristor).

Transients during acceleration with 0.2 Nm step load at 1s show both control systems capable of acceptable mechanical performance (Fig. 11.77).

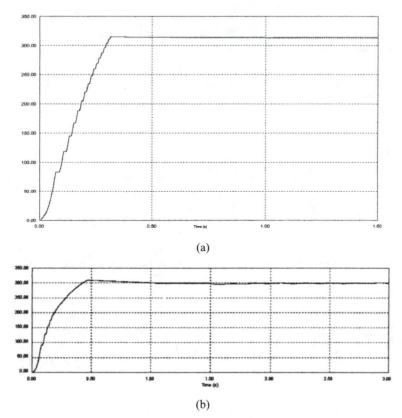

(a)

(b)

Figure 11.77. Start-up speed (rad/s) transient: a) single phase full converter,

b) bipolar winding, 2 thyristors + d.c.-d.c. converter

On the other hand, the steady state no load and step load transients in torque and current, show notable differences between the two solutions (Figure 11.78).

The presence of the C12 capacitor between the bifilar windings (Figure 11.71d) leads to adequate polarity currents in both bifilar windings all the time.

This way the disadvantage of partial use of copper is attenuated and, also, the torque pulsations are reduced notably for the case of two thyristor + d.c.-d.c. converter.

In terms of total system costs, the additional passive components in the two thyristor plus d.c.-d.c. converter seem to offset the count reduction of fast switches (IGBTs) of the 4 switch PWM converter; the filter capacitor C_f (Figure 11.74d) is reduced in the former, however.

The torque pulsations are reduced (peak torque is reduced from 0.6 Nm to 0.35 Nm for an average torque of 0.2 Nm – Figure 11.78). Also the phase current is reduced from 1.1 A (Figure 11.78a) to 0.55 A (Figure 11.78b).

The a.c. source current waveform is about the same for both cases.

Finally, in the two thyristors plus d.c.-d.c. converter case the current sensor and its control are placed in the d.c. link.

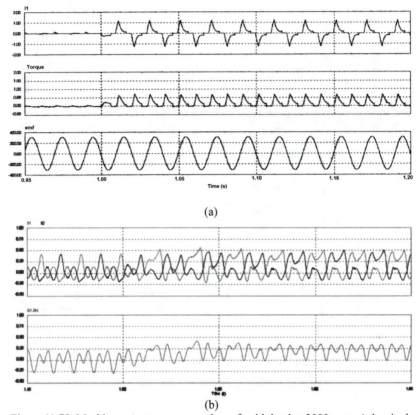

(a)

(b)

Figure 11.78. Machine currents, torque and e.m.f. with load at 3000 rpm: a) the single phase full converter, b) bipolar winding, 2 thyristors + d.c.-d.c. converter.

11.10. SUMMARY

- PM or reluctance rotor a.c. motors are used in high performance drives with PWM voltage source inverters. The higher efficiency and torque density of these motors are their main merits.
- PM a.c. motors may be controlled with rectangular (trapezoidal) current or with sinusoidal current.
- Sinusoidal current control is performed on PM-SMs with distributed windings (q ≥ 2 slots/pole/phase) and with RSMs (reluctance SMs) without or with PMs in axis q ($L_d > L_q$).
- Tooth-wound PMSM with sinusoidal or trapezoidal e.m.f. leads to tower motor costs and copper losses.

- Rectangular current control (for PM a.c. motors with q = 1, called brushless d.c. motors) tends to be simpler as proximity position sensors suffice, if a speed sensor is available. For low speed, 120° current conduction per polarity per phase is adequate; for high speeds, 180° current conduction is better in terms of available torque.
- PM-SM motor drives with motion sensors and vector control are particularly suitable for high performance position or wide speed range (from 100 to 1 to 10000 to 1) applications. Fast torque response is obtained mainly with surface magnets on the rotor.
- Special optimization i_d / i_q (λ_d / λ_q) relationships to torque are required to obtain high performance for steady-state and transients.
- RSMs tend to retain the good performance of PM-SMs at lower costs and for similar control systems.
- Speed control ratios up to 100 to 1 in sensorless drives have been proposed. Involved flux, torque, speed and position observers are needed for sensorless vector control.
- DTFC may be used for PM-SMs and RSMs for high performance for simpler and highly robust control systems, both with motion sensors and sensorless.
- Wide constant power speed range has been demonstrated with PM-SMs or RSMs with PMs.
- The unity power level constant power speed range K_{cpr} is [5, 20]

$$K_{cpr} \approx 0.7\omega_c - 1$$
$$\omega_c = \frac{1}{e_0 - x_d} \tag{11.124}$$

where ω_c is the critical (zero torque) speed in relative units (per base speed ω_b), e_0 is the PM (no-load) voltage in relative units (per rated (maximum) voltage), x_d is the d axis reactance in relative units.
- If we consider that for RSM the d axis is along the low permeance path as for the PM-SMs (change of places of d and q axis) ω_c is valid also for RSM with PMs.
- A theoretically infinite constant power zone (11.124) is obtained for

$$e_0 - x_d = 0, \tag{11.125}$$

however, at lower levels of power.
- Condition (11.125) is met in conventional PM-SMs with $e_0 > 0.5$, in general, to retain some acceptable power factor and with smaller e_0 ($e_0 < 0.3$) for RSMs with low costs PM (in axis q) and good power factor.
- Constant power zone requirements are in some conflict with high torque requirements in steady-state and transients below base speed ω_b.
- Super high speed PMSM drives are characterized by fundamental frequency above 600 Hz, up to 2.4 kHz. In such cases, with limited switching frequency of IGBT converters, synchronous current vector

control or V/f control with stabilizing loop(s) and open loop PWM are to be preferred. Also the machine inductance has to be increased to reduce phase current time harmonics. The simplified control is also imposed by the limited online computation cycle that should correspond to a few electrical degrees, to maintain tight control.

- At low power levels (below 100 W in general), beside three phase, single phase PMSM (with self-starting rotor parking provision) drives may become practical in terms of system costs.
- Home appliance and automotive actuators are typical such low power electric drives.
- Reduced active power switch count converter systems are required to reduce costs at low power.
- The two-thyristors with d.c.-d.c. converter (with one IGBT) and bifilar winding single-phase PMSM drives, investigated in this chapter through digital simulations, is believed to be a good example of a practical lower cost solution.
- Motion-sensorless open loop V/f or I/f control with stabilizing loop are to be used for the scope to reduce the costs of DSP and of control electronics in very small power electric drives.

11.11. PROBLEMS

11.1. A brushless d.c. motor has the data: peak PM flux per phase λ_{PM} = 1.25 Wb, p = 2 - pole pairs, stator resistance r_s = 2 Ω, inverter d.c. input voltage V_d = 500 V. For zero advancing angle and rectangular current i_{dc} = 10 A, determine:
  a. The airgap torque with two phases conducting at any time;
  b. The ideal no-load speed;
  c. Draw the speed versus torque curve obtained.

11.2. For a PM - SM with interior PM - rotor with the data L_d = 0.06 H, L_q = 0.10 H, V_{sn} = 300 V, p = 2, $r_s \approx 0$ and I_{sn} = 10 A, λ_{PM} = 1 Wb, determine:
  a. The current components I_d, I_q and the torque at base speed ω_b (voltage V_{sn} and current I_{sn}) for maximum torque/ current criterion.
  b. Determine the stator flux level, λ_s^*, i_d, i_q and the torque and power for maximum torque / flux condition (11.42) for $3\omega_b$.

11.3. For the indirect vector current control of RSM presented in Section 11.7.2 write a computer program in PSPICE or MATLAB/ SIMULINK and reproduce the results given there.

11.4. For the DTFC of RSM presented in Section 11.7.3. write a computer program in MATLAB/SIMULINK or PSPICE and reproduce the results given there, adding other output data.

11.12. SELECTED REFERENCES

1. **I. Boldea, S.A. Nasar,** "Torque vector control (TVC) - a class of fast and robust torque, speed and position digital controllers for electric drives", EMPS, vol.15, 1988, pp.135-148.

2. **B.V. Murty,** "Fast response reversible brushless d.c. drive with regenerative braking", Record of IEEE-IAS-1984 Annual Meeting, pp.475-480.

3. **T.M. Jahns,** "Torque production in permanent magnet synchronous motor drives with rectangular current excitation", IEEE Trans.vol.IA-20, n.4, 1984, pp.803-813.

4. **S.K. Safi, P.P. Acarnley, A.G. Jack,** "Analysis and simulation of the high - speed torque performance of brushless d.c. motor drives", IEE Proc. Vol. EPA-142, no.3, 1995, pp.191-200.

5. **S. Morimoto, M. Sonoda, Y. Takeda,** "Wide speed operation of interior permanent magnet synchronous motors with high performance current regulator", IEEE Trans vol.IA-30, no.4, 1994, pp.920-925.

6. **T. Jahns,** "Flux weakening regime operation of an interior permanent magnet synchronous motor drive", IEEE Trans vol.IA-23, 1987, pp.681-687.

7. **G. Pfaff et al.,** "Design and experimental results of a brushless a.c. drive", IEEE Trans. Vol.IA-20, no.2, 1984, pp.814-821.

8. **I. Boldea, S.A. Nasar,** "Vector control of a.c. drives", Chapter 10, CRC Press, Florida, USA, 1992.

9. **G.D. Andreescu,** "Robust direct torque vector control system with stator flux observer for PM-SM drives", Record of OPTIM-1996, Brasov, Romania, vol.5.

10. **K. Rajashekara, A.Kawamura, K.Matsuse (editors),** "Sensorless control of a.c. motor drives", IEEE Press, 1996 Section 2, pp.259-379.

11. **L. Cardoletti, A.Cassat, M.Jufer,** "Sensorless position and speed control of a brushless d.c. motor from start-up to nominal speed", EPE Journal, vol.2, no.1, 1988, pp.25-34.

12. **I. Boldea,** Reluctance synchronous machines and drives, book, OUP, Oxford, U.K., 1996.

13. **R.E. Betz,** Theoretical aspects in control of synchronous reluctance machines, Proc IEE, vol.B-139, no.4, 1992, pp.355-364.

14. **A. Fratta, A. Vagati, F. Villata,** "Design criteria of an IPM machine suitable for field weakening operation", Record of ICEM-1990, Cambridge MA, 1990, Part III, pp.1059-1065.

15. **L. Xu, X. Xu, T.A. Lipo, D.W. Novotny,** "Vector control of a reluctance synchronous motor including saturation and iron losses", IEEE Trans. vol.IA-27, no.5, 1991, pp.977-985.

16. **I. Boldea, Z.X. Fu, S.A. Nasar,** "Digital simulation of a vector controlled ALA rotor synchronous motor servodrive", EMPS, vol.19, 1991, pp.415-424.

17. **A. Fratta, A. Vagati, F. Villata,** "Permanent magnet assisted synchronous reluctance drives for constant power applications: Comparative analysis and control requirements", PCIM Europe, 1992, Nurnberg, Germany, pp.187-203.

18. **I. Boldea, N. Munteanu, S.A. Nasar,** "Robust low cost implementation of vector control for reluctance synchronous machines", IEE Proc EPA vol.141, no.1., 1994, pp.1-6.

19. **I. Boldea, Z. Fu, S.A. Nasar,** "Torque vector control (TVC) of ALA rotor reluctance synchronous motors", EMPS vol.19, 1991, pp.381-398.

20. **S. Morimoto, M. Sonoda, Y. Takeda,** Inverter driven synchronous motors for constant power, IA Magazine, vol.2, no.6, 1996, pp.18-24.

21. **H. Kim, K.K. Huh, R.D. Lorenz, T.M. Jahns,** "A novel method for initial rotor position estimation of IPM synchronous machine drives", IEEE Trans. Vol. IA-40, no. 5, 2004, pp. 1369-1378.

22. **Y. Yeong, R.D. Lorenz, T.M. Jahns, S. Sul,** "Initial rotor position estimation of IPM motor", Record of IEEE – IEMDC – 2003, pp. 1218-1223.

23. **S. Shinnaka,** "New mirror - phase vector control for sensorless drive of PMSM with pole saliency", IEEE Trans. Vol. IA – 40, no. 2, 2004, pp. 599-606.

24. **M. Linke, R. Kennel, J. Holtz**, "Sensorless speed and position control of synchronous machines using alternating carrier signal injection", Record of IEEE – IEMDC – 2003, vol. 2, pp. 1211-1217.

25. **I. Boldea, L. Janosi, F. Blaabjerg**, "A modified direct torque control (DTC) of reluctance synchronous motor sensorless drive", EMPS Journal, Taylor & Francis, Vol. 28, no. 2, 2000, pp. 115-128.

26. **K.-H. Kim, M.J. Youn,** "DSP – based high speed sensorless control of a brushless d.c. motor using d.c. link voltage control", EMPS Journal, Taylor & Francis, Vol. 30, no. 2, 2002, pp. 889-906.

27. **P.D. Chandana Perera, F. Blaabjerg, J.K. Pedersen, P. Thorgersen,** "A sensorless, stable V/f control method for PMSM drives", IEEE Trans. Vol. IA – 39, no. 3, 2003, pp. 783-791.

28. **B.-H. Bae, S-.K. Sul, J-.H. Kwon, Ji –S. Byeon,** "Implementation of sensorless vector control of superhigh speed PMSM for turbo-compresor", IEEE Trans. Vol. IA – 39, no. 3, 2003, pp. 811-818.

Chapter 12

SWITCHED RELUCTANCE MOTOR (SRM) DRIVES

12.1. INTRODUCTION

SRMs are doubly salient, singly excited electric motors with passive (windingless) rotors. Their concentrated coil phases are turned-on sequentially, to produce torque, through d.c. voltage pulses which result in unipolar controlled current.

Due to their simple and rugged topology, SRMs have been given considerable attention in the last two decades in the hope of producing alternative low grade and high grade brushless motor electric drives at a lower cost and equivalent performance when compared with a.c. (induction or synchronous) motor drives.

At the time of this writing the SRM drive enjoys only an incipient market but an aggressive penetration of worldwide markets is expected in the near future. Power ratings of SRMs range from a few watts to practically MW units for low speed control range low dynamics, as well as high grade (servo) applications, especially in thermally and chemically harsh environments.

Given the topological peculiarities of the SRM, a few details are considered in order. Also, the operation of SRM with an active phase only at a time, in the presence of magnetic saturation, with open-loop (voltage PWM) or closed-loop (current control) PWM, requires elaborate digital simulation methods.

The basic strategies for low grade and high grade speed and position control with and without motion sensors are also treated in some detail. Some numerical examples are provided to facilitate quick access to magnitudes for a realistic assessment of performance.

12.2. CONSTRUCTION AND FUNCTIONAL ASPECTS

SRMs are made of laminated stator and rotor cores with $N_s = 2mq$ poles on the stator and N_r poles on the rotor. The number of phases is m and each phase is made of concentrated coils placed on 2q stator poles.

Most favored configurations — amongst many more options — are the 6/4 three-phase and the 8/6 four-phase SRMs (Figure 12.1a, b).

These two configurations correspond to q = 1 (one pair of stator poles (and coils) per phase) but q may be equal to 2, 3 when, for the three-phase machine, we obtain 12/8 or 18/12 topologies applied either for low speed high torque direct drives or for high speed stator-generator systems for aircraft [2]. The stator and rotor pole angles β_s and β_r are, in general, almost equal to each other to avoid zero torque zones.

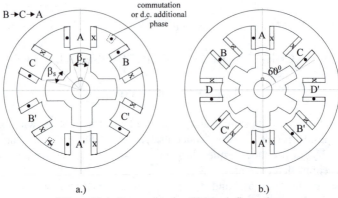

a.)   b.)

Figure 12.1. Representative SRM configurations

The symmetry of the magnetic circuit leads to the almost zero mutual flux linkage in the SRM phases even under saturated conditions. This means that the SRM may work with m–1 phases since no induced voltage or current will appear in the short-circuited phase. Hence the SRM is more fault tolerant than any a.c. motor where the interaction between phases is at the core of their principle of operation. The self-inductance of each phase alone thus plays the key role in torque production.

In the absence of magnetic saturation, the phase self-inductance varies linearly with rotor position, while, in presence of saturation, the respective dependence is nonlinear (Figure 12.2).

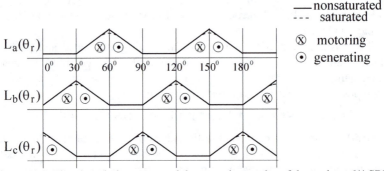

Figure 12.2. The phase inductances and the operation modes of three-phase 6/4 SRM

If the phase flux linkage λ is calculated and plotted versus current for various rotor positions, the $\lambda(\theta_r, i)$ curve family is obtained (Figure 12.3).

The influence of magnetic saturation is evident from Figure 12.3 and is a practical reality in well-designed SRMs, as will be shown later in this section.

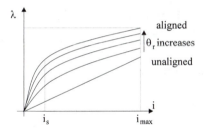

Figure 12.3. Flux / current / position curve family

The instantaneous torque $T_e(i)$ per phase may be obtained through the known coenergy, $W_{mc}(\theta_r)$, formula

$$T_e(i) = \left(\frac{\partial W_{mc}(\theta_r)}{\partial \theta_r}\right)_{i=cons.} \quad ; \quad W_{mc} = \int_0^i \lambda(\theta_r,i)di \qquad (12.1)$$

Equation (12.1) demonstrates the necessity of knowing, through calculations or test, the family of curves $\lambda(\theta_r,i)$.

The total instantaneous torque is

$$T_e = \sum_{i=1}^m T_e(i) \qquad (12.2)$$

Only in the absence of saturation, the instantaneous torque is

$$T_e = \sum_{i=1}^m \frac{1}{2}i_i^2 \frac{\partial \lambda_i(\theta_r)}{\partial \theta_r} \qquad (12.3)$$

Ideally, a phase is turned-on when rotor poles, along the direction of motion, lay between neighboring stator poles $\theta_{on} = 0$ (Figure 12.4), for the motoring operation mode of the respective phase. Only one voltage pulse is applied for a conduction (dwell) angle $\theta_w = \theta_c - \theta_{on}$ in Figure 12.4. During this period, neglecting the resistive voltage drop, the maximum phase flux linkage λ_{max}, for constant speed ω_r, is

$$\lambda_{max} = \int_0^t V_d dt = V_d \frac{\theta_w}{\omega_r} \qquad (12.4)$$

The maximum value of θ_w, for $\theta_{on} = 0$ (zero advance angle), is given by motor design

$$\theta_{w\,max} = \theta_m = \frac{\pi}{N_r} \qquad (12.5)$$

The base speed ω_b corresponds to θ_{wmax} and single voltage pulse V_d with maximum flux linkage λ_{max}, which is dependent on machine design and on

the level of saturation. Thus, to increase the base speed, we have to saturate
the magnetic circuit of that machine.

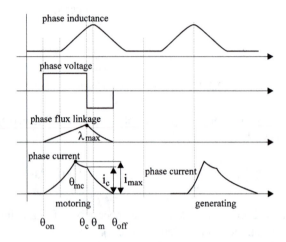

Figure 12.4. Phase inductance, voltage, flux linkage and current

For higher speeds ω_r (above ω_b), the value of θ_w may be slightly reduced
but, definitely, the maximum λ_{max} has to be reduced. This is called flux
weakening. Above base speed ω_b, also, the turn-on angle θ_{on} may be
advanced to reach the available maximum flux λ_{max} at a smaller angle θ_c and
thus allow the current to reach its maximum (at θ_{mc}) sooner and at a higher
level and thus produce more torque. As a consequence, the speed/torque
envelope may be enlarged. On the other hand, the turn-off process of a phase
starts at $\theta_c \leq \theta_m$ and terminates at θ_{off} in the "generating" zone.

The smaller the angle $\theta_{off} - \theta_m$, the smaller the negative torque
"contribution" of the phase going off. In reality, at $\theta_r = \theta_m$ (aligned position)
if the current i_m is already less than (25-30%) of the peak current, the
negative (generating) torque influence becomes small.

Once one phase is turned off at θ_c, another one is turned on, eventually
to contribute positive torque so as to lower the total torque pulsations caused
by the reduction of torque in the phase going off.

It is now evident that the entire magnetic energy of each phase is
"pumped" in and out for each conduction cycle. There are mN_r cycles per
mechanical revolution. A part of this energy is passed over to the incoming
phase through the power electronic converter (PEC) and the rest to the d.c.
bus filtering capacitor of PEC. Below base speed ω_b the current is limited
(and controlled) through PWM (Figure 12.5).

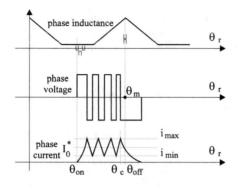

Figure 12.5. PWM below base speed

It should be noted that the interval of conduction is prolonged close to θ_m where the phase inductance is maximum. As mentioned above, at high speeds the phase turn-on angle θ_{on} is advanced and so is the turn-off angle θ_c.

12.3. AVERAGE TORQUE AND ENERGY CONVERSION RATIO

From the energy cycle point of view, based on the family of curves $\lambda(\theta_r,i)$, the two situations presented in Figures 12.4-12.5 (single voltage pulse) and, respectively, current chopping, are shown in Figure 12.6a, b, c.

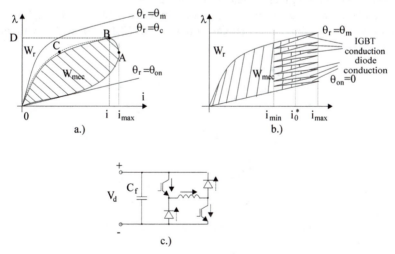

Figure 12.6. Energy cycle per phase: a.) for high speeds (one voltage pulse, Figure 12.4); b.) for low speeds (current chopping, Figure 12.5); c.) phase PEC for unipolar current per phase

The average torque T_{eav} is proportional to the hatched area in Figures 12.6 a and b, W_{mec}. Consequently, for constant speed, single phase conduction, with m phases and N_r rotor poles, the SRM average torque T_{eav} is

$$T_{eav} = \frac{W_{mec} m N_r}{\theta_c - \theta_m} \qquad (12.6)$$

In parallel, an energy conversion ratio E.C. (kW / kVA) is defined as

$$E.C. = \frac{\text{Area of } \overline{OABCO}}{\text{Area of } \overline{OABDO}} = \frac{W_{mec}}{W_{mec} + W_r} \qquad (12.7)$$

As the energy returned during diode conduction, W_r, is reduced by magnetic saturation, the E.C. ratio increases with saturation. Consequently, the kVA rating of the PEC (which depends on $W_{mec} + W_r$) is reduced by magnetic saturation. The E.C. ratio is only 0.5 for linear flux/current curves — in the absence of magnetic saturation — but it can reach values of 0.65-0.67 with magnetic saturation. Also, to keep the W_{mec} high, in comparison with the linear case, the airgap is reduced such that adequate saturation, be obtained at the same level of currents for most rotor positions. This way the level of average torque can be maintained.

Note that the maximum flux per phase and the maximum inductance are reduced by magnetic saturation. Current commutation becomes faster, while the base speed for a given voltage is also increased.

Further increase in the E.C. ratio may be obtained through d.c. premagnetization by a diametrical coil (Figure 12.1a) in series with all phases [3].

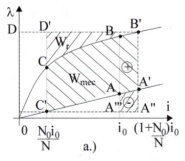

Figure 12.7. Energy cycle with d.c. premagnetization

The energy cycle is moved to the right (Figure 12.7) with the equivalent current $i_0 N_0 / N$ (N_0 - turns in the premagnetization coil, N - turns in the phase coil). The mechanical energy is increased by the area AA′B′B minus area AA′A″A‴, while the energy retrieved during phase demagnetization, W_r, is reduced by the area OCD′D. Consequently, the energy conversion ratio E.C. is notably increased (from 0.64 to 0.74, [3]) at the price of the losses in the premagnetization coil (19% additional copper losses, [3]).

12.4. THE PEAK kW/kVA RATIO

The peak kW/kVA ratio [4] is

$$\text{peak kW/kVA} = \frac{\alpha_s N_r Q}{8\pi} \tag{12.8}$$

where α_s is the stator pole pitch ratio ($\alpha_s = 0.4-0.5$), and Q is

$$Q \approx C\left(2 - \frac{C}{C_s}\right) \tag{12.9}$$

where C is the ratio between the turn-on angle below the stator pole and the stator pole angle β_s.

In general, C = 1 at zero speed and decreases down to C = 0.65 at base speed.

Also, $C_s = \dfrac{\lambda_u - 1}{\lambda_u \sigma - 1}$; $\lambda_u = \dfrac{L_a^{\;u}}{L_u} \approx (4 \div 10)$; $\sigma = \dfrac{L_a^{\;s}}{L_a^{\;u}} \approx (0.3 \div 0.4)$ (12.10)

L_u is the unaligned inductance, and $L_a^{\;u}$ and $L_a^{\;s}$ are the aligned unsaturated and saturated values of phase inductance, respectively.

The peak apparent power, S, of switches in the converter is

$$S = 2mV_d I_{peak} \tag{12.11}$$

where V_d is the d.c. source voltage and I_{peak} is the peak current value.

For an inverter-fed IM drive the peak kW / kVA is

$$\left(\text{peak KW/KVA}\right)_{IM} = \frac{3}{\pi} \frac{V_d I \cdot PF}{K(6V_d I)} = \frac{3}{\pi} \frac{PF}{6K} \tag{12.12}$$

where K is the ratio between the peak current I_{peak} and the peak value of the current fundamental in IMs when inverter-fed (for the 6 pulse mode K = 1.1-1.15); PF is the power factor.

Example 12.1.

For a 6 / 4 SRM with $\sigma = L_a^{\;s} / L_a^{\;u} = 0.4$, $\lambda_u = L_a^{\;u} / L_u = 6$, C = 1, $\alpha_s = 0.4$, calculate the peak kW / kVA ratio.

Solution:

As $\sigma = 0.4$ and $\lambda_u = 6$, the coefficient C_s (from 12.10) is

$$C_s = \frac{\lambda_u - 1}{\lambda_u \sigma - 1} = \frac{6 - 1}{6 \cdot 0.4 - 1} = \frac{5}{1.4} = 3.57 \tag{12.13}$$

Further on from (12.9) Q is

$$Q = C\left(2 - \frac{C}{C_s}\right) = 1\left(2 - \frac{1}{3.57}\right) = 1.72 \tag{12.14}$$

Consequently, from (12.8)

$$\text{peak kW/kVA} = \frac{\alpha_s N_r Q}{8\pi} = \frac{0.4 \cdot 4 \cdot 1.72}{8\pi} = 0.1095 \tag{12.15}$$

For an induction motor with P.F. = 0.85 and K = 1.12

$$\left(\text{peak kW/kVA}\right)_{IM} = \frac{3}{\pi} \frac{0.85}{6 \cdot 1.12} = 0.1208 \tag{12.16}$$

So, in terms of the PWM inverter rating, the IM (with rather good power factor) is only 10-20% better than the PEC for the equivalent SRM which is supposed to produce the same power at the same efficiency.

12.5. THE COMMUTATION WINDINGS

The turn-off process angle (time) $\theta_c - \theta_{off}$ has to be limited. In general, $\theta_c - \theta_{off} < \beta_s$ (β_s is the stator pole angle) is not sufficient, as a strong negative (generating) torque may occur.

Conversely, in the generator mode, the turn-on process has to be advanced within motoring regime (positive inductance slope), but not too much, in order to avoid strong motoring torque influences.

For both cases and all speeds $\theta_{off} - \theta_m < 3-4$ degrees. Through advanced saturation (reduced airgap) and the advancing of turn-on and turn-off angles θ_{on} and θ_{off}, the turn-off interval angle $\theta_{off} - \theta_m$ may be kept (up to a certain speed) below the challenging limits at the price of higher torque pulsations.

If the diametrical coil (proposed for demagnetization) - Figure 12.1a - is turned on whenever an active phase is de-energized, the magnetic energy will be passed to this commutation phase, partly. After the current is zero in the going-off phase, the commutation winding is turned off. Notably smaller turn-off times are obtained this way (Figure 12.8) [5].

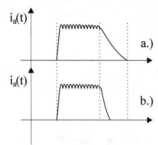

Figure 12.8. Current profile: a.) conventional,
b.) with commutation coil (winding)

The effect of the commutation current turn-off process is evident but we have to consider that an additional coil with an additional phase in the PEC and the corresponding losses make the price for the improvement thus obtained. The energy conversion rate CE is also increased by a few percent. Finally, the interaction of the commutation winding with, say, a faulty (short-circuited) phase notably reduces the fault tolerance claims of conventional SRM. Other distributed winding arrangements for SRMs have been proposed [6,7] for increasing the torque density (Nm/kg of rotor), but they all seem to impede on the fault tolerance claims of conventional SRMs, while the efficiency is not increased either.

In what follows we will deal with modeling, digital simulation and the control of the basic 6 / 4 and 8 / 6 SRMs as they are most representative.

12.6. SRM MODELING

The mathematical model of SRM is highly nonlinear due to magnetic saturation influence on the $\lambda(\theta_r, i)$ curve family, but it allows for phase-torque superposition as the interaction between phases is minimal.

As SRM has double saliency, stator (phase) coordinates are mandatory.

The phase equations are

$$V_{a,b,c,d} = r_s i_{a,b,c,d} + \frac{d\lambda_{a,b,c,d}\left(\theta_r, i_{a,b,c,d}\right)}{dt} \tag{12.17}$$

with the family of curves $\lambda_{a,b,c,d}(\theta_r, i_{a,b,c,d})$ obtained for one phase only (as the periodicity is π / N_s). These curves may be obtained either through theory or through tests. Analytical or finite element methods are used for the scope. Accounting for magnetic saturation and airgap flux fringing is mandatory in all cases.

The motion equations are:

$$J\frac{d\omega_r}{dt} = T_e - T_{load}; \quad \frac{d\theta_r}{dt} = \omega_r \tag{12.18}$$

with
$$T_e = \sum_{a,b,c,d} T_{e_{a,b,c,d}}; \quad T_{e_{a,b,c,d}} = \frac{\partial}{\partial\theta_r} \int_0^{i_{a,b,c,d}} \lambda_{a,b,c,d}\left(\theta_r, i_{a,b,c,d}\right) di_{a,b,c,d} \tag{12.19}$$

Let us use the subscript i for one dominating phase.
Equation (12.16) may be written as

$$V_i = r_s i_i + \frac{\partial\lambda_i}{\partial i_i}\frac{di_i}{dt} + \frac{\partial\lambda_i}{\partial\theta_r}\frac{d\theta_r}{dt} \tag{12.20}$$

Denoting $\dfrac{\partial\lambda_i}{\partial i_i}$ as the transient inductance L_t:

$$L_t\left(\theta_r,i_i\right) = \frac{\partial \lambda_i\left(\theta_r,i_i\right)}{\partial i_i} \qquad (12.21)$$

The last term in (12.20) represents the back e.m.f. E_i

$$E_i = \frac{\partial \lambda_i}{\partial \theta_r} \cdot \omega_r \qquad (12.22)$$

So (12.20) becomes

$$V_i = r_s i_i + L_t\left(\theta_r,i_i\right)\frac{di_i}{dt} + E_i\left(\omega_r,\theta_r,i_i\right) \qquad (12.23)$$

An equivalent circuit with time-dependent parameters may be defined based on (12.23) - (Figure 12.9).

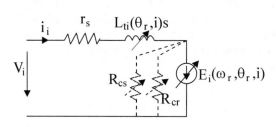

Figure 12.9. Equivalent circuit of SRM with core losses accounted for

The core losses are represented by the variable resistances in parallel with the e.m.f. E_i, based on the assumption that only the main flux produces core losses. Core losses occur both in the stator and in the rotor core as this machine does not operate on the traveling field principle [9,10].

Especially in high speed applications (above 6000 rpm) core loss has to be considered not only for efficiency calculations but also in the transient current response assessment.

Note: Only for the linear case (no magnetic saturation) the instantaneous torque $T_e(t)$ is

$$T_e(t) = \sum \frac{1}{2}\frac{E_i\left(\omega_r,\theta_r,i_i\right)i_i}{\omega_r} \qquad (12.24)$$

As usually, heavy magnetic saturation is present, $E_i\left(\omega_r,\theta_r,i_i\right)$ as in (12.22) is a pseudo e.m.f. as it also includes a part related to the magnetic energy storage. Consequently, the torque is to be calculated only from the coenergy formula.

For more details on this aspect, see [11].

12.7. THE FLUX-CURRENT-POSITION CURVE FITTING

For digital simulations and control purposes the $\lambda_i(\theta_r,i)$ curves family has to be known. The safest way is to use measurements — at standstill or with the machine in rotation. Finite element calculations represent the second best approach. Once this is done what still remains is to determine the inverse function $i(\lambda_i,\theta_r)$ and, eventually, $\theta_r(\lambda_i,i_i)$. There are two main ways to invert the $\lambda_i(\theta_r,i)$ to find either $i(\lambda_i,\theta_r)$ or $\theta_r(\lambda_i,i_i)$.

One way is to use analytical functions (polynomials or exponentials) [12,13]. The second way is to use direct approximations, for example, fuzzy logic [14] or other curve fitting methods. For digital simulations the computation time is not so important. In the case of control, torque calculation, $T_{ei}(\lambda_i,\theta_r,i_i)$, or position estimation, $\theta_r(\lambda_i,i_i)$, is done on line.

Using exponential approximations [12,13] the phase flux linkage $\lambda_i(\theta_r,i)$ is

$$\lambda(\theta_r,i) = a_1(\theta_r)\left(1 - e^{-a_2(\theta_r)i}\right) + a_3(\theta_r)i \qquad (12.25)$$

The periodicity of λ is built in the $a_{1,2,3}(\theta_r)$ functions expressed as Fourier series

$$a_m = \sum_{k=0}^{\infty} A_{mk} \cos\left(k\alpha\theta_r\right) \qquad (12.26)$$

$\alpha = 4$ for a 6/4 machine and $\alpha = 6$ for an 8/6 machine. A_{mk} is the Fourier coefficient of kth order in a_m.

Other rather cumbersome analytical functions of current, based on linear position dependence approximations, are also feasible [15].

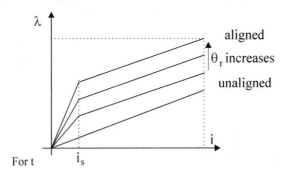

Figure 12.10. Linear $\lambda(\theta_r,i)$ approximations

It is supposed that magnetic saturation occurs at the constant current level i_s irrespective of rotor position. As the airgap is small, in heavily saturated motors, the local overlapping pole zone may saturate quickly. Thus the flux linkage varies linearly with rotor position. It is known, however, that close to the aligned position such an approximation model seems less adequate

$$\lambda = i\left(L_u + \frac{K_s(\theta_r - \theta_0)}{i_s}\right); \quad i \le i_s \tag{12.27}$$

$$\lambda = L_u i + K_s(\theta_r - \theta_0); \quad i \ge i_s \tag{12.28}$$

where L_u is the unaligned inductance value, K_s is the only coefficient to be found from the family of $\lambda(\theta_r, i)$ curves — eventually one intermediate position information suffices; θ_0 is the position angle of the pole entry end; $\theta_{nmax} = \theta_m$ — the pole exit end position angle.

12.8. SRM DRIVES

SRM drives may be classified with respect to a few criteria:
- with motion (position, speed) sensors;
- without motion sensors (sensorless).
  They also may be:
- general — for low dynamics applications (moderate speed range and costs);
- high grade (performance) — for servos.
  They differ in complexity, cost and performance. Performance is defined by the energy conversion ratio, speed control range, precision and quickness of torque response.

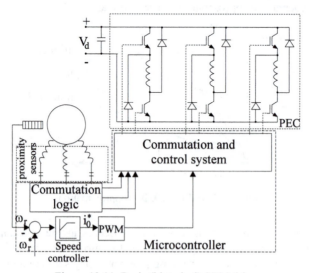

Figure 12.11. Basic (historical) SRM drive

Servo applications require precise speed or position control and fast torque response (in the milliseconds range). Historically, for low cost SRM drives, proximity (Hall) position sensors have been used to trigger the turn-off conducting phase and at the same time, turn-on of the incoming phase. For constant speed only, even the position angle between two proximity sensors signals may be estimated.

Consequently, advancing the turn-on and turn-off angle may be performed in such drives only for steady-state or low dynamics.

Also, starting under load poses problems because, from the three (four) proximity sensor signals, we may only infer what phase has to be turned on but not exactly the initial position. A separate speed signal is necessary for safe starting under load.

In the following section we will discuss an up-to-date general drive, a high grade drive with precision (encoder) position feedback and an advanced motion-sensorless drive.

12.9. GENERAL PURPOSE DRIVE WITH POSITION SENSOR

For applications with wide speed range but moderate energy conversion or dynamic performance, a precision position sensor is used to commutate the phases at speeds down to a few rpm. No speed sensor is used (Figure 12.12).

The core of the general purpose SRM drive in Figure 12.12 is the dependence of the turn-on θ_{on} and turn-off θ_{off} angles (for motoring and generating) on speed. A linear dependence is the obvious choice.

We may distinguish three regions:
- the low speed - constant torque - zone: θ_{on} = cons., θ_{off} = cons., $\omega_r < \omega_b$;
- the constant power zone, $T_e\omega_r$ = cons.: θ_{on} and θ_{off} decrease with speed $\omega_r > \omega_b$;
- the $T_e\omega_r^2$ = cons. zone, above ω_{m1} (Figure 12.13a)

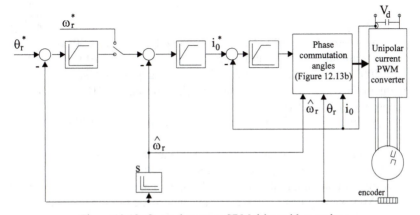

Figure 12.12. General purpose SRM drive with encoder

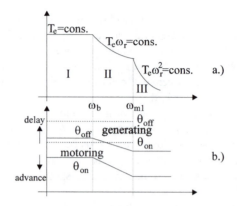

Figure 12.13. a.) Torque/speed zones, b.) turn-on and off angles θ_{on} and θ_{off} for

motoring and generating

The above control philosophy assumes flat top currents below base speed, obtained through closed-loop (current control) PWM and single-voltage pulses above base speed when (and if) current control is no longer feasible due to lack of voltage reserve (the e.m.f. overcomes the d.c. link voltage V_d).

As is easy to see, no complicated data on machine flux/current/position curves is necessary as $(\theta_{on})_{min}$ and $(\theta_{off})_{min}$ are to be obtained through trial and error. For more details on such a drive see [16].

The drive could produce moderate performance over a wide speed range but quick torque response optimum energy conversion or maximum torque/speed envelope is not guaranteed.

However, there are many applications when the general purpose SRM drive suffices.

Example 12.2.

A 6 / 4 three-phase SRM has the following data: $\beta_s = \beta_r = 30°$ (stator and rotor pole angle), $J = 0.002 kgm^2$, maximum current $i_{omax} = 10A$, d.c. voltage $V_d = 300V$, $r_s = 1.5\Omega$, maximum flux (aligned position) $\lambda_{max} = 0.8Wb$, minimum flux $\lambda_{min} = 0.16Wb$, linear flux/current curves (12.27-12.28) are supposed, with $i_s = 2A$ (Figure 12.10).

A position sensor with 1024 pulses per revolution is available.
Calculate:
• the unaligned inductance, K_s coefficient in (12.27-12.28) and the maximum average torque available for $i_{max} = 10A$ at zero speed.
• using Matlab/Simulink run digital simulations on a general purpose SRM drive (Figures 12.14-12.17) and choose the turn-on and -off angles to explore starting transients, step load and step-speed responses.

Solution:

At zero speed the conduction angle $\theta_{rmax} - \theta_0 = \beta_s = 30^0 = \pi / 6$ with $\theta_0 = 0$, for the aligned position $\theta_{rmax} = 30°$.

In the unaligned position $\theta_r = \theta_0$ and from (12.27)

$$\lambda_{min} = i_{max} L_u \qquad (12.29)$$

So the unaligned inductance L_u is

$$L_u = \frac{\lambda_{min}}{i_{max}} = \frac{0.16}{10} = 0.016 \, H \qquad (12.30)$$

Also, from (12.28) λ_{max} is

$$\lambda_{max} = i_{max} L_u + K_s \beta_s \qquad (12.31)$$

$$K_s = \frac{0.8 - 0.16}{\pi / 6} = 1.223 \, Wb/rad \qquad (12.32)$$

Simulation results on a SRM motor drive with PEC are presented. The motor model was integrated in a block (SRM in Figure 12.14).

Changing motor parameters is done by clicking on this block. A dialog box appears and you may change them by modifying their default values.

The drive system consists of PI speed controller ($K_i = 10$, $T_i = 0.05s$), motor block (voltage equations for each phase, motion equations, (12.18-12.20), angle selection and advancing block, and three blocks (A, B, C) for matrix calculation (as functions of θ_r and phase current): Equations (12.33-12.35).

$$\frac{\partial \lambda_i}{\partial i_i} = \begin{bmatrix} L_u & & L_u \\ L_u + \dfrac{K_s}{i_s}(\theta_r - \theta_0) & & L_u \\ L_u + \dfrac{K_s}{i_s}(\theta_0 + \pi/3 - \theta_r) & & L_u \end{bmatrix} \qquad (12.33)$$

$$\frac{\partial \lambda_i}{\partial \theta_r} = \begin{bmatrix} 0 & 0 \\ \dfrac{K_s}{i_s}i_i & K_s \\ -\dfrac{K_s}{i_s}i_i & -K_s \end{bmatrix} \qquad (12.34)$$

$$\frac{\partial\left(\int \lambda_i di_i\right)}{\partial \theta_r} = \begin{bmatrix} 0 & 0 \\ \dfrac{K_s}{2i_s} i_i^{\ 2} & K_s i_i \\ -\dfrac{K_s}{2i_s} i_i^{\ 2} & -K_s i_i \end{bmatrix} \tag{12.35}$$

Each line of these matrices represents a case function of phase inductance for 3 operation modes (see Figure 12.2) based on (12.27 and 12.28). The columns represent the two cases for $i \le i_s$ and $i \ge i_s$.

The angle selection and advancing block produce the voltage PWM between θ_{on} and θ_c angles and, respectively, negative voltage supply until θ_{off} (where the current for the working phase becomes zero); see Figure 12.5. In this block the firing angle is selected for each phase as a function of rotor position information.

The study examines the system's behavior for starting, load perturbation, motoring and generating of the SRM drive.

The integration step (50μs) can be modified from the Simulink's *Simulation/Parameters*.

To find out the structure of each block presented above, unmask them (*Options/Unmask*). Each masked block contains a short help screen describing that block (inputs/outputs/parameters).

The block diagram of the electric drive system is presented in the Figure 12.14.

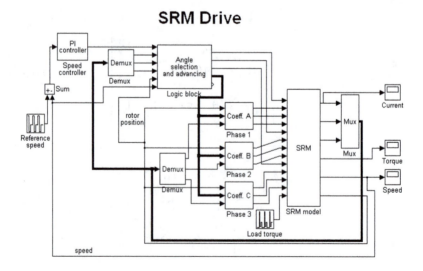

Figure 12.14. Block diagram of simulated SRM drive

The following Figures represent the speed (Figure 12.15), torque (Figure 12.16) responses, and current waveforms (Figure 12.17) for speed step response from 0 to 700 rad/s (advancing angle is 5^0 after 500rad/s), changing the speed reference at 0.3s (from 700rad/s to 150rad/s; the machine is working as a generator until it reaches the reference speed), load torque is applied at 0.5s (8Nm), unloading is done at 0.6s; and, at the same time, the speed reference is changed to 400rad/s and loading is done again at 0.8s (3Nm).

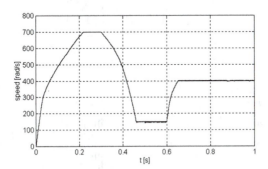

Figure 12.15. Speed response

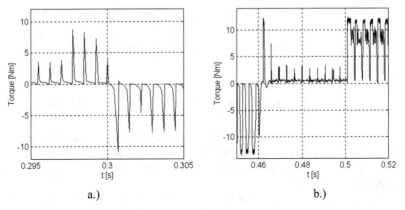

a.)   b.)

Figure 12.16. Torque response

12.10. HIGH GRADE (SERVO) DRIVES

High grade (servo) drives have a strong capability for position, speed or torque control, characterized by high energy conversion ratios, precision, robustness and quickness of torque control and rather wide speed range control. Traditionally, d.c. brush, brushless (PMSM) or (recently) advanced control IM drives are used as servos.

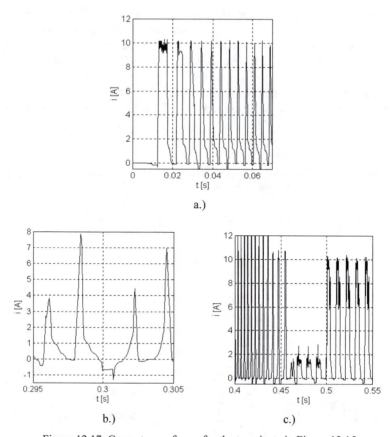

Figure 12.17. Current waveforms for the transients in Figure 12.15

The strong nonlinearity in the flux/current/position and torque/current/position curves makes, in principle, the SRM apparently less attractive for servos. However the motor simplicity and ruggedness in such applications may not be overlooked.

An attempt to build a SRM servo drive may start with a revisitation of torque production in the SRM.

a. Torque sharing and current shaping

In servos torque pulsations should be reduced to less than 1% of rated torque if smooth (precise) torque control is to be obtained. If only one phase can produce torque (positive inductance slope) at any time, chances to reduce the torque pulsations are low, especially if winding losses are to be limited.

A 4-phase (8/6) SRM has two phases capable of producing torque at the same time, while, in a three-phase machine, only one phase can produce torque at any time. Consequently, 4 phases seem necessary for servo performance. In this case, single-phase torque production will be alternated with two-phase torque action for low losses or low torque pulsations. The total torque T_e is the sum of the torques per phase T_{ei} (Figure 12.18)

$$T_e = \sum_{i=1}^{m} T_{ei}(i_i, \theta_{ri}); \quad i = 1,...,m; \quad m = 4 \tag{12.36}$$

The torque response time is basically equal to the time to reach the maximum flux level in the machine and it is in the order of milliseconds for the kW power range.

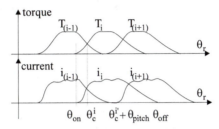

Figure 12.18. Phase torque components and phase current shaping

How to turn on and off various phases and profile the current waveforms to minimize torque ripple is not a problem with a unique solution.

A second condition is required. To determine the flux/torque relationships for SRMs we may apply two optimization criteria for torque sharing:

• maximum torque per unit winding losses (below base speed);
• maximum torque per unit admissible flux level (above base speed)

For both criteria the crucial problem is when to switch from one phase to the next. In [17] the switch over angle θ_c^i (for the first criterion) and θ_c^λ (for the second) are chosen so that in these conditions the values of torque are equal to each other (half the reference torque: $T_e^* / 2$) and so are the currents (Figure 12.18), and the flux levels.

The θ_{on} angle is calculated to allow the current rise required in the machine characteristics. Before θ_c^i (θ_c^λ), the phase with stronger torque was in charge, while after that the incoming phase predominates in torque production.

The problem is how to find the current versus position profiles for various speeds and reference torques. This task definitely requires a thorough knowledge of the $\lambda(i,\theta_r)$ curves which have to be inverted to produce $i_i(\omega_r, T_i, \theta_r)$ functions. Various mathematical approximations of $\lambda_i(i_i,\theta_r)$ functions determined through tests may be used for the scope [17]. Here is an example

$$\lambda(i,\theta_k) = a_0(\theta_k)\tan^{-1}(a_1 i) + a_2(\theta_k) i \tag{12.37}$$

where θ_k are discrete rotor positions.

From the co-energy formula the corresponding torque is computed. The torque sharing functions may be implemented in a feedforward manner in the sense that four reference currents and current controllers make sure that the reference currents are closely tracked (Figure 12.19).

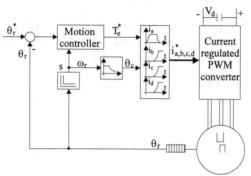

Figure 12.19. High grade SRM drive with
feedforward torque sharing optimization

To store all the waveforms for various speeds would require much memory and thus, in [17], only the waveforms for maximum speed are stored. For lower speeds the phase voltage (flux) available is not surpassed.

It may be seen from Figure 12.19 that the switch over angle θ_c (θ_c^i at low speeds and θ_c^λ at high speeds) has to be precalculated based on the optimization criteria mentioned above. In general, $\theta_c^i > \theta_c^\lambda$ as expected.

Torque pulsations of less than 4% from 0.5rpm to 5000rpm and good positioning performance have been reported in [17, 18]. Also, it has been shown that the maximum torque/loss criterion produces more power at low speeds while the maximum torque/flux enlarges the constant power speed/torque envelope.

Still not solved satisfactorily is the self-commissioning mode when the drive itself, before calibration, should gather and process all the data for the $\lambda(\theta_r,i)$ curve family within one minute as is done currently for a.c motor drives.

12.11. SENSORLESS SRM DRIVES

Sensorless SRM drives may be open-loop or closed-loop with respect to speed (or position) control. Open-loop sensorless driving implies fixed dwell angle $\theta_w = (\theta_c-\theta_{on})$ control with synchronization as is done in stepper motors [18] (Figure 12.20). Increasing the dwell angle tends to increase the stability but with lower efficiency. To solve this problem the dwell angle is increased with d.c. link current (load in general). Though simple, such a method has stability problems, produces low dynamics and allows for a limited speed control range.

Consequently, open-loop sensorless drive solutions seem today of some historical importance but hardly practical.

Closed-loop sensorless SRM drives have been proposed to compete with the, by now, commercial sensorless a.c. drives capable of at least 100 to 1 speed range and torque response in the milliseconds range with good speed precision (less than 0.2-0.3% of rated speed).

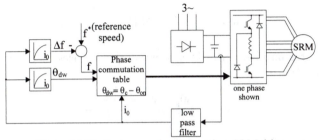

Figure 12.20. Basic open-loop sensorless SRM drive

Competitive closed-loop sensorless SRM drives should be high grade and, in addition, capable of producing, through estimation (or observation) the rotor position $\hat{\theta}_r$ and speed $\hat{\omega}_r$ with good precision, robustness and on time (fast). Among the closed-loop strategies for position and speed estimation we mention the incremental inductance estimator in the active phase, diagnostic pulse injection in the activated or passive phase, mutual inductance estimation, observers, etc. For a collection of papers on these solutions see [19] section 4.

To the best of our knowledge, no close loop sensorless SRM drive has reached the markets as of this writing. We will dwell only a little on two promising solutions: the combined voltage-current-model-based [20] and the fuzzy logic (rule-based) observers [14].

12.12. THE VOLTAGE-CURRENT MODEL-BASED POSITION SPEED OBSERVER

Basically, the phase voltages and currents are measured with wide-band frequency sensors with optoisolation and analog-digital fast (2μs or so) converters. In any case, the $\lambda(\theta_r, i)$ family of curves has to be *a priori* measured or FEM calculated and curve fitted through derivable functions.

The voltage model is represented by the stator phase equations

$$\frac{d\lambda_i}{dt} = V_i - r_s i_i \qquad (12.38)$$

This model may be used to estimate the flux, but for low speeds, the model is erratic due to integration drift, voltage and current measurements noise and stator resistance variation with temperature.

From the $\lambda(\theta_r,i)$ curves, with current measured and $\hat{\theta}_r$ estimated in a previous time step, we may estimate back the current $\hat{i}_i$ with the current error $\Delta i_i(k)$. A position correction $\Delta\theta_r(k)$ is operated to drive this error to zero. Thus the voltage and current models have already been used together once [20]

$$\Delta\theta_i(k) = -\left(\frac{\partial\lambda_i}{\partial i_i}\right)\Delta i_i(k) / \frac{\partial\lambda_i}{\partial\theta_r(k)} \qquad (12.39)$$

The three position corrections (from all three-phases) may be averaged (or weighted) to give

$$\Delta\theta_r = \frac{\left(\Delta\theta_a + \Delta\theta_b + \Delta\theta_c + ...\right)}{m} \qquad (12.40)$$

Thus the estimated position $\theta^e(k)$ is

$$\theta^e(k) = \theta_p(k) + \Delta\theta_r \qquad (12.41)$$

where $\theta_p(k)$ is the value estimated in the previous time step.

Now, with a new position $\theta^e(k)$ and the predicted flux values from the voltage model, once again new currents are estimated and new current errors $\Delta'i_i$ appear. Flux corrections $\Delta\lambda_i(k)$, based on $\lambda(\theta_r,i)$ curves, are

$$\Delta\lambda_i = \frac{\partial\lambda_i}{\partial i_i}\Delta'i_i \qquad (12.42)$$

$$\Delta\lambda_i(k) = \lambda_i(k) + \Delta\lambda_i \qquad (12.43)$$

Finally, the rotor position is predicted using its estimated values at three time steps K–2, K–1, K and a quadratic prediction [20]

$$\theta^p(k) = 3\theta^e(k) - 3\theta^e(k-1) + \theta^e(k-2) \qquad (12.44)$$

The flux correction reduces the integrator drift. Mechanical load parameters are not required and the magnetic saturation is included through the $\lambda(\theta_r,i)$ curves. Still, the speed estimator (from position) is to be added. It has to be noted that a great number of calculations is required. A TMS310C31 DSP may provide for a 100μs computing (and control) cycle. Good results, for position control only, have been reported, down to 30rpm, in [20].

The control system for the high grade SRM drive has to be added to the voltage-current model position (and speed) observer (Figure 12.21) to produce a high performance SRM drive.

Fuzzy logic may also be used to realize a voltage-current observer similar to that presented in this section.

In essence, the $\lambda(\theta_r,i)$ curve family is approximated (fitted) through the fuzzy logic approach, avoiding analytical approximations as above. Digital simulations have shown [14] such a system capable of position estimation with an error less than 0.4 degree.

Alternatively, sliding mode rotor position and speed estimators based on given $L_s(\lambda_s, \theta_r)$ curves, with flux estimator, has been also proven to produce good motion sensorless control [21] for a 5:1 constant power range by switching from maximum torque/current to maximum torque/flux optimal turn on θ_{on} and turn off θ_{off} phase angle estimation. High torque dynamic response has been also demonstrated with direct average torque control and on-line correction of off-line reference current and reference θ^*_{on} and θ^*_{off}, based on PI torque error output [22].

For more on position-sensorless control see Refs [23 - 25].

The first sensorless 100 to 1 speed range (and 0.3% of rated speed error and quick torque response SRM drive) is apparently still due.

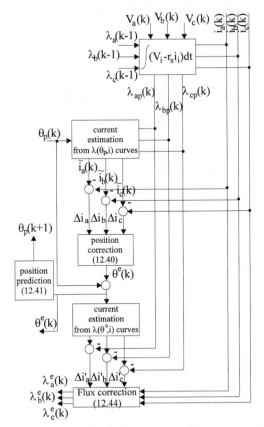

Figure 12.21. Voltage-current model position observer

12.13. SINGLE-PHASE SRM CONTROL

The single phase SRM with equal count of stator and rotor poles (2/2 or 4/4) and self starting parking position provision seems very attractive for home appliances or automotive actuators below 1Nm of torque because of the possibility of utilizing only one active power switch; that is, a lower cost drive is obtained.

In addition, with 6/6, 4/4, and even 2/2 stator/rotor pole count, the single phase SRM may be designed in some conditions for more torque density than a three phase SRM, which is characterized, anyway, by one phase acting at any time.

The self-starting parking position may be obtained by:

- Parking PMs on top of special poles (Figure 12.21a)
- Step airgap under the stator poles $g_2 / g_1 = 1.5 - 2.5$ (Fig. 12.21b) or snail cam rotor shape.
- Magnetic saturation of roughly one half of the rotor poles with the other half providing a double airgap $g_2 \sim 2g_1$ (g_1 airgap under the saturable zone – Figure 12.21c) [26].

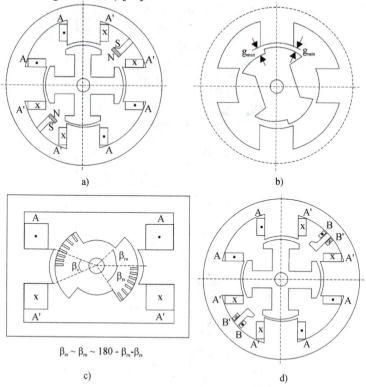

Figure 12.21. Self starting parking position methods: a) parking magnet, b) stepped airgap, c) saturated half rotor pole plus stepped airgap, d) energy-drain starting phase.

- With an additional starting phase that is automatically turned on when the active phase turns off, some additional torque is produced by the auxiliary phase even during operation at nonzero speed (Figure 12.21d) [27, 28].

Basically all methods of providing self starting position allow for unidirectional motion from standstill.

However once the motor starts in one direction, in presence of two Hall sensors (in general), it is feasible to brake the rotor regeneratively and then rotate it in the reverse direction of motion. This way bidirectional (or four quadrant) motion is claimed with single phase SRM [28].

A basic power electronics converter for the single phase SRM without and with starting phase is shown in Figure 12.22.

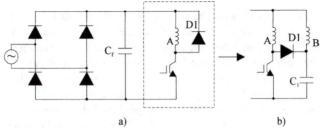

a)   b)

Figure 12.22. Single IGBT (MOSFET) converter: a) for single phase SRM,
b) for single phase SRM with energy-drain starting phase

The single phase SRM main winding A works only during about half the time as only then the torque is unidirectional (positive or negative).

However, the saturated rotor half – pole with stepped airgap configuration (Figure 12.21c) claims that the motoring cycle (negative torque) may be wider than the generating one (Figure 12.23).

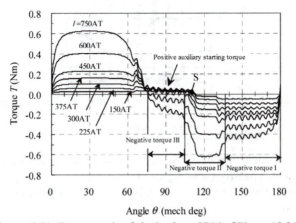

Figure 12.23. Torque cycle of single phase SRM of Figure 12.21c

There are two zero torque positions at 75° and, respectively, 110° besides the zero and 180° ones. Moreover, for low current the torque between 75° and 110° is positive, while for large current it is negative (motoring).

If the rotor is positioned for $\theta_r=75°$, then, with a small current, the rotor is moved to $\theta_r=110°$ in the opposite direction of desired motion.

Further on, with high current (at 110° rotor position) the large negative torque should be able to move the rotor to the left till $\theta_r=0$, when again negative torque is available, to continue motor acceleration.

Playing with magnetic saturation zone (made of slots and teeth on the rotor half-pole), the starting torque at low current may be increased to 10% of rated torque, to handle easy starts. Heavy starts are, apparently, hardly possible.

On the other hand, with the starting phase B present (Figure 12.21d), if the rotor is initially aligned to the main phase A poles, a current pulse does not move the rotor, but, when the phase A is turned off, the energy is drained through the starting phase B, tending to align the rotor to its axis. This way, when the second voltage pulse is applied to the main phase, the latter should provide for motor self-starting from the starting phase axis towards main phase axis. Once the motor starts in the preferred direction of motion, and two Hall proximity sensors are available, the motor may be regeneratively braked and then controlled (on the fly) to accelerate in the opposite direction of motion (Figure 12.24). Four quadrant operation is thus claimed [28].

A nonhesitant heavy (direct) bidirectional motor start does not seem possible even with this solution.

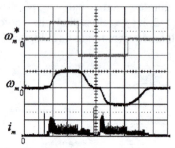

Figure 12.24. Four quadrant operation 15000 rpm/div and 5s/div

12.14. SUMMARY

- SRM is a single or multiphase doubly salient simple and rugged electric motor with a passive (windingless) rotor.
- The SRM phases are fed sequentially with voltage pulses through a sort of PWM.
- Phase commutation is rotor position dependent.
- The torque has to be calculated from the co-energy derivative method due to magnetic saturation.

- In general, the motor phases have almost zero interaction and thus SRM is claimed to be fault tolerant.
- The SRM is a motor totally dependent on the unipolar current two-quadrant multiphase chopper.
- One or, at most, two-phases may conduct at a time.
- Torque pulsations are reduced by torque sharing and current profiling, when a precision position sensor (estimator) is required.
- While low grade - general purpose - SRM drives are not difficult to control, high grade (servo) SRM drives require notable on-line (and off-line) computation efforts.
- The flux/current/position curve family is required for high grade (servo) SRM drives with motion sensors or sensorless.
- These curves are either FEM calculated or measured at standstill or in running (dynamic) tests. They are also curve fitted through analytical functions or mapped by a fuzzy logic approach.
- SRMs are simple and rugged, may withstand thermally harsh environments (like avionics, metalurgy, etc.) and thus, despite their more complicated control (for high grade performance), they seem to have a future. An incipient market has surfaced and strong marketing strides seem close.
- Single controlled power switch single phase SRM drives at low power seem feasible apparently for 4 quadrant operation in light start residential and automotive applications.

12.15. PROBLEMS

12.1. A 6/4 three-phase SRM has the data: maximum current i_{omax} = 10A, unaligned inductance L_u = 10mH, aligned inductance L_a^s = $6L_u$; flux/current/position curves are considered piece-wise linear (Figure 12.22). The stator pole width angle β_s = 30°. The interpole width angle is β_i = 30°. The d.c. link voltage V_d = 300V for the unipolar current PWM converter.

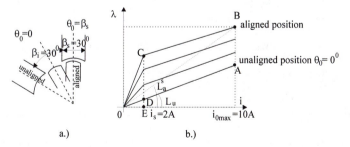

Figure 12.25. Linear piece-wise flux / current curve and unaligned and aligned position

Determine:

    a.   the maximum flux linkage;

    b.   the base speed ω_b for zero advance angle ($\theta_0 = 0°$);

    c.   the maximum average torque at zero speed;

    d.   torque versus position for constant current $i_{av} = 5A$;

    e.   for $\theta_c = -5°$ and $\theta_{off} = 22°$ calculate the maximum flux level in the phase at $2\omega_b$; determine the corresponding current from flux / current / position curves in Figure 12.22.

12.2.   For the SRM of problem 12.1 and constant $\theta_{on} = \theta_0 = 0°$ and $\theta_{off} = 24°$ use Simulink to investigate the starting transients from zero to base speed with a load of 50% of maximum average torque at zero speed and the controller in example 12.5.

12.16. SELECTED REFERENCES

1.   **T.J.E. Miller,** Switched reluctance motors and their control, OUP, 1993.

2.   **C. Ferreira, E. Richter,** About channel independence for multichannel switching reluctance generating systems, Record of IEEE-IAS, Annual Meeting, 1996, pp.816-821.

3.   **D.A. Philips,** Switched reluctance drive, new aspects, IEE Trans., vol.PE-5, no.4, 1990, pp.454-458.

4.   **T.J.E. Miller,** Converter volt-ampere requirements of the switched reluctance motor drive, IEEE Trans., vol.IA-21, no.5, 1985, pp.1136-1144.

5.   **W.F. Ray, P.L. Lawrenson et al.,** High performance switched reluctance brushless drives, IBID, vol.IA-22, no.4, 1986, pp.722-730.

6.   **F. Liang, Y. Liao and T.A. Lipo,** A new variable reluctance motor utilizing an auxiliary commutation winding, IBID, 1993.

7.   **H.Y. Li, Y.Zhao, T.A. Lipo and F. Liang,** A doubly salient doubly excited variable reluctance motor, Record of IEEE-IAS, Annual Meeting, 1992.

8.   **B.C. Mecrow,** New winding configurations of doubly salient reluctance machines, Record of IEEE-IAS, Annual Meeting, 1992, Part.I, pp.249-256.

9.   **P. Materu, R. Krishnan,** Estimation of switched reluctance motor losses, Record of IEEE-IAS, Annual Meeting, 1988, vol.I, pp.176-187.

10.   **Y. Hayashi, T.J.E. Miller,** A new approach to calculating core losses in the SRM, IBID, 1994, vol.I, pp.322-328.

11.   **V.V. Athani, V.N. Walivadeker,** Equivalent circuit for switched reluctance motor, EMPS vol.22, no.4, 1994, pp.533-543.

12.   **I.A. Byrne, J.B. O'Dwyer,** Saturable variable reluctance machine simulation using exponential functions, Proc. of Int. Conf. on Stepping motors and systems, University of Leeds, July 1976, pp.11-16.

13.   **D. Torrey,** An experimentally verified variable reluctance machine model implemented in the Saber ^{TM} circuit simulation, EMPS Journal, vol.23, 1995.

14.   **A. Cheok, N. Ertugrul,** A model-free fuzzy-logic-based rotor position sensorless switched reluctance motor drive, Record of IEEE-IAS, Annual Meeting, 1996, vol.I, pp.76-83.

15.   **A. Radun,** Design considerations for the switched reluctance motor, Record of IEEE-IAS, Annual Meeting, 1994, vol.I, pp.290-297.

16.   **B.K. Bose, T.J.E. Miller,** Microcomputer control of switched reluctance motor, Record of

IEEE-IAS, Annual Meeting, 1985, pp.542-547

17. **P.C. Kjaer, J.J. Gribble, T.J.E. Miller**, High grade control of switched reluctance machines, Record of IEEE-IAS, Annual Meeting, 1996, vol.I, pp.92-100.

18. **J.T. Buss, M. Eshani, T.J.E. Miller**, Robust torque control of SRM without a shaft-position sensor, IEEE Trans., vol.IA-33, no.3, 1996, pp.212-216.

19. **K. Rajashekara, A. Kawamura, K. Matsuse** (editors), Sensorless control of a.c. drives, IEEE Press, 1996, pp.433-485.

20. **P.P. Acarnley, C.D. French, IH. Al-Bahadly**, Position estimation in switched-reluctance drives, Record of EPE-95, Sevilla, Spain, pp.3765-3770.

21. **M.S. Islam, M.M. Anwar, I. Husain,** "A sensorless wide-speed range SRM drive with optimally designed critical rotor angles", Record of IEEE – IAS – 2000 Annual Meeting.

22. **R.B. Inderca, R.W. De Doncker,** "High dynamic direct torque control for switched reluctance drives", Record of IEEE – IAS – 2001 Annual Meeting.

23. **M. Eshani, B. Fahimi,** "Position sensorless control of switched reluctance motor drives", IEEE Trans. Vol. IE – 49, no. 1, 2002, pp.40-48.

24. **B. Fahimi, A Emadi, R.B. Sepe,** "Position sensorless control", IEEE – IAS Applications Magazine, vol. 10, 2004, pp. 40-47.

25. **R. Krishnan,** Switched reluctance motor drives, book, CRC Press, Florida, 2002.

26. **T. Higuchi, J.O. Fiedler, R.W. De Doncker,** "On the design of a single-phase switched reluctance motor", Record of IEEE – IEMDC – 2003, pp. 561-567.

27. **R. Krishnan, A.M. Staley, K. Sitapati,** "A novel single phase switched reluctance motor drive", Record of IEEE – IECON – 2001, pp. 1488-1493.

28. **R. Krishnan, S.T. Park, K. Ha,** "Theory and operation of a four quadrant SRM drive with a single controllable switch – the lowest cost four quadrant brushless motor drive", Record of IEEE –IAS – 2004 Annual Meeting.

Chapter 13

PRACTICAL ISSUES WITH PWM CONVERTER MOTOR DRIVES

13.1. INTRODUCTION

Induction (IM), permanent magnet a.c. (PM-SM), reluctance synchronous (RSM) and switched reluctance (SRM) motors are brushless multiphase motors fed through PWM voltage-source converters (PECs) with bipolar and, respectively (for SRM), unipolar current capability. In all these drives, the line side converter is (still), in general, a diode rectifier with a d.c. link filter capacitor.

Also, the motors are fed through a cable of notable length (sometimes up to some hundreds of meters), with ultrafast voltage pulses (1-3μs or less in most cases). What are the effects of such drives on the motor itself and on the environment?

Let us enumerate a few of them:

- additional motor losses due to current and flux harmonics;
- current harmonics injected in the a.c. power supply (supply filter is necessary);
- electromagnetic interference due to the rather high switching frequency (up to 20kHz with IGBTs, in the hundreds of kW power range per unit);
- high frequency leakage currents' influence on motor current control and on circuit breakers;
- overvoltages along the stator coils first turns due to wave reflections of the PWM converter (special filters are required); these overvoltages are magnified by long cables;
- the steep-front voltage pulses produce both electromagnetic and electrostatic stray currents (voltages) in the bearings causing wear unless special measures are taken.

All these issues are called practical as they occur in industry where these drives are used every day for many hours.

13.2. THE BASIC PWM CONVERTER DRIVE

A basic PWM inverter drive with a.c. motors (or with SRMs) is made of a PEC, a power cable and the motor (Figure 13.1).

As variable speed drives are applied in various industries, the environment (thermal or chemical), the distance between the motor and the PEC, and the load cycle varies widely with the application. Variable speed drives may be introduced as new systems (the PEC and the motor) or the old motor (IM or SM) — designed for sine wave supply — is not changed but rerated for the PEC supply and the new speed control range.

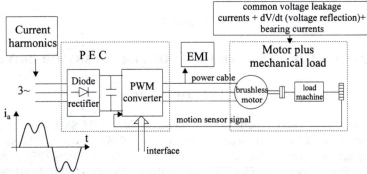

Figure 13.1. Basic PWM converter drive

Current harmonics in the power supply in the presence of the diode rectifier, long cable effects on voltage reflection (and dV/dt) at motor terminals, bearing currents and leakage currents are all consequences of the short rise time of voltage pulses in the PWM converter.

13.3. LINE CURRENT HARMONICS

Increasing the use of PWM converter drives for pumps, blowers, fans, and other industrial applications may lead to a notable infusion of current harmonics in the industrial systems.

A few questions arise:

- are harmonics filters needed?
- what type of filters (5th, 7th, 11th, 13th or combinations of these) are required?
- how to design (size) such filters?

Six-pulse PWM converters with diode rectifiers on the power system side draw harmonic currents of orders 5, 7, 11, 13, 17. In principle, each PEC has its own distinct harmonic signature.

The harmonic factor HF to define this signature is

$$HF = \frac{\sqrt{\sum_{5,7,11,...} v^2 I_v^2}}{I_1} \tag{13.1}$$

The HF depends on the fact that the PWM converter is fed from a single-phase or three-phase supply, with or without line reactors, with or without an isolation transformer. HF varies in the interval 2 to 4% for three-phase supplies and up to 12% for single-phase a.c. supplies, for powers up to 400kW or more and IGBTs as SRCs.

The harmonic factor HF is related to the total harmonic distortion (THD) by

$$THD = HF \cdot \frac{Drive\ kVA}{SC\ kVA} \cdot 100\% \tag{13.2}$$

where drive kVA is the drive rated kVA and SC kVA is the short-circuit kVA of the power system at the point of drive connection.

In general, a 5% THD is required (IEEE-519-1992 and IEC 1000-3-2(3) standard). If this constraint is met, no line filter is needed. However, in general, this is not the case and a filter is necessary.

Harmonics filters consist of several sections of single (or double) tuned series L-C resonant circuits (Figure 13.2).

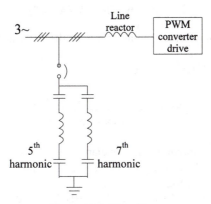

Figure 13.2. Line filters

Separate filters for each drive above 15kW are advisable but aggregate filters for multidrive nodes may also be used, especially if not all drives are working at full load most of the time.

Slightly below the resonance frequency (0.95), series LC filters behave like a capacitor and their design for this situation avoids the possibility of parallel resonance due to component parameter change because of temperature or aging.

The resonance frequency of a single tuned filter f_r is

$$f_r = \frac{1}{2\pi\sqrt{LC}}; \quad \alpha = \frac{f_r}{f_n} \tag{13.3}$$

α - detuning factor, f_n - nominal resonance frequency of the filter $f_n = vf_1$.

The filter impedance for v^{th} harmonic, $Z_f(v)$, is,

$$\underline{Z}_f(v) = R_f + j\left[2\pi f_1 vL - 1/(2\pi f_1 vC)\right] \tag{13.4}$$

The filter capacitor reactive power kVAF (three-phase) for 60Hz is

$$kVAF = 2\pi f_1 10^{-3} CV_{LL}^2 = 0.3777 CV_{LL}^2 \tag{13.5}$$

where V_{LL} is the line voltage (rms) and C is in farads. The filter capacitor kVA is about (25-30%) of drive kVA rating. A harmonic filter attenuates all harmonic voltages at the point of connection but maximum attenuation occurs for frequencies close to its resonance frequency.

The attenuation factor $a_v(h)$ is

$$a_v(h) = \frac{V(h)}{V_f(h)} \qquad (13.6)$$

$$V(h) = \%I(h) \cdot h \cdot (\text{Drive kVA} / \text{SC kVA}) \qquad (13.7)$$

$V(h)$ is the h harmonics voltage without the filter and $V_f(h)$ is with the vth order filter.

Neglecting the filter resistance

$$a_v(h) = \left(1 + \frac{(v\alpha)^2}{1-(v\alpha/h)^2}\right) \frac{\text{kVAF}}{\text{SC kVA}} \qquad (13.8)$$

In general, $a_v(h)$ should be higher than 1.0 to provide some attenuation.

The THD has to be limited for the voltage harmonics making use of $V(h)$ of (13.7) for finding $V_f(h)$.

The filter is designed such that the THD is less than 5%. The allowed current THD may increase with the short-circuit ratio R_{sc} at the point of connection [2] from 5% for ($R_{sc} \leq 20$) to 20% for $R_{sc} > 1000$

$$R_{sc} = \frac{\text{utility maximum short - circuit current}}{\text{maximum demand fundamental current}} \qquad (13.9)$$

For a 5th-order harmonic current filter, the drive current waveform changes notably (Figure 13.3d).

The peak current is increased but continuous current conduction is obtained by eliminating the 5th current harmonic.

Designing adequate line filters for a local power industrial system with more than one PWM converter drive is a rather complex problem whose solution also depends on the power source impedance, through the short-circuit current ratio Rsc or through the drive kVA / SC kVA.

As the relative power of PWM converter drives is increasing and the power quality standards are more and more challenging, it seems that the trend is to provide each drive with its line filter to comply with some strict regulations.

Common point of connection filter additions may be made in "hot" PWM converter-drive zones with weak local power supplies.

In PWM converters we included both, the voltage source inverters for a.c. motor drives, and the multiphase choppers for SRM drives.

Line voltage instabilities in the presence of line filters with some fast (direct torque and flux - DTFC) control systems of a.c. motor drives [3] and with d.c. chopper-fed motor drives [4] have been reported. This aspect has to

be taken care of when the drive with incorporated line filter is designed and
built.

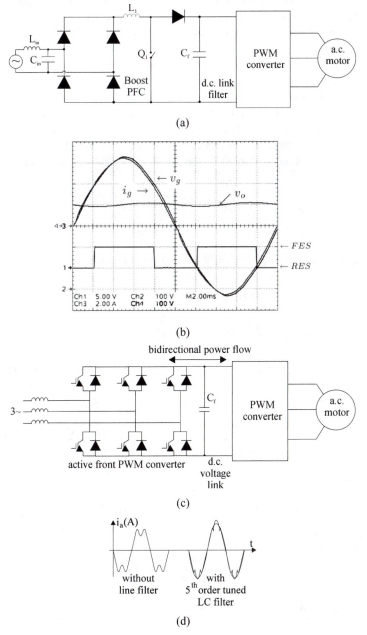

(a)

(b)

(c)

(d)

Figure 13.3. The boost PFC: a) the configuration, b) the obtained input voltage and
current, c) the active front rectifier, d) 5th harmonic filtered in input current

Example 13.1. Line filter sizing

Let us consider that a PWM converter IM drive of 100kVA is connected to a local power system with a short-circuit power SC kVA = 1000kVA. The line current 5th, 7th and 11th harmonics are $i_5 = 0.15I_1$, $I_7 = 0.03I_1$, $I_{11} = 0.01I_1$; line frequency $f_1 = 60Hz$, line voltage $V_{LL} = 440V$ (rms).

Calculate:

- the line current harmonic factor HF;
- the total distortion factor THD for the current;
- the filter capacitor C tuned to the 5th harmonic with $\alpha = 0.95$ (detuning factor), and the corresponding inductance L_f;
- the voltage attenuation factors $a_5(h)$ for the 5th, 7th and 11th harmonics;
- the voltage total harmonic distortion factor THD with the filter off and on.

Solution:

According to (13.1), HF is

$$HF = \frac{\sqrt{\sum_{5,7,11} v^2 I_v^2}}{I_1} = \sqrt{(5\cdot0.15)^2 + (7\cdot0.03)^2 + (11\cdot0.01)^2} = 0.7865 \quad (13.10)$$

From (13.2) the THD current is

$$THD = HF\cdot\frac{Drive\ kVA}{SC\ kVA}\cdot100\% = 0.7865\cdot\frac{100}{1000}\cdot100\% = 7.865\% \quad (13.11)$$

As THD is higher than the 5% recommended for drive kVA / SC kVA = 1/10, a line filter is required.

The filter is tuned to the 5th harmonic with $\alpha = 0.95$, so, from (13.3)

$$LC = \frac{1}{4\pi^2(f_n\alpha v)^2} = \frac{1}{4\pi^2(60\cdot0.95\cdot5)^2} = \frac{0.1308\cdot10^{-6}}{25}s^2 \quad (13.12)$$

Also, the capacitor kVA is

$$kVAF = 0.3\cdot Drive\ kVA = 0.3\cdot100 = 30\,kVA \quad (13.13)$$

From (13.5) C is

$$C = \frac{kVAF}{0.377\cdot V_{LL}^2} = \frac{30}{0.377\cdot440^2} = 411\cdot10^{-6}F = 411\mu F \quad (13.14)$$

The harmonic voltage V(h) (with filter off) in the power source (in %) is found from (13.7)

$$V(5) = \%(I(5)) \cdot 5 \cdot \frac{100}{1000} = 15 \cdot 5 \cdot \frac{100}{1000} = 7.5 \%$$

$$V(5) = \%(I(7)) \cdot 7 \cdot \frac{100}{1000} = 3 \cdot 7 \cdot \frac{100}{1000} = 2.1 \% \qquad (13.15)$$

$$V(5) = \%(I(11)) \cdot 11 \cdot \frac{100}{1000} = 1 \cdot 11 \cdot \frac{100}{1000} = 1.1 \%$$

We may calculate the voltage THD before filtering

$$THD(\%) = \sqrt{(5 \cdot V(5))^2 + (7 \cdot V(7))^2 + (11 \cdot V(11))^2} =$$
$$= \sqrt{(5 \cdot 7.5)^2 + (7 \cdot 2.1)^2 + (11 \cdot 1.1)^2} = 42.056 \% \qquad (13.16)$$

The voltage attenuation factors are

$$a_s(5) = \left(1 + \frac{(5 \cdot 0.95)^2}{1 - (5 \cdot 0.95 / 5)^2}\right) \frac{0.3}{10} = 6.942$$

$$a_s(7) = \left(1 + \frac{(5 \cdot 0.95)^2}{1 - (5 \cdot 0.95 / 7)^2}\right) \frac{0.3}{10} = 1.2845 \qquad (13.17)$$

$$a_s(11) = \left(1 + \frac{(5 \cdot 0.95)^2}{1 - (5 \cdot 0.95 / 11)^2}\right) \frac{0.3}{10} = 0.862$$

Based on the definition of $a_v(h)$, (13.6), we may calculate $V_f(5)$, $V_f(7)$, $V_f(11)$ and thus, from (13.1), the new voltage, THD_f (with the filter on) is

$$THD_f = \sqrt{(5 \cdot V(5) / a_s(5))^2 + (7 \cdot V(7) / a_s(7))^2 + (11 \cdot V(11) / a_s(11))^2} =$$
$$= \sqrt{(5 \cdot 7.5 / 6.942)^2 + (7 \cdot 2.1 / 1.284)^2 + (11 \cdot 1.1 / 0.862)^2} = 18.66 \%$$

$$(13.18)$$

The beneficial effect of the 5th-order harmonic filter is only partly evident: a reduction of voltage THD from 42% to 18.66%. Note that in our example the 7th and 11th current harmonics were exaggerated and this explains why the final value of THD_f is not much smaller (eventually below 8%). Also, the short-circuit power of the local grid was unusually small. An increase in the SC kVA of the local grid would improve the results.

Apart from passive input filters, hybrid (active-passive) filters may be used to compensate the input current and voltage harmonics and eventually deal with short voltage sags in the a.c. source.

The boost power factor correcter (PFC) is a typical such device (Figure 13.3a), applied for single phase a.c. source. [18, 20]

The boost PFC converter operates in continuous conduction mode. This way, with a proper control, the sinusoidal input current is reconstructed in phase with the a.c. source voltage (Fig. 13.3b). [18]

Also for input current filtering and unity power factor, the active front end rectifier-inverter (Fig. 13.3c) may replace the diode rectifier plus the passive input filter [21 - 23].

The active front end PWM converter provides quite a few functions:

- Stabilizes the a.c. link voltage "against" limited overload or against a.c. source short-lived voltage sags (three phase, two phase, single phase);
- Produces rather sinusoidal current input with limited current, controllable, power factor angle.
- Provides for fully bidirectional power flow, so needed for fast braking drives (such as in elevators).

13.4. LONG MOTOR POWER CABLES: VOLTAGE REFLECTION AND ATTENUATION

In many new and retrofit industrial applications the PWM converter and motor must be placed in separate locations, and thus long motor cables are required. The high rate of voltage rise (dV/dt) of up to 6000V/µs, typical for IGBT inverters, has adverse effects on motor insulation and produces electrostatic-caused currents in the bearings. The distributed nature of the long cable L-C parameters results in overvoltages which further stress the motor insulation. In addition, voltage reflection of the long cable is dependent on inverter pulse rise time ($t_r = 0.1$ to 5µs) and on the cable length which behaves like a transmission line.

PWM voltage pulses travel along the power cable at about half light speed ($U^* = 150$-200m/µs) and, if the pulses take longer than one third the rise time to travel from converter to motor, full reflection occurs; the voltage pulse level is doubled at motor terminals.

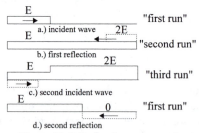

Figure 13.4. Reflection wave cycle

Let us consider a finite length cable with an infinite dV/dt voltage wave (pulse) traveling from converter to motor. The motor represents a high impedance with respect to such fast voltage pulses. The motor behaves like an equivalent uncharged capacitor in this case. Upon arrival, the incident wave is reflected (the capacitor charges). The reflected plus incident wave doubles the voltage at motor terminals (2E). The line is thus charged at 2E while at the sending end, the inverter output voltage is still E. Hence a −E (negative reflection) wave now travels from inverter to motor and thus, finally, the motor voltage becomes again E after three "runs" along the cable length (Figure 13.4). This is why it is said that, if the pulses take longer than one third of rise time to travel from inverter to motor, the reflected voltage is 2E at the motor terminals.

We may thus calculate the ideal peak line to line voltage V_{LL} at the motor terminals as [5,6]

$$V_{LL} = \begin{cases} \dfrac{3l_c \cdot E}{U^* \cdot t_r} \cdot \Gamma_m + E; & \text{for } t_{travel} < t_r / 3 \\ E \cdot \Gamma_m + E; & \text{for } t_{travel} > t_r / 3 \end{cases} \qquad (13.19)$$

where Γ_m is the motor realistic reflection coefficient derived from transmission line theory

$$\Gamma_m = \frac{Z_{motor} - Z_{0c}}{Z_{motor} + Z_{0c}} \qquad (13.20)$$

where Z_{motor} is the motor equivalent impedance and Z_{0c} is the surge impedance of the cable (of length l_c)

$$Z_{0c} = \sqrt{\frac{L_c}{C_c}} \qquad (13.21)$$

A similar reflection coefficient Γ_c may be defined on the PWM converter output side. In the qualitative analysis in Figure 13.4, $Z_{motor} = \infty$ and, thus, $\Gamma_m = 1$. In reality, the motor impedance is more than 10-100 times larger than Z_{0c} and thus $\Gamma_m \cong 1$.

For $\Gamma_m = 1$, to reduce the overvoltage to almost zero, $t_{travel} < t_r / 3$ and thus

$$\frac{3l_c \cdot E}{U^* \cdot t_r} \cdot \Gamma_m \leq 0.2E \qquad (13.22)$$

Equation (13.22) provides a condition to calculate the critical cable length l_c above which virtual voltage doubling occurs. For E = 650V d.c. bus (480V a.c. system), $\Gamma_m = 0.9$, $U^* = 165$m/μs, from (13.22) we obtain

$$\frac{3l_c \cdot E}{U^* \cdot t_r} \cdot \Gamma_m = E \qquad (13.23)$$

$$l_c = \frac{U^* \cdot t_r}{3\Gamma_m} = \frac{165}{3 \cdot 0.9} \cdot t_r = \left(61.11 \cdot t_r [\mu s]\right) \text{ meters} \qquad (13.24)$$

Five times shorter cables in comparison with the critical length l_c (Figure 13.5) are required to reduce the motor overvoltage to 20% (13.22). This latter condition may not be met in many applications.

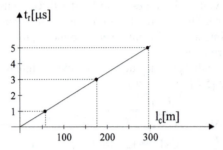

Figure 13.5. Voltage rising time t_r versus critical cable length

For long cables, the solution is to reduce the reflection coefficients at motor side $\Gamma_m(Z_m)$ (13.20) and on converter output side $\Gamma_c(Z_c)$. This may be done by adding a filter at motor terminals, for example, to reduce Z_m (Figure 13.6). Consequently, a low pass filter at converter output will increase the voltage pulse rise time which has the effect of allowing cables longer than l_c (Figure 13.6b).

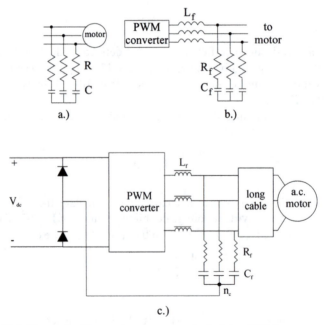

Figure 13.6. First-order filters: a.) at motor terminals, b.) at PWM converter output, c) LRC filter with diode clamp

According to [6] the total voltage across the R-C motor terminal filter, E_r, is

$$E_r = 2E\left(1 - \frac{Z_{0c}}{R + Z_{0c}} e^{-\frac{t}{(R+Z_{0c})C}}\right); \quad R = Z_{0c} \tag{13.25}$$

To design the filter, we may limit E_r to $E_r = 1.2E_0$ (20% overvoltage) for $t = t_r$ (rising time)

$$1.2E = 2E\left(1 - \frac{Z_{0c}}{R + Z_{0c}} e^{-\frac{t_r}{2Z_{0c}C}}\right) \tag{13.26}$$

Thus we may calculate the filter capacitor C while the resistance $R = Z_{0c}$ (Z_{0c} — the cable surge impedance).

On the other hand, the influence of inverter output low pass filter is materialized by a voltage delay

$$V(t) = E\left(1 - e^{-\frac{t_r}{\tau}}\right) \tag{13.27}$$

$$\tau = \sqrt{L_f C_f} \tag{13.28}$$

where τ is the filter time constant. Evidently,

$$\tau \geq t_r \tag{13.29}$$

The capacitor has to be chosen as $C_f > 1_c.10^{-10}$ (F): L_f is obtained from (13.28) with τ imposed and the resistance R_f is calculated assuming an overdamped circuit

$$R_f \geq \sqrt{\frac{4L_f}{C_f}} \tag{13.30}$$

Typical results with a motor terminal filter and inverter output filter are given in Figure 13.7a and b, respectively.

Note: Increase in the rise time of the terminal voltage may impede the control system performance in the motor parameter identification (for tuning) mode. Care must be exercised when designing the drive to account for this aspect.

Recently, overvoltages (above the doubling effect of ideal reflection) of up to 3 times the d.c. link voltage E have been reported in connection with some PWM patterns which, once modified adequately by pulse elimination techniques, lead to lowering the ideal overvoltage to twice the value of E [7].

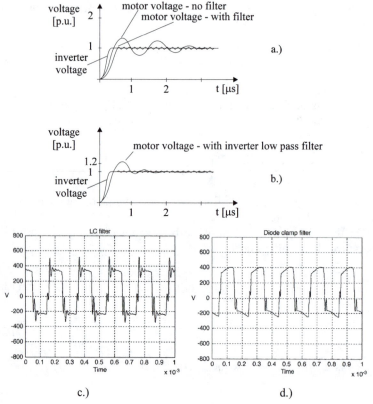

Figure 13.7. Motor terminal voltage: a.) with motor terminal RC filter,

b.) with inverter low pass filter, c) 2.2 kW induction motor with low pass filter $L_f =$

2.6 mH, $C_f = 1.4\ \mu F$, d) 2.2 kW induction motor with low pass filter $L_f = 2.6$ mH, C_f

$= 1.4\ \mu F$ and diode clamp

A rather low cost solution to reduce drastically motor terminal overvoltage due to high dV/dt and their propagation along motor cable (as shown earlier in this paragraph) adds to the differential LRC filter (Fig. 13.6c) two diodes in the d.c. link whose middle point is connected to the capacitors null point (Fig. 13.6c) [24].

The LC filter still operates in the differential mode to reduce dV/dt but the diode-clamp occurs whenever an overvoltage due to ringing is present. When the capacitor null point n_c voltage becomes larger than $\pm V_{dc}/2$ the corresponding diode (upper or lower) starts conducting; this way the d.c. link voltage is clamped. For a 2.2 kW drive, $L_f = 2.6$ mH, $C_f = 1.4\ \mu F$ would be suitable [24].

The simulated motor terminal voltage at 8 kHz fundamental frequency with $f_{sw} = 4.5$ kHz switching frequency (Fig. 13.7c, d) for the LRC filter without and with diode-clamping shows a 15 to 20 % further reduction of motor terminal overvoltage with the diode-clamp. Experiments fully confirmed these results [24].

Medium and large IM still experience failures in their windings due to voltage sources (high dV/dt and overvoltage) [25]. Attributing these failures to winding insulation weakness or to cable length and type (fully shielded or only bundled) or to PWM strategy in the converter is not always straightforward and a comprehensive study is required to find the actual cause.

Adding an LC filter with diode clamp looks like a fairly practical solution to the problem, especially when the existing motor is kept and only the inverter drive is added anew.

13.5. MOTOR MODEL FOR ULTRAHIGH FREQUENCY

For the fast rising voltage pulses produced by the PWM at the motor terminals, the motor behaves like a complex net of self and mutual inductances and capacitors between winding turns C_s and between them and earth C_p (series parallel) — Figure 13.8.

Besides the stator model (Figure 13.8), the rotor windings (if any) have to be added. Owing to the presence of parallel capacitors C_p and $C_{p1,2}$ the voltage pulse at the motor terminals (a, b, c), in the first nanoseconds, is not uniformly distributed along the winding length. Thus, even if neutral n of the stator windings is isolated, the first coils of windings have to withstand more than (60-70%) of the voltage pulse. Consequently, they are heavily stressed and their insulation may wear out quickly. Increasing the voltage rise time at motor terminals reduces this electrostatic stress and thus even existing line-start motors may work with PWM converter drives for a long time.

It is also possible to use special thin insulation for motor windings and/or use random windings to distribute the initial voltage pulse with high dV/dt randomly through the winding.

13.6. COMMON MODE VOLTAGE: MOTOR MODEL AND CONSEQUENCES

The structure of a PWM inverter drive is shown in Figure 13.9.

Assume that the zero sequence impedance of the load (motor) is Z_0. The zero sequence voltage V_0 is

$$V_0 = \frac{V_{an} + V_{bn} + V_{cn}}{3} = \frac{V_a + V_b + V_c}{3} - V_n \qquad (13.31)$$

and
$$i_0 = \frac{i_a + i_b + i_c}{3} \qquad (13.32)$$

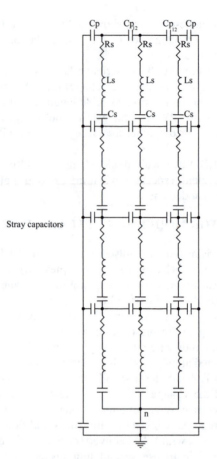

Figure 13.8. Motor RLC equivalent circuit for superhigh frequencies

(stator only considered)

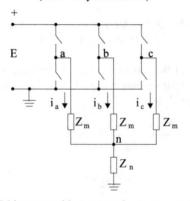

Figure 13.9. PWM inverter with motor and common mode impedance Z_n

(isolated null point)

The relationship between V_0 and i_0 is:

$$V_0 = Z_0 i_0 \qquad (13.33)$$

Consequently, common mode current i_n and common mode voltage at neutral point V_n are:

$$i_n = i_a + i_b + i_c = \frac{3}{Z_0 + 3Z_n} \frac{V_a + V_b + V_c}{3} \qquad (13.34)$$

$$V_n = \frac{3Z_n}{Z_0 + 3Z_n} \frac{V_a + V_b + V_c}{3} \qquad (13.35)$$

As known, the common current mode is decoupled from the differential current mode (the balanced phase impedance Z_m is not involved).

Consequently, the equivalent common mode input voltage V_{0in} is

$$V_{0in} = \frac{V_a + V_b + V_c}{3} \qquad (13.36)$$

and the machine impedance is as in (13.34) and Figure 13.10

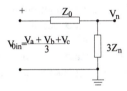

Figure 13.10. Motor model for common mode input voltage

For each commutation sequence in the converter, the common mode input voltage V_{0in} varies with $E/3$ approximately (E – d.c. link voltage). To the quick common voltage pulses, the motor behaves like stray capacitors between stator windings and stator laminations in parallel with capacitors through the airgap to the shaft and in parallel with capacitors from the null point through bearings to the common grounding point (Figure 13.11) [8,9].

Let us note that the power cable, common mode choke (to be detailed later) and line reactors may be represented as common voltage series and parallel impedances Z_s, Z_p between the inverter and the motor.

Also, the $3Z_n$ impedance may be replaced by capacitors and by the equivalent circuit of the voltage of the rotor shaft to ground V_{rg} (Figure 13.11).

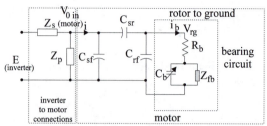

Figure 13.11. Common mode motor model for high frequencies

The bearing is represented by the bearing resistance (R_b), in series with a capacitor C_b which is in parallel with a nonlinear impedance Z_{fb} representing the intermittent shorting of capacitance C_b through bearing film breaking by the voltage pulses.

C_{sr} - equivalent capacitor from stator to rotor;
C_{sf} - equivalent capacitor between stator and frame;
C_{rf} - equivalent capacitor between rotor laminations and frame.

It now becomes evident that any voltage pulse from the inverter, when a switching in the inverter takes place, also produces a common mode voltage V_{0in} of E/3. This common voltage pulse results in a high common mode leakage current i_n. If large, this common mode (leakage) current can influence the null protection system. Reducing i_n means implicitly to reduce the shaft voltage V_{rg} and the bearing noncirculating current i_b [9]. Though i_b << i_n, it manages to deteriorate the bearings through the breakdown of lubricant film.

We will first deal with the stator common mode (leakage) current i_n which, in some cases, may be as large as the rated current.

13.7. COMMON MODE (LEAKAGE) STATOR CURRENT REDUCTION

As we are, for the time being, interested only in the leakage current $i_l(t)$, we may lump the motor and cable through a series RLC circuit whose components may be determined experimentally (Figure 13.12) by applying a pulse from the converter and measuring the leakage current i_l, which looks like that in Figure 13.12.b.

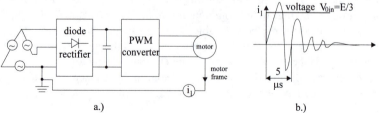

a.)   b.)

Figure 13.12. Measuring the leakage current: a.) test circuit,
b.) current response to step voltage

The leakage current response resembles that of an RLC circuit. L is related to the cable inductance, C is the stray capacitance between windings and frame and R is related to the motor only.

After adding a common mode choke L_c at motor terminals (Figure 13.13.a), we obtain an equivalent circuit as in Figure 13.13b.

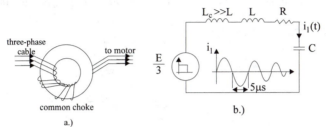

Figure 13.13. Common mode choke: a.) equivalent circuit, b.) leakage current $i_l(t)$

The peak leakage current is notably reduced (Figure 13.13b) by the presence of the common mode choke but the rms value is still not drastically reduced.

Placing a secondary coil (connected to a resistor R_t) on the common choke core [10] (Figure 13.14a.) has proved to drastically reduce both peak and rms leakage current values (Figure 13.14.b).

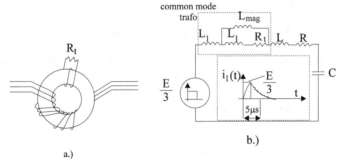

Figure 13.14. a.) common mode trafo, b.) equivalent circuit and leakage current response

The common mode trafo with secondary resistor R_t (L_l – leakage inductance, L_{mag} – magnetizing inductance) reduces the main flux in comparison with the common mode choke. Consequently, the ferrite core size may be reduced notably [10].

Reducing drastically the rms value of the leakage current leads to lowering the risk for incorrect operation of residual current-operated circuit breakers. Implicitly, it reduces, to some extent, the shaft voltage V_{rg} and the bearing currents. It appears, however, that bearing currents still have to be further reduced to avoid premature bearing damage through lubrication film electrostatic breakdown.

13.8. CIRCULATING BEARING CURRENTS

Noncirculating common mode bearing currents occur, as shown in the previous section, as a result of common mode voltage pulses (Figure 13.11).

Recently, an additional bearing current — the circulating component — due to the electrostatic current leaks to the laminated core, due to stator coil sides currents, along the machine stack length: $2\Delta i_a$, $2\Delta i_b$, $2\Delta i_c$, has been discovered (Figure 13.15a [9]).

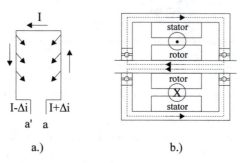

Figure 13.15. a.) current unbalance, b.) net flux and bearing circulating current

These unbalanced longitudinal currents produce a circular net flux that is responsible for an additional, circulating, frame-bearing shaft current.

13.9. REDUCING THE BEARING CURRENTS

Examining the bearing failure in some PWM converter drives after only a few months of operation, bearing fluting (induced by electrical discharge machining (EDM)) has been noted. Bearing fluting manifests itself by the appearance of transverse grooves or pits in the bearing race. The shaft voltage V_{rg} (Figure 13.11) is a strong indication for the presence of bearing currents.

EDM may occur if the electrical field in the thin lubricant film is in excess of 15V (peak)/μm which, for films between 0.2 to 2μm, means 3 to 30V_{pk} of shaft voltage, V_{rg}.

Among measures proposed to reduce the shaft voltage V_{rg} and, implicitly, bearing wear we mention here:
- outer-race insulation layer (Figure 13.16a);
- dielectric-metallic Faraday airgap or complete foil (Figure 13.16b) - or paint along full stator length;
- copper-plated slot stick covers (Figure 13.16c);
- one insulated bearing and shaft grounding brush and good motor grounding (to high frequency) from motor to mechanical load and from motor to inverter (drive) — Figure 13.16d [25].

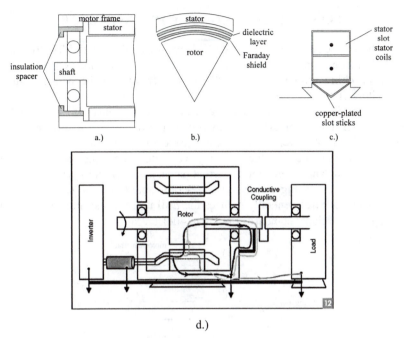

Figure 13.16. Methods to reduce bearing currents: a.) bearing insulation spacer, b.) conduction foil or paint, c.) conduction slot sticks, d.) one insulated and shaft grounding brush and motor, good high-frequency grounding from motor to load and from motor to drive

In fact, all methods introduce a small capacitor in series with the bearing circuit (Figure 13.11) and this is expected to ultimately reduce the bearing currents. The bearing insulation spacer seems to be a must. Only when stator and winding shielding is added - either through conductive foil or paint - is the shaft voltage reduced to less than 5% of its initial value (of the conventional machine) to values of 1-1.2 V(peak). Also, the stator temperature does not change notably due to these conductive shields (in which eddy currents are induced) and thus the methods proposed seem safe to apply [8]. Finally, when common mode chokes are used (see Section 13.8), the shaft voltage was initially around 30V. After using the full Faraday shield (Figure 13.16b) the shaft voltage was further reduced to less than 1.5V [8].

It is yet to be seen, if the common mode trafo is capable of reducing enough the danger of bearing wear occurrence so as to eliminate or at least reduce the demand for Faraday shields.

13.10. ELECTROMAGNETIC INTERFERENCE

The step change in voltage caused by fast switching IGBTs produces high frequency common mode (Section 13.7) and normal mode currents after each switching sequence in the PWM inverter, due to the parasitic stray capacitors of electric motors.

The oscillatory currents with frequencies from 100kHz to several MHz radiate EMI fields (noises) in the environment, influencing various electronic devices such as AM radio receivers, medical equipments, etc.

A motor model, including stray capacitors and the corresponding common mode and normal mode current paths, is shown in Figure 13.17.

As expected, using a shielded three-core cable (with the shield used as the grounding wire) generally reduces the EMI noise to the limits prescribed through international standards (less than 40dBμV/m).

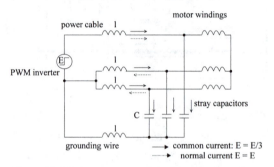

Figure 13.17. A.c. motor model with stray capacitors

Alternatively, making use of a common mode transformer (Section 13.7) and $R_F L_F$ normal mode filters (Figure 13.18) for the normal mode currents, produces acceptable EMI noise limits with a conventional power cable made through bundling the three feeding wires and the ground wire [11].

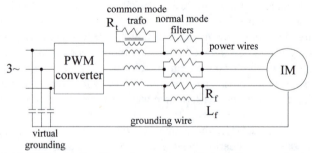

Figure 13.18. EMI reduction through common mode transformer and natural mode filters mounted at inverter output terminals

13.11. THE AUDIBLE NOISE

The audible noise, in fact a part of the EMI spectrum (15 Hz to about 18 kHz), is a main performance index of an electric drive. Audible noise standards, measures to reduce the audible noise in PWM electric drives and methods to measure it (ISO 3340 - 46 standard) constitute a subject of wide interest and debate [12]. In general, the PWM supply in electric drives increases the level of noise related to grid connected electric motors.

Approaches to reduce PWM excited noise include: using an LC filter at motor terminals, changing the switching frequency, operation above 18kHz, use of random PWM [13] or making the PWM drive play music with its noise.

Standards for noise emission of induction motors, for example, are: ISOR 1680, IEC-34-9, NEMA M61-12.49.

13.12. LOSSES IN PWM CONVERTER DRIVES

Losses in PWM converter drives occur both in the motor and in the PEC. The PEC produces current and flux harmonics in the motor and thus additional winding losses and core losses are present in PWM converter drives [14]. The PWM converter losses are of conduction and commutation (switching) types [15].

Though there is a rich literature related to loss modeling in motors and PWM converters, experimental findings are still crucial in view of the complexity of the phenomena. As the switching frequency increases, the PEC losses increase while the motor losses only slightly decrease.

Total drive losses only slightly increase when switching frequency varies from 2 – 10 kHz in a 2 kW motor, for example. Higher power units mean lower switching frequencies; also, it has been shown that if the motor efficiency is high so is PEC efficiency.

A maximum efficiency drive is obtained if the flux increases with torque and there are quite a few control methods to implement it. For vector or DTFC control, adequate flux/torque reference correlation solves the problem rather satisfactorily. However, the speed response, in the first few milliseconds after a step torque reference increase occurs, is rather slow until the flux in the motor reaches the higher (according to torque reference) levels.

13.13. SUMMARY

- The introduction of a large scale of PWM converter drives with line-side diode rectifiers for powers now up to 500kW per unit and more, has produced considerable progress in energy usage and productivity increases through variable speed.
- However, there are some side effects related to the ultrafast rise time voltage pulses produced by the PWM converter with switching

frequency up to 20 kHz sent to the motor often through long power cables, as imposed by application.

- Electromagnetic interference (EMI) is related to conductive (through power source) and electromagnetic voltage radiations at frequencies from 100 kHz to megahertz and produces communication (radio, etc.) damage around or crosstalk in neighboring digital equipment. This aspect is treated extensively in power electronics books although it is by now a field in itself.

- Line current harmonics of 5th, 7th, 11th, ... order occur. Limited by demanding international or national standards, they are defined by the total harmonic distortion (THD) factor of harmonic voltages produced in the power source. Passive line filters may solve the problem but care must be exercised to check that filter voltage instabilities do not occur in presence of high performance PWM converter drives.

- If power cables between PWM converter and motor are longer than a critical l_c, which corresponds to one third of voltage pulse rise time t_r (t_r = 0.1-5μs) travel at about half light speed, the voltage is doubled, by reflection, at the motor terminal. This fast high voltage pulse may damage the motor insulation as it stresses initially only the first coils of the stator windings. So it has to be reduced — for cable length $l > l_c$ — by inverter-side low pass filters or by motor terminal filters.

- Increasing the voltage rise time might impede some self-commissioning strategies for motor parameter identification based on step voltage response standstill tests, which are in use for some commercial drives.

- Some PWM strategies may lead to tripling the inverter voltage pulse at motor terminals but special pulse elimination methods may reduce them to the classical double voltage, taken care of as explained above.

- Common mode (zero sequence) voltage pulses occur in the machine and they produce electrostatic leakage currents in the motor — to the frame structure. For such high frequencies a distributed capacitor model of the machine is required. The motor bearing appears also as an elaborate circuit in this case. The common mode stator leakage current may reach the level of the rated current and may impede the action of residual current circuit breakers and produce notable shaft voltage and bearing currents.

- Common mode chokes and common mode transformers are proposed to reduce the leakage current, shaft voltage and bearing currents. Apparently they are not enough to secure a long life for bearings.

- Current bearing existence is related to rather high shaft voltages (in the order of 5 to 30V peak values); reducing this voltage to 1-2V would reduce the chances for bearing damage caused by electrostatic breakdown of the bearings' 2μm thick film lubricant. This is done through Faraday complete shields (paints) placed along the stator bore and winding end connections, or similar methods.

- The future? It seems to be either to improve existing solutions for curing the side effects of PWM converters or to produce close to sinusoidal input-output converters where all these problems are inherently solved at the price of a more sophisticated converter-motor control and higher costs.

13.14. PROBLEMS

13.1. <u>The line filters</u>: For a PWM converter drive of drive kVA = 25kVA, connected to a grid point with a short-circuit capacity SC kVA = 500kVA, the measured line current harmonics are i_5 = 25%, i_7 = 5%, i_{11} = 1.5%, line harmonics; f_1 = 60Hz and line voltage V_{LL} = 220V (rms).
Calculate:
    a.   the line current harmonic factor HF;
    b.   the filter capacitor C and inductance L tuned to the 5th harmonic with a detuning factor α = 0.95;
    c.   the voltage total harmonic distortion factor THD_f after the filter is introduced.

13.2. <u>The voltage reflection</u>: A PWM converter drive with a rated kVA of 10kVA for maximum phase voltage V_n = 120V(rms) at 60Hz has the motor supplied through a power cable with a surge impedance Z_0 = $Z_n/10$ where Z_n is the rated load impedance of the motor. The IGBTs in the PEC have a rising time t_r = 0.5μs. The electromagnetic wave speed along the cable is U^* = 160m/μs.
Calculate:
    a.   the reflection coefficient Γ_m at motor terminals and motor load impedance Z_n for which the reflected voltage is only 20% of d.c. link voltage;
    b.   the cable critical length l_c;
    c.   for a 20% reflected wave, the required RC filter located at motor terminals.

13.15. SELECTED REFERENCES

1. **S.M. Peeran, C.W.P. Cascadden,** "Application, design and specification of harmonic filters for variable frequency drives", IEEE Trans.vol.IA-31, no.4, 1995, pp.841-847.

2. **J.W. Gray, F.J. Haydock,** "Industrial power quality considerations when installing adjustable speed drive systems", IBID vol.32, no.3, 1996, pp.646-652.

3. **A.M. Walczyna, K. Hasse, R. Czarnecki,** "Input filter stability of drives fed from voltage inverters controlled by direct flux and torque control methods", Proc. IEE, vol.EPA-143, no.5, 1996, pp.396-401.

4. **B. Mollit, J. Allan,** "Stability characteristics of a constant power chopper controller for traction drives", Proc IEE.vol.B, no.1(3), 1978, pp.100-104.

5. **A. von Jouanne, D.A. Rendusora, P.M. Enjeti, J.W. Gray,** "Filtering tehniques to minimize the effect of long motor leads on PWM inverter fed a.c. motor drive systems",

IEEE Trans.vol.IA-32, 1996, pp.919-925.

6. **A.von Jouanne, P. Enjeti, W. Gray,** "Application issues for PWM adjustable speed a.c. motor drives", IEEE-I.A. magazine, vol.2, no.5, 1996, pp.10-18.

7. **R. Kerkman, D. Leggate, G. Skibinski,** "Interaction of drive modulation and cable parameters on a.c. motor transients", Record of IEEE-IAS, Annual meeting, vol.1, 1996, pp.143-152.

8. **D. Busse et al,** "An evaluation of electrostatic shielded induction motor: a solution for rotor shaft voltage build-up and bearing current", IBID, vol.I, pp.610-617.

9. **S. Chen, T.A. Lipo, D.W. Novotny,** "Circulating type bearing current in inverter drives", IBID, vol.1, pp.162-167.

10. **S. Ogasawara, H. Akagi,** "Modeling and damping of high frequency leakage current in PWM inverter fed AC motor drive systems", IEEE Trans.vol.IA-32, no.5, 1996, pp.1105-1114.

11. **S. Ogasawara, H. Ayano, H. Akagi,** "Measurement and reduction of EMI radiated by a PWM inverter-fed AC motor drive system", IEEE Trans., vol.IA-33, no.4, 1997, pp.1019-1026.

12. **P. Enjeti, F. Blaabjerg, J.K. Pedersen,** Adjustable speed AC motor drives, application workbook, EPE 1997, Trondheim, Norway.

13. **F. Blaabjerg, J.K. Pedersen, L. Oesterguard, R.L. Kirlin, A.M. Trzynadlowski, S.Logowski,** "Optimized and nonoptimized random modulation techniques for VSI drives", Proc of EPE-95, vol.1, 1995, pp.19-26.

14. **A. Boglietti, P. Ferraris, M. Lazzani, M. Pastorelli,** "Energetic behavior of induction motors fed by inverter supply", Record of IEEE-IAS, Annual Meeting, 1993, pp.331-335.

15. **F. Blaabjerg, U. Jaeger, S. Munk-Nielsen, J.K. Pedersen,** "Power losses in PWM-VSI inverter using NPT or PT IGBT devices", IEEE Trans.vol.IE-10, no.3, 1995, pp.358-367.

16. **I. Abrahamsen, J.K. Pedersen, F. Blaaberg,** "State of the art of optimal efficiency control of induction motor drives", Proc. of PEMC-96, vol.2, 1996, pp.163-170.

17. **F. Abrahamsen, F. Blaabjerg, J.K. Pedersen,** "On the energy optimized control of standard and high-efficiency induction motors in CT and HVAC applications", Record of IEEE-IAS, Annual Meeting, 1997, vol.1., pp.612-628.

18. **D.M. Van de Sype, K. De Gusseme, A.P. Van den Bossche, J.A.A. Melkebeek,** "Sampling algorithm for digitally controlled boost PFC converters", IEEE Trans. Vol. PE – 19, no. 3, 2004, pp. 649-657.

19. **S. Chattopadhyay, V. Ramanarayanan, V. Iayashankan,** "A predictive switching modulator for current mode control of high power factor boost rectifier", IEEE Trans. Vol. PE-18, no. 1, 2003, pp. 114-123.

20. **I. Zhang. F.C. Lee, M.M. Iovanovic,** "An improved CCM single phase PFC converter with a low frequency auxiliary switch", IEEE Trans. Vol. PE – 18, no.1, 2003, pp. 44-50.

21. **S. Chattopadhyay, V. Ramanarayanan,** "Digital implementation of a line current shapping algorithm for three phase high power factor boost rectifier without voltage input sensing", IEEE Trans. Vol. PE – 19, no. 3, 2004, pp. 709-721.

22. **M. Kazmierkovski, R. Krishnan, F. Blaabjerg,** (editors), Control in Power Electronics, book, Academic Press, New York, 2003.

23. **K. Stockman, M. Didden, F. D'Hulster, R. Belmans,** "Bag the sags" IEEE – IA Magazine, vol. 10, no. 5, 2004, pp. 59-65.

24. **N.Hanigovszki, J. Poulsen, F. Blaabjerg,** "Novel output filter topology to reduce motor overvoltage", IEEE Trans. Vol. IA – 40, no. 3, 2004, pp. 845-852.

25. **M. Fenger, S.R. Campbell, J. Pedersen,** "Motor winding problems caused by inverter drives", IEEE – IA Magazine, Vol. 9, no. 4, 2003, pp. 22-31.

Chapter 14

LARGE POWER DRIVES

14.1. POWER AND SPEED LIMITS: MOVING UP

Large power drives are defined as those drives beyond the reach of low voltage (up to 660V a.c. source) PWM voltage source inverters with IGBTs (and high switching frequency). The two level IGBT-inverter power limit per unit has increased steadily in the last decade to about 2MW (at the time of this writing) by paralleling four 500kW units. With respect to voltage level, we have noted the introduction of 3 to 5 IGBT power cells (diode rectifier plus single phase PWM inverter) in series per phase — properly insulated from each other and to ground — to supply 1MW and more a.c. motors at line voltages up to 4.5kV (rms) [1].

So powers above 1-2MW, where GTO or thyristor power electronic converters are a must, may be termed as large powers. High powers per unit have been achieved with thyristor rectifier-current source inverter synchronous motors: 100MW for speeds up to 3000rpm, 30MW at 6000rpm, 3MW at 18000rpm. In low speed applications, such as cement mills (up to 20rpm, 5-6Hz) and up to 11MW power [2], thyristor cycloconverter-synchronous motor drives are predominant.

Recently, 3 level GTO voltage source inverters (at 15MW, 60Hz and 6kV line voltage [rms]) for synchronous motor drives have been introduced for steel main rolling mills [3]. It seems that this new breakthrough will drastically change the large power drives' technologies spectrum. For limited speed control ($\pm$20% around rated speed) doubly fed induction motor drives with a step down transformer-cycloconverter supplying the rotor circuit have been built for powers up to 400MW for pumped-storage hydropower plants. In such applications, the motoring is used during off-peak energy consumption hours for pumped storage and generating is used during peak hours [4]. Variable speed is useful to reduce energy consumption during motoring and extract energy at maximum available hydroturbine efficiency during generating mode.

The rotor converter power is proportional to speed control ratio (up to 20%) and so are the costs. Resistive starting for motoring is still required.

Note: By now, electric or diesel-electric multimotor propulsion systems for powers up to 6MW are provided with 2 level GTO (IGBT in the future) voltage-source inverters and induction motors or with cycloconverters and induction (or synchronous motors) or with thyristor controlled rectifier current inverters. For ship electric propulsion cycloconverters-synchronous motor drives at low speed prevail. Note that all these solutions are similar to those mentioned above for large power drives.

A recent newcomer to the field is the diode rectifier-voltage source inverter with insulated gate commutated thyristor (IGCT) for powers up to

5MW at 4.16kV. Finally, a multilevel active front bidirectional PWM converter synchronous motor drives at 50 MW/unit and at 60 KV (with cable made high voltage stator winding) has been dedicated very recently in a gas compressor submarine application. Six main large power drive technologies are summarized in Table 14.1.

Table 14.1. Large power drives

Converter type	Motor type	Power range	Speed range	Power quality	Power flow
3 level GTO rectifier – voltage source inverter	synchronous (without rotor damper winding)	10MVA (15VA / 1 minute)	6 Hz	unity power factor, low line harmonics, high efficiency through regenerative GTO snubbers	bidirectional
Thyristor cycloconverter	synchronous (without rotor damper winding)	10MW	6 Hz	rather low power factor	bidirectional
Rectifier – current source inverter a.) with thyristors	synchronous (with rotor damper winding)	30 MW 100 MW 3 MW	120 Hz(6000 rpm) 60 Hz(3600 rpm) 300 Hz(18000 rpm)	low power factor and high line current harmonics at low speed	bidirectional
b.) with GTOs	induction synchronous (with rotor damper winding)	1 - 4 MW	Up to 300 Hz	unity power factor and low line current harmonics	
Step down transformer and cycloconverter in the rotor circuit	doubly fed induction motor	Up to 100 - 200 MVA	Frequency: ± 20Hz variable, in the rotor; 50 (60)Hz constant in the stator	rather low power factor though stator active and reactive power control is possible	bidirectional
Diode rectifier current source inverter and transformer in the rotor circuit	doubly fed induction motor	Up to 10 MW or more	60 Hz	low power factor	unidirectional
Diode rectifier current source inverter with integrated gate commutated thyristors	cage rotor induction motors	5 MW 4 KV	3000 rpm	Close to unity source side power factor: it needs strong input current + filter (transformer)	unidirectional

14.2. VOLTAGE SOURCE CONVERTER SYNCHRONOUS MOTOR DRIVES

Both the 3-level GTO inverters and the cycloconverters are, in fact, voltage source type PECs with load current control and they drive the excited rotor synchronous motor at the unity power factor, above 5% of rated speed. Being both a voltage source type, the SM rotor may be cageless which is a notable simplification with sizeable motor costs reduction.

Also, the vector control system of cycloconverter is similar to that of PWM-IGBT converter. There are major differences related to the fundamental output frequency limit, which is $f_1/3$ for the standard cycloconverter.

In contrast, the voltage source inverter is limited in frequency only by the switching frequency in the converter which, for large power GTOs, is, today, around 300Hz.

Also, the cycloconverter is notably less expensive but the power factor, especially at low motor frequencies (well below the rated one) is rather low.

A 3-level GTO inverter system is shown on Figure 14.1.

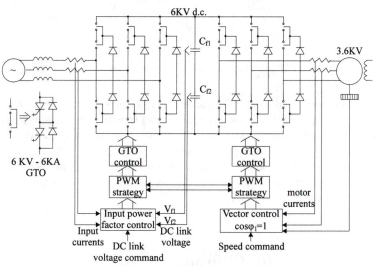

Figure 14.1. A 3-level GTO inverter - SM drive

To reduce the losses in the GTOs a regenerative snubber circuit topology is used. Space vector voltage PWM techniques are applied to improve the motor current waveform, taking advantage of the three-level voltage available in the d.c. link circuit.

Unity power factor control for the motor side and on the source side may be performed through DC link voltage (filter capacitor voltage) control.

A high power cycloconverter SM drive configuration is shown in Figure 14.2.

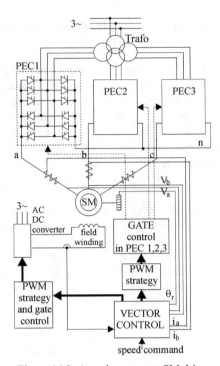

Figure 14.2. A cycloconverter SM drive.

Each phase is fed from a double rectifier bridge — one for each current polarity — to produce operation in 4 quadrants (positive and negative output voltage and current) with phase angle control. The output voltage waveform is "carved" from a sequence of adequate (same polarity, highest value) sections of the three-phase voltage waveforms of the input source and constant frequency f_1 (Figure 14.3).This is why the output frequency is a fraction of the input frequency.

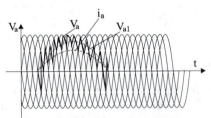

Figure 14.3. Output voltage V_a, and ideal current i_a of the cycloconverter for unity power factor operation of SM

The motor current waveform is rather close to a sinusoid, while the voltage waveform (consequently the motor flux) has harmonics whose order is related to the ratio between the output frequency f_2 and input frequency f_1.

Theoretically, $f_{2max} = f_1 / 2$. Unfortunately, the input currents are closer to rectangular waveforms (Figure 14.4) and are not even fully symmetric.

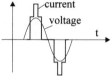

Figure 14.4. Input phase voltage and current of cycloconverters

A lagging power factor is obtained as the thyristors are line (input source) commutated.

To decouple the thyristor commutation process of three phases, the leakage inductance of the source side transformer has to be concentrated mostly in its secondary.

The line current harmonics and the input lagging power factor (lower at lower motor speeds) are serious demerits of the otherwise rather simple and rugged cycloconverter configuration.

Now since the cycloconverter is also a voltage source, as the 3-level GTO inverter system is, vector control of both is similar, though, in general, the first one is applied for low frequency drives ($f_2 < 6Hz$) and the second one for up to 60Hz (or more, as required).

14.3. VECTOR CONTROL IN VOLTAGE SOURCE CONVERTER SM DRIVES

In essence, vector control of SMs is performed either in d-q rotor coordinates or in stator flux coordinates. The d-q model equations for a salient pole cageless rotor SM is (Chapter 10)

$$\overline{V}_s = r_s \overline{i}_s + \frac{d\overline{\lambda}_s}{dt} + j\omega_r \overline{\lambda}_s \qquad (14.1)$$

$$\overline{\lambda}_s = \lambda_d + j\lambda_q; \quad \overline{i}_s = i_d + ji_q; \quad \overline{V}_s = V_d + jV_q; \qquad (14.2)$$

$$\lambda_d = L_d i_d + L_{dm}(i_d + i_F); \quad L_d = L_{dm} + L_{sl} \qquad (14.3)$$

$$\lambda_q = L_q i_d; \quad L_q = L_{qm} + L_{sl} \qquad (14.4)$$

$$\frac{d\lambda_F}{dt} = V_f - r_f i_F \qquad (14.5)$$

$$T_e = \frac{3}{2}p(\lambda_d i_q - \lambda_q i_d) \qquad (14.6)$$

$$\frac{J}{p}\frac{d\omega_r}{dt} = T_e - T_{load} \tag{14.7}$$

For steady-state ($d/dt=0$) and unity power factor the space vector diagram (from equations (14.1)-(14.4)) is as shown in Figure 14.5.

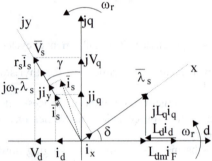

Figure 14.5. Space vector diagram of SM at steady-state and unity power factor
($\varphi_1 = 0$)

During the vector control process the stator current space vector $\bar{i}_s$ differs from the reference space vector $\bar{i}_s^*$. Consequently, the actual current $\bar{i}_s$ may be written in flux (M, T) coordinates

$$\bar{i}_s = i_M + ji_T \tag{14.8}$$

Evidently, for unity power factor (zero angle, φ_1, between $\overline{V}_s^*$ and $\bar{i}_s^*$) i_M should be zero. In reality, during transients at least, as the field current i_F correction is rather slow (to maintain constant stator flux), the unity power factor condition is abandoned for a short time and a nonzero value of i_M is referenced.

How to calculate, approximately, a reference flux current i_M^* is not simple unless the motor has a nonsalient rotor: $L_d = L_q$. Let us calculate the torque from (14.6) and (14.3)-(14.4)

$$T_e = \frac{3}{2}p\left[L_{dm}i_F + \left(L_d - L_q\right)\cdot i_d\right]\cdot i_q \tag{14.9}$$

As $L_d \geq L_q$ and $i_d < 0$ (demagnetizing effect, for unity power factor) the reluctance torque (small as it may be) will be negative. *So it would be adequate to use a cageless nonsalient pole rotor SM.* This is possible in 2(4)-pole rotors used in high speed applications, but it seems still possible for a 5-6Hz, 4(6) multi-pole rotor, 11MW SM where the rotor pole pitch is 0.66m.

The stator flux may be written as:

$$\lambda_s^* = L_{dm}i_M^* \tag{14.10}$$

where

$$i_M^* \approx i_F \cos\delta + i_x^* \cdot \frac{L_d}{L_{dm}} \qquad (14.11)$$

is strictly valid for $L_d = L_q$.

For $L_d = L_q$ the torque from (14.6) is

$$T_e = \frac{3}{2}p\left(\lambda_d i_q - \lambda_q i_d\right) \qquad (14.12)$$

with

$$i_T^* = \frac{T_e^*}{\frac{3}{2}p\lambda_s^*} = i_y^* \qquad (14.13)$$

From (14.11) we may calculate

$$i_x^* = \left(i_M^* - i_F \cos\delta\right)\frac{L_{dm}}{L_d} \qquad (14.14)$$

The stator flux has to be controlled directly through the field current control. But first the stator flux has to be estimated.

We again resort to the combined voltage (in stator coordinates) and current model (in rotor coordinates)

$$\hat{\overline{\lambda}}_{sv}{}^s = \int\left(\overline{V}_s - r_s \overline{i}_s\right)dt + \overline{\lambda}_{s0} \qquad (14.15)$$

$$\hat{\overline{\lambda}}_{si}{}^r = L_{sl}i_d + L_{dm}\left(i_d + i_F\right) + jL_q i_q \qquad (14.16)$$

$$\hat{\overline{\lambda}}_s = \hat{\overline{\lambda}}_{sv}\cdot\frac{T_c}{1+sT_c} + \hat{\overline{\lambda}}_{si}{}^r e^{j\theta_{er}}\frac{1}{1+sT_c} \qquad (14.17)$$

For low frequencies, the current model prevails while at high frequency, the voltage model takes over. If only short periods of low speed operation are required, the current model might be avoided, by using instead the reference flux λ_s^* in stator coordinates

$$\hat{\overline{\lambda}}_s = \hat{\overline{\lambda}}_{sv}\cdot\frac{T_c}{1+sT_c} + \lambda_s^* e^{j\theta_{er}}\frac{1}{1+sT_c} \qquad (14.18)$$

$$\sin\left(\theta_{er}+\delta\right) = \frac{\text{Imag}\,\hat{\overline{\lambda}}_s}{\left|\hat{\overline{\lambda}}_s\right|}; \quad \cos\left(\theta_{er}+\delta\right) = \frac{\text{Re}\,\hat{\overline{\lambda}}_s}{\left|\hat{\overline{\lambda}}_s\right|}; \qquad (14.19)$$

with θ_{er} measured, δ may be found (Figure 14.5).

Finally, the vector current control in stator flux coordinates is shown in Figure 14.6.

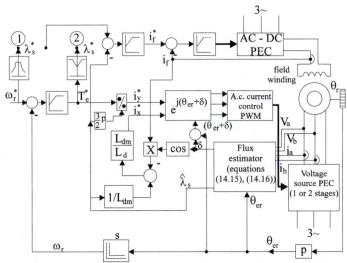

Figure 14.6. Vector current control of SM fed from voltage source type (1 or 2 stages) PECs and unity power factor

The kind of direct vector current control system shown in Figure 14.6 "moves" most of parameter dependence problems into the flux estimator but relies heavily on the rotor position feedback (a resolver is used generally).

Voltage decoupling is not considered but it could be handled as for PM-SM drives when d.c. current (i_x, i_y) controllers are used and the motion induced voltage is added only along axis y

$$V_x^* = 0 \tag{14.20}$$

$$V_y^* = \omega_r \lambda_s \tag{14.21}$$

In this case, open-loop (voltage) PWM is performed (Figure 14.7).

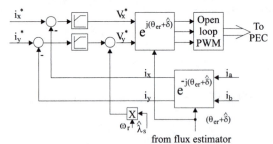

from flux estimator

Figure 14.7. Voltage decoupler in stator flux coordinates and d.c. current controllers

The voltage decoupler with d.c. current controllers is known to give better performance around and above rated frequency (speed) and to be less sensitive to motor parameter detuning. Flux control as done in Figure 14.6.

proved to provide fast response in MW power motors due to the delay compensation by nonzero transient flux current reference i_x^*.

Also, fast speed control and reversal has been proved practical with vector current control.

In drives where the load is dependent only on speed the reference flux switch is on 1 (Figure 14.6), while it is on position 2 where the torque may vary at any speed. The flux level has to increase with torque to preserve the unity power factor with variable load.

Response frequency of more than 600rad/s in torque and speed control has been reported for a 11MW drive with a 3-level GTO inverter system [3]. Similar results have been obtained with a 2500kW cycloconverter SM drive [6].

Besides vector current control with voltage decoupler, direct torque and flux control (DTFC) for the unity power factor may be performed to obtain a simpler and more robust control while preserving quick response.

14.4. DIRECT TORQUE AND FLUX CONTROL (DTFC)

Vector current control, as discussed in the previous paragraph, is performing direct flux control but it does not do direct torque control. Also it uses current vector rotators.

The stator flux estimator equations (14.15-14.19) could easily be augmented to also perform torque estimation (Figure 14.8).

For the unity power factor, the angle between stator flux and current is 90°. Also, a reactive torque RT is defined as

$$RT = Q_1 / \omega_1 \qquad (14.22)$$

where Q_1 is the reactive power and ω_1 the primary frequency.

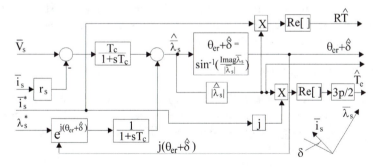

Figure 14.8. Flux and torque observer

The reactive torque RT may be estimated from (14.15) and works well for steady-state and above 5% of rated speed. To make sure that the power factor is unity, the $\hat{RT}$ should be driven to zero through adding or subtracting from reference field current (Figure 14.9). Finally, we should remember that the torque and flux error and stator flux position in one of the 60° wide sectors solely lead to a voltage vector (or a combination of) in the PEC, to form the table of optimal switching (and timings) as in any DTFC system. The table of optimal switchings (TOS) remains the same as for IMs or PM-SMs. Consequently, the DTFC system is as in Figure 14.9.

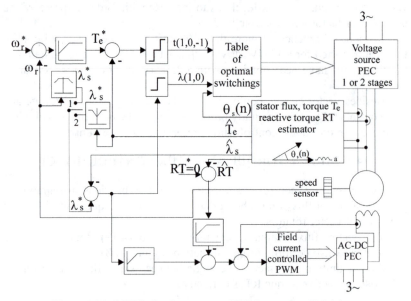

Figure 14.9. DTFC of voltage source PWM converter SM drives

A prompt start up process requires the flux to be present in the machine. To do so, first we initiate the field current control while the stator is fed through a zero voltage vector until the stator flux λ_s surpasses the reference flux for the first time. After that, the motor may start.

14.4.1. Sensorless control

We should note that the rotor position is not present directly in the DTFC scheme but rotor speed is. Consequently, for sensorless control, a rotor speed estimator is required. We may define rotor speed $\hat{\omega}_r$ as:

$$\hat{\omega}_r = \hat{\omega}_1 - \frac{d\hat{\delta}}{dt}; \quad \hat{\omega}_1 = \frac{d\hat{\theta}_s}{dt}; \quad \hat{\theta}_s = \sin^{-1}\left(\frac{\operatorname{Imag}\hat{\overline{\lambda}}_s}{\left|\hat{\overline{\lambda}}_s\right|}\right) \tag{14.23}$$

where $\hat{\bar{\lambda}}_s$ is the stator flux; $\hat{\omega}_1$, space vector speed, and $\hat{\delta}$ is flux vector angle with respect to rotor d axis.

To a first approximation $(L_d = L_q)$ for $\cos\varphi_1 = 1$ the torque is (14.9)

$$\hat{T}_e = \frac{3}{2}pL_{dm}i_Fi_q \approx \frac{3}{2}pL_{dm}i_Fi_s\cos\delta \qquad (14.24)$$

with
$$sign\left(\cos\hat{\delta}\right) = sign\left(\hat{T}_e\right) \qquad (14.25)$$

So
$$\cos\hat{\delta} \approx \frac{2\hat{T}_e}{3pL_{dm}i_Fi_s}; \quad \sin\hat{\delta} = \sqrt{1-\cos^2\hat{\delta}} \qquad (14.26)$$

i_F and i_s are measured and L_{dm} has to be known.

Consequently,
$$\frac{d\hat{\delta}}{dt} = \cos\hat{\delta}\frac{d\left(\sin\hat{\delta}\right)}{dt} - \sin\hat{\delta}\frac{d\left(\cos\hat{\delta}\right)}{dt} \qquad (14.27)$$

Equation (14.27) may represent an estimator for $\frac{d\hat{\delta}}{dt}$. On the other hand, the stator flux speed $\hat{\omega}_1$ is as shown in Chapter 11

$$\hat{\omega}_1 = \frac{\hat{\lambda}_{s\alpha}(k-1)\cdot\hat{\lambda}_{s\beta}(k) - \hat{\lambda}_{s\beta}(k-1)\cdot\hat{\lambda}_{s\alpha}(k)}{\left|\hat{\bar{\lambda}}_s\right|^2 T} \qquad (14.28)$$

where T is the sampling time.

Using (14.28) in (14.27)

$$\frac{d\hat{\delta}}{dt} = \left(\cos\hat{\delta}(k-1)\sin\hat{\delta}(k) - \sin\hat{\delta}(k-1)\cos\hat{\delta}(k)\right)/T \qquad (14.29)$$

As expected, for steady-state the angle $\hat{\delta}$ is constant and thus $\frac{d\hat{\delta}}{dt} = 0$.

14.5. LARGE MOTOR DRIVES: WORKING LESS TIME PER DAY IS BETTER

Large power drives do not work, in general, 24 hours a day, but many times they come close to this figure. Let us suppose that a cement mill has a 5MW ball mill drive that works about 8000 hours/year, that is, about 22 hours/day (average) for 365 days and uses 45000MWh of energy.

A 16MW drive would work for 4000 hours to consume about 44000MWh (not much less). However, the 16MW drive may work only off peak hours.

The energy tariffs are high for 6 hours (0.12$/kWh), normal for 8 hours (0.09$/kWh) and low for 10 hours (0.04$/kWh).

As the high power drive works an average of 11hours/day/365 days (or 12 hours/day excluding national holidays), it may take advantage of low tariff for 10 hours and work only 2 hours for normal tariff..

The lower power drive, however, has to work at least 4 peak tariff hours. Though the energy consumption is about the same, the cost of energy may be reduced from $3,700,000 to $1,700,000 per year, that is more than 50%.

As investment costs are not double for doubling the power (60% more) and the maintenance costs are about the same, the revenue time is favorable for the higher power drive working less time per day.

14.6. RECTIFIER-CURRENT SOURCE INVERTER SM DRIVES - BASIC SCHEME

Standard high speed (50(60)Hz or more) large power variable speed SM drives, traditionally above only 750kW, use rectifier-current source inverters (CSI). Load commutation of thyristors is required. The SMs are capable of leading power factor angle, $\varphi_1 = -(6°-8°)$, and thus, above 5% of rated speed, load commutation is feasible.

Both the phase-delay rectifier (PDR) and the CSI make use of thyristors and thus they may be built at low costs.

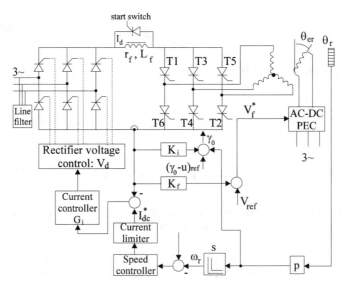

Figure 14.10. Basic scheme of Rectifier - CSI-SM drive

To reduce torque pulsations at low speed, PWM (current notches) of the otherwise rectangular (120° wide) currents may be used.

Here we introduce the basic scheme (Figure 14.10), steady-state equations and performance with load commutation and an advanced control scheme which provides constant leading power factor with load and speed response without overshooting.

The basic scheme of the rectifier-CSI-SM drive in Figure 14.10 uses an indirect vector current control scheme and contains:

- a fully controlled (phase-delay) rectifier;
- a current choke (r_f, L_f);
- a CSI-load commutated;
- a rotor position sensor (γ_0) or a terminal voltage zero crossing sensor (γ);
- advance angle (γ and γ_0 increasing with load (I_{dc})) controller for constant leading power factor angle and safe commutation (limited overlapping angle u);
- speed controller with current limiter;
- d.c. current controller;
- an a.c.-d.c. PEC for the field winding.

The CSI provides only the synchronization between the rotor position and stator current 120°-wide blocks through the advance angle $\gamma_0(\gamma)$.

The controlled rectifier modifies the d.c. link (and stator) current level, that is the torque, in order to perform speed control.

The d.c. link voltage produced by the rectifier V_d is positive for motoring and negative for generating. Through proper changing of the inverter firing sequence, regenerative braking and speed reversal are obtained. Sufficient e.m.f. (field current motion-induced stator voltage) to turn off the CSI thyristors is available only above 5% of rated speed.

14.7. RECTIFIER-CSI-SM DRIVE - STEADY-STATE WITH LOAD COMMUTATION

Let us consider that the machine is running above five percent of rated speed. Therefore, sufficient e.m.f. is provided by the field current to produce safe load commutation. By commutation, we mean the turning on and off process of current in the motor phases. Ideally, the d.c. link current is constant in time (because of a high filter inductance L_f). Thus the inverter distributes the d.c. link current as 120° current blocks of alternate polarity between the stator phases in pairs. With instantaneous (ideal) commutation, the ideal currents are as shown in Figure 14.11.

As the SM contains inductances, a sudden change in phase currents, as implied by instantaneous commutation (Figure 14.11) is, in fact, impossible.

Intuitively, it follows that the machine will impose some exponential (or almost linear) variation of currents throughout the commutation process. So the actual currents are trapezoidal rather than rectangular.

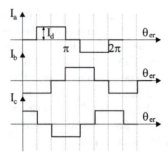

Figure 14.11. Ideal stator current waveforms

As the commutation process is rather fast — in comparison with the current period — the machine behaves approximately according to the voltage behind the subtransient inductance principle (Figure 14.12).

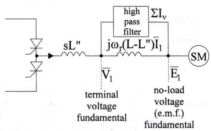

Figure 14.12. Equivalent circuit under mechanical steady-state

The voltage behind the subtransient inductance L'', V_1, is, in fact, the terminal voltage fundamental if phase resistance is neglected. The no-load (e.m.f.) voltage E_1 is also sinusoidal. The space vector (or phasor) diagram may be used for the current fundamental I_1 for $L–L''$ with L as synchronous inductance and L'' the subtransient inductance along the d-q axes (Figure 14.13) [7].

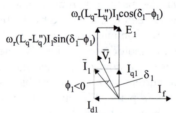

Figure 14.13. Space vector (or phasor) diagram for the voltage behind subtransient inductance

$$V_1 \cos \delta_1 = E_1 - \omega_r \left(L_d - L_d{''}\right) I_1 \sin\left(\delta_1 - \phi_1\right) \qquad (14.30)$$

$$V_1 \sin \delta_1 = \omega_r \left(L_q - L_q{''}\right) \cdot I_1 \cos\left(\delta_1 - \phi_1\right) \qquad (14.31)$$

In general, the machine may have salient poles ($L_d \neq L_q$) and, definitely, damper windings ($L_d'' \ll L_d$, $L_q'' \ll L_q$). The currents induced in the damper windings due to the nonsinusoidal character of stator currents are neglected here.

During commutation intervals the machine reacts with the subtransient inductance. An average of d-q subtransient inductances L_d'' and L_q'' is considered to represent the so-called commutation inductance L_c given by

$$L'' = L_c \approx \frac{1}{2}\left(L_d'' + L_q''\right) \tag{14.32}$$

14.7.1. Commutation and steady-state equations

To study the commutation process let us suppose that, at time zero, phases a and c are conducting (Figure 14.14) (T_1T_2 in conduction)

$$\text{at } t = 0 \; I_a = +I_d, \; I_c = -I_d \text{ and } I_b = 0 \tag{14.33}$$

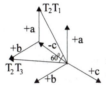

Figure 14.14. Stator m.m.f. (60° jump from T_1T_2 to T_2T_3 conduction)

The advance angle control requires a jump of 60° counterclockwise in the stator current space vector. That is, leave −c conducting and switch phase +a for +b with T_2T_3 (after T_1T_2).

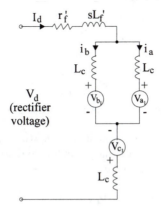

Figure 14.15. Equivalent circuit for commutation

During this commutation process phases +a and +b are both conducting (in parallel) with phase −c continuing conduction (Figure 14.15).

The loop made of phases a, b in parallel (Figure 14.15) has the equation

$$L_c \frac{di_a}{dt} + V_{a1} = L_c \frac{di_b}{dt} + V_{b1} \tag{14.34}$$

$$i_d = i_a + i_b \tag{14.35}$$

At the end of the commutation interval $(t = t_c)$, phase b is conducting

$$i_a = 0, \quad i_b = I_d \quad \text{for} \quad t = t_c \tag{14.36}$$

The fundamental voltages V_{a1} and V_{b1}, represent, as explained above, voltages behind transient inductances and are thus considered sinusoidal in time.

Eliminating i_b from (14.34)-(14.35) yields

$$2L_c \frac{di_a}{dt} = -\left(V_{a1} - V_{b1}\right) \tag{14.37}$$

For successful commutation, the current in phase a should decrease to zero during the commutation interval $(di_a/dt < 0)$. Consequently, the line voltage V_{ab1} should necessarily be positive when the current switches from phase +a to phase +b (Figure 14.16).

If the commutation process is triggered sufficiently before V_{ab1} goes to zero, from positive values, the phase a current i_a will be driven to zero.

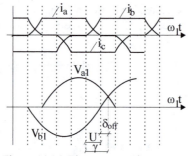

Figure 14.16. The commutation process

Before V_{ab1} goes to zero, a time interval t_{off} $(\delta_{off} / \omega_r)$ is required with $i_a = 0$ to allow the recombination of charges in the thyristor T_1.

According to Figure 14.16 the phasing of V_{ab1} is

$$V_{ab1} = V_{a1} - V_{b1} = -V_1\sqrt{6}\sin(\omega_r t - \gamma); \quad 0 < \omega_r t < \gamma \tag{14.38}$$

with V_1 the rms value of phase voltage fundamental. Integrating (14.38) from $t = 0$ to $t_c = u / \omega_r$, we obtain

$$V_1\sqrt{6}\left(\cos(\gamma - u) - \cos\gamma\right) = 2L_cI_d\omega_r \qquad (14.39)$$

The angle u corresponds to the overlapping time of currents i_a and i_b. For $I_d = 0$, $u = 0$ and γ should be greater than u for successful commutation. Ideally, $\gamma - u = \delta_{off}$ should be kept constant so both γ and u should increase with current.

It is now clear (as $\gamma_{max} = 60°$) that the maximum current commutated safely is inversely proportional to the commutation inductance L_c. The lower L_c the better, *so a damper winding in the rotor is necessary.*

The rms phase current fundamental is related to the d.c. current I_d by

$$I_1 \approx \frac{\sqrt{6}}{\pi}\frac{\sin(u/2)}{(u/2)}I_d \approx \frac{\sqrt{6}}{\pi}I_d \qquad (14.40)$$

and was obtained by using the trapezoidal current shape of Figure 14.16. The power factor angle ϕ_1 between V_1 and I_1 (Figure 14.17) is approximately [7]

$$\phi_1 \approx \gamma - u/2 \qquad (14.41)$$

The inverter voltage V_1 is made of line voltage segments disturbed only during the commutation process

$$V_1(t) = V_{a1}(t) - V_{c1}(t); \ 0 \le \omega_r t \le (\pi/3 - u) \qquad (14.42)$$

$$V_1(t) = V_{a1}(t) - V_{c1}(t) + L_c\frac{di_a}{dt}; \ (\pi/3 - u) \le \omega_r t \le \pi/3 \qquad (14.43)$$

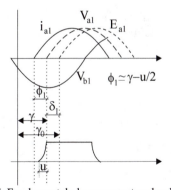

Figure 14.17. Fundamental phase current and voltage phasing

Finally, making use of (14.37)-(14.43), we obtain the average inverter voltage V_{1av}

$$V_{1av} = \frac{3}{\pi}\int_0^{\pi/3}V_1(t)d(\omega_r t) = \frac{3}{\pi}\left(V_1\sqrt{6}\cos\gamma - L_cI_d\omega_r\right) \qquad (14.44)$$

So the commutation process produces a reduction in the motor terminal voltage V_1 by $L_c I_d \omega_r$ for a given V_{1av}.

Neglecting the power losses in the CSI and in the motor

$$I_{dc} V_{1av} = 3 V_1 I_1 \cos \varphi_1 = T_{eav} \cdot \frac{\omega_r}{p}; \quad I_1 = I_d \frac{\sqrt{6}}{\pi} \qquad (14.45)$$

As expected, the simultaneous torque $T_e(t)$ is

$$T_e(t) = \frac{I_{dc} V_1(t)}{\omega_r / p} \qquad (14.46)$$

Consequently, the torque pulsates as does the inverter voltage (Figure 14.18).

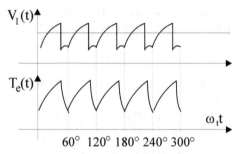

Figure 14.18. Inverter voltage $V_1(t)$ and motor torque pulsations during steady-state

The torque pulsations in Figure 14.18 are different from voltage pulsations during commutation periods especially due to the filtering effect of the motor rotor damper winding.

14.7.2. Ideal no-load speed

To calculate the ideal no-load speed (zero I_d, zero torque), we have to use all equations in this section (from (14.30)-(14.31)) to obtain

$$\delta_1 = 0, \ V_1 = E_1, \ u = 0, \ \gamma = \gamma_0;$$
$$V_1 = L_{dm} I_F \omega_{r0} / \sqrt{2} \qquad (14.47)$$

Finally, from (14.39)-(14.44)

$$\omega_{r0} = \frac{V_{1av0} \pi}{3 \sqrt{3} L_{dm} I_F \cos \gamma_0} \qquad (14.48)$$

Also, the rectifier voltage V_r is related to inverter voltage V_1

$$V_{rav} = V_{1av} + r_f I_d = V_{1av}; \quad (I_d = 0) \qquad (14.49)$$

14.7.3. Speed control options

To vary speed, we need to vary the ideal no-load speed. The available options are:
- inverter voltage variation through rectifier voltage variation up to maximum voltage available;
- reducing the field current i_F (flux weakening);
- modify the control angle γ_0. As γ_0 may lay in the 0 to 60° interval, this method is not expected to produce significant speed variations. However varying γ (or γ_0) is used to keep the power factor angle ϕ_1 rather constant ($\phi_1 \approx \gamma - u/2$) or $\delta_{off} =$ cons. for safe commutation.

Example 14.1. A rectifier-CSI-SM drive is fed from a V_L = 4.8 kVa.c. (line to line, rms), power source, the magnetization inductance L_{dm} = 0.05H and rated field current (reduced to the stator) i_{Fn} = 200A.
Determine:
- The maximum average rectifier voltage available;
- The ideal no-load speed ω_{r0} for $\gamma_0 = 0$;
- For L_c / L_{dm} = 0.3, $\gamma = 45^0$ and I_d = 100A, ω_r / ω_{r0} = 0.95, calculate the fundamental voltage V_1 and the overlapping angle u.
Solution:
The fully controlled rectifier produces an average voltage V_r (Chapter 5, equations (5.54)).

$$V_r = \frac{3V_L\sqrt{2}}{\pi}\cos\alpha \qquad (14.50)$$

So
$$V_{r\,max} = \frac{3 \cdot 4800\sqrt{2}}{\pi} \cdot 1 = 6466\ V \qquad (14.51)$$

The ideal no-load speed (14.48) is

$$\omega_{r0} = \frac{6466 \cdot \pi}{3\sqrt{3}0.05 \cdot 200 \cdot 1} = 390\ rad/s \qquad (14.52)$$

The voltage fundamental V_1 becomes (14.44)

$$V_1 = \frac{6646 \cdot \pi/3 + 0.95 \cdot 390 \cdot 0.3 \cdot 0.05 \cdot 100}{\sqrt{6}\cdot\cos 45^0} = 4336\ V \qquad (14.53)$$

Now, from (14.39) the overlapping angle u is found

$$4336\sqrt{6}\left(\cos(\gamma - u) - \cos\gamma\right) = 2 \cdot 0.3 \cdot 0.05 \cdot 390 \cdot 0.95 \cdot 100 = 1111.5 \qquad (14.54)$$
$$\cos(\gamma - u) = 0.807; \quad u = 4.8^\circ$$

The overlapping angle u is small, indicating an unusually strong damper winding. Higher values of u are practical.

14.7.4. Steady-state speed/torque curves

By now we know that the rectifier-CSI-SM drives exhibit a finite ideal no-load speed much like d.c.-brush motors with separate excitation.

However, it is important if the speed increases or decreases with torque for given inverter voltage V_{Iav}. An increase of speed with torque means an unstable speed/torque characteristic while speed decreasing with torque means a statically stable characteristic.

While successful commutation has to be provided up to the maximum load torque considered, the drive may work with γ_0 = cons. (direct position sensor required) or with ϕ_1 = cons. = γ–u / 2.

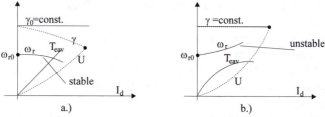

Figure 14.19. Steady-state curves (V_{Iav} = cons., i_F = cons.): a.) for constant γ_0 (position sensor); b.) for constant γ (terminal voltage zero crossing angle).

The above relationships suffice to calculate the ideal motor speed/torque (I_d) curves and the variation of overlapping angle u with load (I_d).

For γ_0 = cons. the speed / torque curve proves to be stable (Figure 14.19a). This is not the case for γ = cons. (Figure 14.19b).

So far, the field current was considered constant with load (I_d). If the field current increases with load to keep the power factor constant (ϕ_1 = cons. = γ–u/2), it means that we should increase γ with load as (anyway) the overlapping angle u does so (Figure 14.19). This way the speed will decrease with load and thus, as a bonus, static stability is restored (Figure 14.20).

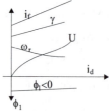

Figure 14.20. Steady-state curves (V_{Iav} = cons.), i_F - variable, ϕ_1 = cons. = γ–u / 2

Results in Figure 14.20 lead naturally to the conclusion that, for high performance variable speed control, both the angle γ and field current i_F

should increase with load to secure constant power factor and safe commutation. These conclusions will be fully exploited in the dynamics and control section later in this chapter.

14.7.5. Line commutation during starting

As already mentioned, at low speeds (below 5% of rated speed), the e.m.f. is not high enough to produce a leading power factor ($\phi_1 < 0$) and thus secure load commutation (the resistive voltage drop is relatively high). For starting either line commutation or some forced commutation (with additional hardware) [4] is required.

To start, the angle $\gamma(\gamma_0)$ is chosen to be zero to obtain maximum torque per ampere. When the commutation of phases is initiated, both thyristors conducting previously are first turned off by applying a negative voltage at the machine terminals through increasing the rectifier delay angle to about 150°-160°. To speed up the d.c. link current attenuation to zero, the d.c. choke is short-circuited through the starting thyristor (Figure 14.10).

Current notches occur and thus the torque pulsations may also be slightly reduced, while the motor accelerates smoothly.

14.7.6. Drive control loops

While the basic scheme of the control system remains the same as in Figure 14.10, we will briefly discuss some practical solutions for speed, ω_r, current, I_d, advance angle (γ) and field voltage control.

A typical speed controller capable of providing for safe load commutation during the transients is shown in Figure 14.21. The d.c. link current controller, the reference angle γ^* and the field current voltage V_f^* are all included.

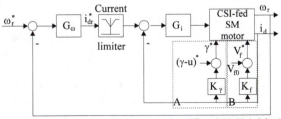

Figure 14.21. Speed control system (rectifier-CSI-SM drive)

A PI speed controller is, in general, used

$$I_d^* = \left(\omega_r^* - \omega_r\right)G_\omega \tag{14.55}$$

with

$$G_\omega = K_{c\omega}\left[1 + 1/T_{i\omega}s\right] \tag{14.56}$$

where $K_{c\omega}$ is the gain and $T_{i\omega}$ the integral time constant. Safe load commutation has to be provided, in general, up to 150% of rated current. A current limiter is required. The integral part of the speed controller is thus not desirable during the current limiter saturation period. This may be done by inhibiting the integral part of the current controller, as long as the current is higher than a limit value.

The current controller may also be of PI type

$$V_r^* = G_i \left(I_d^* - I_d \right) \tag{14.57}$$

with

$$G_i = K_{ci} \left[1 + 1 / T_{ii} s \right] \tag{14.58}$$

As expected, $T_{ii} \ll T_{i\omega}$ and, in general, $T_{i\omega} > 4T_{ii}$. The advance angle γ^* will be increased with the load current as suggested in the previous section

$$\gamma^* = \left(\gamma - u \right)^* + K_\gamma I_d \tag{14.59}$$

The initial angle $\left(\gamma - u \right)^* \approx 15° - 25°$ provides for safety margin as required by slow, standard (low cost) thyristors

$$\left(\gamma - u \right)^* \geq \omega_{r \max} t_{off}; \quad t_{off} = 0.3 - 0.5 \text{ms} \tag{14.60}$$

Though this is only a linear (intuitive) approximation, it has been proved practical. As in most CSI-SM drives, fast drive response is not a high priority, a field voltage referencer V_f^* suffices instead of a field current controller

$$V_f^* = V_{f0} + K_f I_d \tag{14.61}$$

This way, a rather constant safety margin angle $\left(\gamma - u \right)^*$ with increasing loads and a constant power factor (leading power factor angle $\phi_1 = -(8° - 10°)$) are obtained. For control numerical details related to a practical case, see [8]. Such drives are, by now, standard in industry for powers up to 30MW (5500rpm) per unit [9].

To improve (speed up) the current response at high speeds, an e.m.f. V_c compensator may be added to the d.c. link current controller. The e.m.f. may be based on the flux λ'' behind the commutation inductance L_c

$$\overline{\lambda}'' = \int \left(\overline{V}_s - r_s \overline{i}_s \right) \cdot dt - L_c \overline{i}_s \tag{14.62}$$

$$V_c = \left| \overline{V}_c \right| = \left| j\omega_r \lambda'' \right| \tag{14.63}$$

Above 5% of rated speed, the $\overline{\lambda}''$ calculator is good enough even if based only on the voltage model.

Feedforwarding the value of the V_c to the d.c. link current controller produces notable increases in the current response quickness. The above control system - in fact, of indirect vector current type - looks rather simple but relies heavily on known machine data and parameters.

A more direct, robust approach, such as direct vector control, has also been proposed [10].

As an alternative to this concept, the DTFC principle is introduced here.

14.7.7. Direct torque and flux control (DTFC) of rectifier-CSI-SM drives

As the stator current is rather trapezoidal, the stator flux is nonsinusoidal even under steady-state. The subtransient flux linkage λ'', defined above, is however, very close to a traveling wave during steady-state. For DTFC, we thus make use of λ''. A tentative DTFC configuration is shown in Figure 14.22.

The CSI provides 6 nonzero current space vectors $\bar{I}_{1,...,6}$ produced by 2 phases conducting at any time outside the commutation zones (Figure 14.23.)

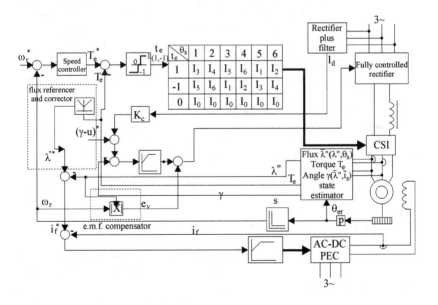

Figure 14.22. Tentative DTFC system for CSI-SM drives

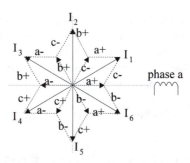

Figure 14.23. CSI-current space-vectors

The subtransient flux λ'' position in one of the 6 sectors (Figure 14.23) is used to select the required current space vector, depending also on the torque error sign. As the machine is overexcited, the forthcoming current vector should be located at an angle greater than 90° with respect to the $\overline{\lambda}''$ position. So for $\overline{\lambda}''$ in the first sector (0° to 60°), I_3 should be applied for positive torque error and I_5 for negative torque error.

A hysteresis angle may be allowed at the boundary between sectors to reduce the switching frequency. The hysteresis band of the torque controller may be increased with speed to avoid PWM of current at high speeds.

For zero torque error the thyristors of upper and lower lag of a phase are applied such that to minimize the switching frequency of the thyristors. In fact the current (I_0) in the machine is zero, but the a.c. choke limits the d.c. link current.

The subtransient flux estimator could be built in many ways. A combination of voltage (14.62) and current model is the obvious way to build a wide speed range subtransient flux estimator. Sensorless control may also be considered but a speed estimator is additionally required.

14.8. SUB- AND HYPER-SYNCHRONOUS IM CASCADE DRIVES

14.8.1. Limited speed control range for lower PECs ratings

Limited speed control range is required in many applications such as high power pumps, fans, etc. Low motor speed range control, (20-30% around rated speed) implies frequency control.

Stator frequency control, in either SMs or IMs, no matter the speed control range, requires full motor power PECs. So it is costly for the job done. It is very well known that the power balance in the wound rotor of IMs is characterized by the slip formula

$$SP_{elm} = P_{Co2} + P_r; \quad S = \frac{\omega_1 - \omega_r}{\omega_1}; \quad \omega_1 = 2\pi f \qquad (14.64)$$

where P_{elm} is the electromagnetic (active) power that "crosses the airgap", or is exchanged between rotor and stator in IMs; P_{Co2} is the rotor winding loss and P_r is the electric active power extracted or introduced (injected) into the rotor.

Neglecting, for the time being, the rotor winding loss ($P_{Co2} = 0$), the electric power injected into the rotor at slip frequency $f_2 = Sf_1$ is

$$P_r \approx SP_{elm} = \frac{\omega_1 - \omega_r}{\omega_1} P_{elm} \qquad (14.65)$$

The speed control range is defined through the minimum speed, ω_{rmin}, or the maximum slip, S_{max},

$$S_{max} = \frac{\omega_1 - \omega_{rmin}}{\omega_1} \qquad (14.66)$$

So the maximum active electric power injected in the wound rotor is

$$P_{rmax} \approx S_{max} P_{elm} \qquad (14.67)$$

For a 20% speed control range $S_{max} = 0.2$ and thus the PEC required to handle the rotor injected power P_{rmax} is rated to S_{max}, that is 20%-30% of motor rated power.

In general, as the rotor voltage required at low slip frequency $f_{2max} = S_{max}f_1$ is low, a step-up transformer is mandatory to exchange energy with the power grid also used to feed the stator windings (Figure 14.24).

In essence, P_r may be positive or negative and thus sub- or hyper-synchronous operation is feasible, but the rotor side PEC has to be able to produce positive and negative sequence voltages as

$$f_1 = f_2 + \frac{\omega_r}{2\pi}; \quad f_2 = Sf_1 \underset{<}{\overset{>}{0}} \qquad (14.68)$$

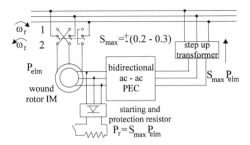

Figure 14.24. Sub- and hyper-synchronous IM cascade system

For $S < 0$, $f_2 < 0$ and thus negative sequence voltages are required. As the machine is reversible, both motoring and generating should be feasible, provided that a bidirectional power flow is allowed for through the rotor side PECs.

Also, a smooth transition through S = 0 is required. For S = 0 (stator produced synchronism) f_2 = 0 and thus d.c. current (m.m.f) should be injected into the rotor. These challenging constraints restrict the types of PECs to be used for direct a.c.-a.c. conversion.

Cyclo or matrix converters and two "back to back" voltage source PWM inverters are adequate for the scope.

Note: The so-called slip-recovery scheme is simple but, allowing only unidirectional power flow (from machine rotor to power grid), may work only as a subsynchronous motor (or as a hypersynchronous generator). On top of that the total power factor of the machine is notably low. This is the reason why we do not treat it here.

Let us note that the sub- and hyper-IM cascades may work both as a motor and a generator above and below f_1 / p and thus are used for pumped storage power plants. In this latter case, for motoring, the turbine will have, in general, to change the direction of motion to act as a pump. This explains the presence of a power switch 1-2 (Figure 14.24). Also, to protect the rotor side converter from overvoltages or overcurrents "induced" by stator faults (short-circuits, etc.) and for starting, a controlled resistor is introduced at the rotor terminals (Figure 14.24).

14.8.2. Sub- and hyper-operation modes

The induction machine equations in stator field (synchronous) coordinates are (Chapter 8, equations (8.50)-(8.52))

$$\overline{V}_s = r_s \overline{i}_s + \frac{d\overline{\lambda}_s}{dt} + j\omega_1 \overline{\lambda}_s \qquad (14.69)$$

$$\overline{V}_r = r_r \overline{i}_r + \frac{d\overline{\lambda}_r}{dt} + jS\omega_1 \overline{\lambda}_r \qquad (14.70)$$

$$\overline{\lambda}_s = L_s \overline{i}_s + L_m \overline{i}_r; \ \overline{\lambda}_r = L_r \overline{i}_r + L_m \overline{i}_s \qquad (14.71)$$

$$T_e = \frac{3}{2} p \operatorname{Re}\left(j\overline{\lambda}_s \overline{i}_s^*\right) = -\frac{3}{2} p \operatorname{Re}\left(j\overline{\lambda}_r \overline{i}_r^*\right) \qquad (14.72)$$

In the cascade configuration, the rotor circuits are fed with three-phase voltages (and currents) represented by the space vector $\overline{V}_r$.

Note that $\overline{V}_r$ and $\overline{i}_r$ in equations (14.69-14.70) are referred to the stator and expressed in synchronous coordinates. During steady-state $\overline{V}_r$, $\overline{i}_r$

represent quantities at stator frequency ω_1, while in the real machine their frequency is $\omega_2 = S\omega_1$.

For steady-state $(d/dt = 0)$ equation (14.70) becomes

$$\overline{V}_r = r_r \overline{i}_r + jS\omega_1 \overline{\lambda}_r \qquad (14.73)$$

The ideal no-load speed ω_{r0} is obtained for zero rotor current

$$\overline{V}_r = j(\omega_1 - \omega_{r0})\overline{\lambda}_r \qquad (14.74)$$

From (14.69) with $\overline{i}_r = 0$ and $d/dt = 0$

$$\overline{V}_s = r_s \overline{i}_{s0} + j\omega_1 L_s \overline{i}_{s0}; \quad \overline{\lambda}_{r0} = L_m \overline{i}_{s0} \qquad (14.75)$$

Neglecting r_s $(r_s = 0)$

$$\overline{\lambda}_{r0} = \overline{\lambda}_{s0} \cdot \frac{L_m}{L_s} = \frac{\overline{V}_s}{j\omega_1} \frac{L_m}{L_s} \qquad (14.76)$$

From (14.74) with (14.76)

$$\overline{V}_{r0} = \overline{V}_{s0} \cdot S\omega_1 \frac{L_m}{L_s} \qquad (14.77)$$

For ideal no-load, the stator and rotor voltages (in synchronous coordinates) are, for steady-state, d.c. quantities. They may be written as

$$V_{r0} = V_{s0} \left(\frac{\omega_1 - \omega_{r0}}{\omega_1} \right) \frac{L_m}{L_s} \qquad (14.78)$$

$$\omega_{r0} = \omega_1 \left(1 - \frac{V_{r0}}{V_{s0}} \frac{L_s}{L_m} \right) \qquad (14.79)$$

For $V_{r0} > 0$ (zero phase shift) and $V_{s0} > 0$, $\omega_{r0} < \omega_1$ and subsynchronous operation is obtained. In contrast, for $V_{r0} < 0$ (180° phase shift) (and $V_{s0} > 0$), $\omega_{r0} > \omega_1$, the hypersynchronous operation is obtained.

On the other hand, multiplying (14.73) by $3/2i_r{}^*$ and extracting the real part, we obtain

$$P_{rotor} = \frac{3}{2} Re(\overline{V}_r \overline{i}_r{}^*) = \frac{3}{2} r_r i_r{}^2 - S \frac{\omega_1}{p_1} T_e \qquad (14.80)$$

As known, $\omega_1 T_e / p_1 = P_{elm}$ is the electromagnetic (airgap) power transferred through the airgap from stator to rotor

$$P_{rotor} = \frac{3}{2} r_r i_r{}^2 - S P_{elm} \qquad (14.81)$$

It is now clear that we may have $P_{elm} > 0$ (motoring) or $P_{elm} < 0$ (generating) for both positive and negative slips, that is subsynchronous and hypersynchronous operation, provided that the active electric power injected in the rotor P_{rotor} may be either positive or negative.

The definition of slip frequency ω_2

$$\omega_2 = \omega_1 - \omega_r = S\omega_1 \qquad (14.82)$$

Equation (14.82) shows that for $S < 0$ ($\omega_r > \omega_1$), $\omega_2 < 0$, that is the phase sequence in the rotor is inverse with respect to the stator.

Also multiplying (14.73) by $3/2i_r^*$, but extracting the imaginary part, we obtain

$$Q_{rotor} = \frac{3}{2}\operatorname{Imag}\left(\overline{V}_r \overline{i}_r^*\right) = S\omega_1 \frac{3}{2}\operatorname{Imag}\left(j\overline{\lambda}_r \overline{i}_r^*\right) \qquad (14.83)$$

Neglecting the stator resistance and replacing $\overline{\lambda}_s$ with $\overline{\lambda}_r$ in (14.69) we may obtain, after multiplication with i_s^* (and $\overline{i}_s = -\overline{i}_r$; $\overline{\lambda}_r \approx \overline{\lambda}_s - L_{sc}\overline{i}_s$; $\overline{V}_s = j\omega_1\overline{\lambda}_s$)

$$Q_1 = \frac{3}{2}\operatorname{Imag}\left(\overline{V}_s \overline{i}_s^*\right) = \frac{3}{2}\omega_1 L_{sc} i_s^{\,2} - \frac{Q_{rotor}}{|S|} \qquad (14.84)$$

So, if reactive power is injected in the rotor, it will be subtracted from the reactive power injected in the stator. Eventually, leading power factor ($Q_1 < 0$) in the stator is obtained at the cost of lagging power factor in the rotor. If the PEC on the rotor side itself is capable — through capacitors — of producing Q_{rotor}, then the total power factor may be close to unity or even slightly leading. However, as the rotor voltage is rather low, this is hardly a practical way to produce reactive power through capacitors.

On the other hand, the rotor side PEC may be of a voltage source or current source type. It seems that rotor voltage control leads to motor-only stable operation in the subsynchronous mode and to generator-only in the hypersynchronous mode. For rotor current control, both motoring and generating are stable in sub- and hyper-synchronous modes [11].

14.8.3. Sub- and hyper-IM cascade control

The power structure of the sub- and hyper-IM cascade drive is as shown in Figure 14.24. The control system is related to controlling the speed and, eventually, the stator reactive power for motoring and active and reactive stator power for generating.

As the electromagnetic power P_{elm} is obtained at fixed stator frequency, the torque T_e is

$$T_e = \frac{P_{elm}}{\omega_1} p \qquad (14.85)$$

But the torque may be estimated from stator flux as

$$T_e = \frac{3}{2}p\,Re\left(j\overline{\lambda}_s \overline{i}_s^{\,*}\right) = \frac{3}{2}p\lambda_s i_{sT} \tag{14.86}$$

Now we may define for the stator a "reactive torque"

$$T_{Q_1} = \frac{Q_1}{\omega_1} \tag{14.87}$$

$$T_{Q_1} = \frac{3}{2}Im\,ag\left(j\overline{\lambda}_s \overline{i}_s^{\,*}\right) = \frac{3}{2}\lambda_s i_{sQ} \tag{14.88}$$

$$\left(\overline{i}_s\right)_{ref} = i_{sT} - ji_{sQ} \tag{14.89}$$

As the value of $f_1(\omega_1)$ is rather large, the stator flux estimation may be performed through the voltage model

$$\overline{\lambda}_s = \int\left(\overline{V}_s - r_s \overline{i}_s\right)dt + \overline{\lambda}_{s0} \tag{14.90}$$

$$\overline{\lambda}_s \approx \left(\overline{V}_s - r_s \overline{i}_s\right)\frac{T_c}{1+sT_c} + \overline{\lambda}_{s0} \tag{14.91}$$

In essence, the reference stator current $\left(\overline{i}_s\right)_{ref}$ has to be reproduced through the control system. What we need to control, in fact, is the rotor current $\left(\overline{i}_r\right)_{ref}$

$$\left(\overline{i}_r\right)_{ref} = -\left(\overline{i}_s\right)_{ref} - j\overline{I}_\mu = -i_{sT} - j\left(I_\mu - I_{sQ}\right) \tag{14.92}$$

where $\overline{I}_\mu$ is the main flux magnetizing current. Only approximately

$$\overline{I}_\mu = \frac{\overline{\lambda}_s}{L_s} \tag{14.93}$$

$\left(\overline{i}_r\right)_{ref}$ is still in synchronous coordinates. It has to be transformed into rotor coordinates

$$\left(\overline{i}_r\right)^r_{ref} = e^{-j\left(\theta_{er} - \omega_1 t\right)}\left|\left(\overline{i}_r\right)_{ref}\right| \tag{14.94}$$

So we need a rotor position sensor. Once the reference rotor current is known, a.c. current controllers PWM may be used in the rotor side PEC to produce sinusoidal currents at slip frequency $S\omega_1$.

Note: The rotor position angle θ_{er} and rotor speed considered as measured here, may be estimated and thus a motion sensorless drive is obtained.

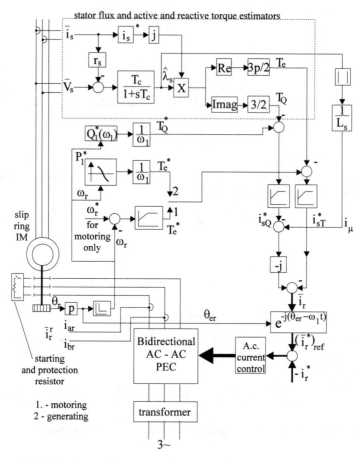

Figure 14.25. Fixed stator frequency motor-generating control with sub- and hyper-synchronous IM cascade

Figure 14.25 summarizes the controller just described for both options — motoring and generating — with independent stator torque or reactive power (torque) control.

For generating, the stator active power input has to be commanded but, again (with ω_1 = cons.), the reference torque T_e^* may be calculated, so the motor and generating control is similar. Regenerative braking is considered in the drive control mode.

The stator reactive power Q_1 is transformed into a reactive power reference torque T_Q^*. In fact, the active (torque) i_{sT} and reactive i_{sQ} stator current components are proportional to P_{elm} and Q_1 as the stator flux λ_s is constant.

The control system requires a rotor position sensor (resolver, for example) or a rotor position observer if sensorless control is targeted.

For starting, the PEC in the rotor has to be separated from the rotor as high voltages and currents occur. Note that the PEC is designed for 20%-30% of machine full power (and voltage). Alternatively, the drive may be started from rotor side with shortcircuited stator. Then the stator is opened and the conditions for selfsynchronization are quickly provided at variable speed (within the slip speed range). After selfsynchronization loading may be performed.

It also has been shown [4] that current faults (short-circuits) in the stator produce overvoltages in the rotor. Stator voltage variations induce rotor overcurrents. In all these situations, the starting resistors may be connected for protection purposes.

The sub- and hyper- IM cascade represents a rather generalized (unified) solution for limited variable speed motoring and generating applications.

While pumped storage hydropower plants of various powers (up to 400MW/unit, in principle) are the obvious applications, there are other industrial plants where such systems are performance-cost competitive.

Transient behaviour for motor operation mode of such a doubly fed induction machine with vector control in a 400 MW pump – storage power plant at Ohkawachi (Japan) [14] is shown in Figure 14.26.

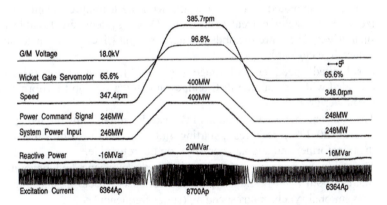

Figure 14.26. 400 MW unit doubly fed induction machine motoring transient with smooth passage through conventional synchronous speed

The smooth passage through the conventional synchronous speed ($n_1=f_1/p$) is evident, together with fast active power (torque) response and small variation of reactive power. This particular case apparently refers to the largest power electric motor drive so far.

14.9. SUMMARY

- High power means power levels beyond the reach of IGBT PWM converters. This limit goes up by the year and it is now (2005) about 6MW.

- GTOs and thyristors are the standard SCRs used in large power industrial drives. But IGCTs are catching on.

- Representative converters for large power may be of voltage source type or current source type.

- 3-level GTO (IGCT) 2-stage PWM a.c.-a.c. voltage source converters for high speeds and cycloconverters (for low speeds) represent the current controlled voltage source type and are used with cageless SMs.

- Phase-delay rectifier current source inverters are of the current source type and are used with cage-rotor SMs.

- Vector current control at unity power factor - indirect or direct version - is used to control the current in the voltage source type a.c.-a.c. converter SM drives.

- E.m.f. compensation or voltage decoupler may be added for better response at high speeds.

- DTFC (direct torque and flux control) may be used for voltage source a.c.-a.c. PECs for a simpler and more robust system.

- Phase delay rectifier-current source inverter SM drives have a low input power factor at low speeds (due to the rectifier) though the CSI is load (e.m.f.) commutated as the SM works with the leading constant power factor. Both vector current control and DTFC systems are feasible for such drives. There are methods to improve input (line) power factor and reduce line current harmonics.

- For limited speed control range ($\pm(20\text{-}30\%)$) wound-rotor IMs are provided with a rotor side PEC and, eventually, a step-up transformer to the power grid.

- The sub- and hyper-synchronous IM cascade, suitable for limited speed control (20-30%), requires a starting and protection resistor, other than the transformer and a.c.-a.c. converter on the rotor side.

- The cyclo or matrix converters can handle motoring and generating modes both sub- and hyper-synchronously (below and above conventional synchronous speed ω_1 (stator frequency)).

- Though the rotor side converter can handle bidirectional reactive power flow, the total power factor of the system is, in general, slightly lagging (unless larger capacitor filter in the d.c. link is added).

- As expected, vector current control or DTFC is to be applied to such systems for quick, precise and robust speed (power) control.

- Dual stator winding stators with single-phase short-circuited (nested) winding or anisotropic rotors of adequate pole number combinations, with only one stator winding converter-fed, have been recently proposed for medium power but are still in the laboratories [12,13].

14.10. PROBLEMS

14.1. Unity power factor SM drive: A nonsalient pole large SM motor is fed from a 3-level PWM voltage source inverter at ω_r = 376.7rad/s, p = 2 with a V_L = 4.6kV line to line voltage (rms). The motor rated phase current is I_n = 1000A (rms), the synchronous inductance is l_d = l_q = 0.75(p.u.). The drive works at the unity power factor and the leakage inductance l_{sl} = 0.17 for star connection. Neglecting the stator resistance determine:

    a.   The stator flux space vector amplitude;

    b.   For a dq flux angle $\delta = 30^0$ and rated current calculate the i_d, i_q current components;

    c.   The field current i_F and the corresponding torque T_e.

14.2. An IM cascade drive is designed for a ±20% speed control around synchronous speed. With L_m / L_s = 0.93 and stator voltage V_{sn} = 5kV, (star connection line voltage (rms)) I_{sn} = 1000A, L_{sc} = $0.05V_{sn} / i_{sn}\omega_1$; $\omega_1 = 2\pi60$rad/s.Determine:

    a.   The rotor voltage V_{r0} (value and sign) - in synchronous coordinates - for $\omega_{r0} = (1\pm0.2)\omega_1$ at no-load.

    b.   The required reactive power injection to the rotor side for $\omega_r = 0.8\omega_1$ (S = 0.2) to produce unity power factor on stator side.

    c.   The cycloconverter and step-up transformer — on rotor side — rating.

14.11. SELECTED REFERENCES

1. **P.H. Hammond,** A new approach to enhance power quality for medium voltage a.c. drives, IEEE Trans.vol.IA-33, no.1, 1997, pp.202-208.

2. **R.A. Errath,** 15000HP gearless ball mill drive in cement - why not, IEEE Trans.vol.IA-32, no.3, 1996, pp.663-669.

3. **H. Okayama,** Large capacity high performance 3 level GTO inverter system for steel main rolling mill drives, Record of IEEE-IAS, Annual Meeting, 1996, volI., pp.174-179.

4. **H. Stemmler,** High power industrial drives, Proc. IEEE vol.82, nr.8, 1994, pp.1266-1286.

5. **S. Nonaka, Y. Ueda,** A PWM GTO current source converter - inverter system with sinusoidal input and output, Record of IEEE-IAS, 1987, Annual Meeeting, vol.I, pp.247-252.

6. **T. Nakama, H. Ohsawa, K. Endoh,** A high performance cycloconverter-fed synchronous machine drive system, IEEE Trans., vol.IA-20, no.5, 1984, pp.1278-1284.

7. **J. Rosa,** Utilization and rating of machine commutated inverter - synchronous motor drives, IEEE Trans., vol.IA-15, no.2, 1979, pp.395-404.

8. **S. Nishikata, T. Kataoka,** Dynamic control of a selfcontrolled synchronous motor drive system, IEEE Trans. Vol.IA-20, no.3, 1984, pp.598-604.

9. **M. Miyazaki, et.al.,** New application of thyristor motor drive system, Record of IEEE-IAS, Annual Meeting, 1984, pp.649-654.

10. **L. Boyter, L. Schmidt,** Microprocessor controlled converter fed synchronous motor using subtransient flux model, Record of IEEE, 1990, MIT, Cambridge, Part 2., pp.427-433.

11. **A. Masmoudi, M.B.H. Kamoun and M. Poloujadoff,** A comparison between voltage and

current controls of doubly fed synchronous machine with emphasis on steady-state characteristics, Record of ICEM-1996, Vigo-Spain, vol. 3., pp.212-217.

12. **W.R. Brassfield, R. Spee, T.G. Habetler,** Direct torque control of brushless doubly-fed machines, IEEE Trans., vol.IA-32, no.5, 1996, pp.1098-1104.

13. **Y. Liao, L. Xu, Li Zhen,** Design of a doubly fed reluctance motor for adjustable speed drives, IBID, pp.1195-1203.

14. **T. Kawabara, A. Shibuya, H. Furata,** "Design and dynamic response characteristics of 400 MW adjustable speed pump storage unit for Ohkawachi power station", IEEE Trans. Vol. EC – 11, no. 2, 1996, pp. 376-394.

Chapter 15

CONTROL OF ELECTRIC GENERATORS

15.1. INTRODUCTION

Practically all electric energy is produced through electric generators driven by prime-movers. The solar power panels are the notable exception.

The prime-movers extract their mechanical energy from the water or wind energy, from steam (by burning oil or coal nuclear or fuel) or gas – burning thermal energy, gasoline or Diesel fuel in internal combustion engines (ICE).

The prime-movers are basically wind, hydro, steam turbines or ICEs.

The standard is that electric energy is produced in power plants with the prime-movers (turbines) operated at controlled, almost constant, speed, through a speed governor that acts on the fuel input rate of the prime-movers.

Synchronous generators (up to 1400 MVA/unit) are used in standard power systems. They provide controlled constant voltage (through their d.c. rotor excitation current control) and constant frequency ($f_n = n_n p$) through prime mover speed control.

Regional, national and continental electric power systems have been built by adding synchronous generators in parallel.

The rather weak coupling between the frequency (active power) and voltage (reactive power) control in synchronous generators works well in stiff power systems with sizeable spinning power reserves.

The recent opening of electrical energy markets leads to the separation of electric energy production, transmission and distribution to customers. Also, decentralization in energy production is accentuating and results in more distributed electric power systems.

The tendency to produce the daily peaks of electric energy closest to their consumption has led to smaller generator units, with variable output.

The tighter environmental requirements have prompted more renewable energy conversion systems. Wind (up to 4 MW/unit) and hydro energy (up to 770 MW/unit) are typical renewable energy sources with a softer effect on the environment.

In distributed electric power systems, the coproduction of heat and electricity is proposed through super high speed gas turbine – electric generators up to 3 MVA/unit at 15000 rpm or 100 KVA/unit at 70000 rpm. Small companies, villages or town sections may profit from such highly efficient decentralized energy production solutions.

Autonomous generators sets, driven mainly by gas turbines or ICE, even by small hydro or wind turbines, are also used for emergency (or standby) power in Telecom, hospitals, banks, etc with power/unit up to a few hundred KW.

Finally, more and more electric energy on board vehicles is needed for better comfort, better gas-mileage in standard ICE vehicles, but also for the newly developed hybrid electric vehicles (HEV). Electric machines on HEV for powers up to 100 kW and more should work both as motors, for vehicle propulsion assistance, and as electric generators for battery recharging and electrical vehicle braking.

Large and medium power synchronous generators in power systems are controlled in terms of voltage (reactive power), while the frequency (speed) or active power is controlled through the mechanical prime-mover's speed governor.

In the distributed power systems of the future, fast active power and voltage control may be harnessed better when variable speed is allowed for.

Variable speed electric generators imply power electronics converter interfaces to produce constant voltage and frequency output.

Wind and small hydro turbine electric generators are typical for almost mandatory variable speed control.

So full rating (100%) power electronic converters with electrically, excited, or PM, synchronous generators, or cage-rotor induction generators up to 4-10MW are to be expected.

Partial rating (20-30%) power electronic converters connected to the wound rotor of induction generators have already been introduced for limited (±20-5%) speed control range up to 400 MVA in pump-storage hydropower plants.

Switched reluctance generators with full rating power electronics control have been proposed for aircraft (up to 250KW) or for HEVs (up to 100 KW).

Also, full rating (100%) power electronics control PM brushless generators/motors are introduced as starter-alternators up to 100 KW and more, on board hybrid (or electric) vehicles.

In what follows we will concentrate on a few essential control aspects of electric generators, recommending the interested reader to go to Ref. 1 and then take it from there.

The main issues treated here are:
- Rather constant frequency (active power) and voltage (reactive power) control of synchronous generators connected to electric power systems.
- Limited variable speed wound-rotor induction generator control.
- Full power electronics control of variable speed PM synchronous generators.
- Full power electronics control of variable speed cage-rotor induction generators.
- The control of claw-pole rotor alternators on automobiles.
- IPM synchronous starter-alternator control on HEV.
- Switched reluctance generator output control at variable speed.

15.2. CONTROL OF SYNCHRONOUS GENERATORS IN POWER SYSTEMS

A.c. power systems operate satisfactorily when their frequency and voltage remain nearly constant and thus vary in a limited and controllable mover, when active and reactive power loads they serve are modified.

Active power flow is controlled by modifying the prime mover fuel flow rate to control the speed of the generator with a few percent variation range.

On the other hand, reactive power flow is related to generator terminal voltage control, within a few percent change range, by varying the d.c. field rotor winding current.

When too large an electric power load occurs, the speed may collapse. Similarly, too large a reactive power load leads to voltage collapse.

When a synchronous generator (SG) acts alone on its loads such as in generator sets or it is the largest by far in the respective power system, the former is controlled for constant (isochronous) speed.

In contrast, when an SG operates in a large power system, the load power is shared by many SGs. In this case it is mandatory to control the generator with a speed droop.

A voltage droop is required for voltage control, or to provide for reactive power load sharing between SGs.

Automatic generation control (AGC) divides the electric power generation between SGs, while the automatic reactive power control (AQC) distributes the reactive power contribution between same SGs.

As there is some coupling between P (active power) and Q (reactive power) controls, especially for weak power systems, power system stabilizers (PSS) are added for decoupling.

All these principles are visible in the general SG control system in Figure 15.1.

The speed governor and its controller and the frequency (speed)/power droop curve serve to control the speed to produce the required active power P^*.

The speed droop curve may be raised or lowered to decrease or increase the power share of the respective generator.

On the generator side, the voltage controller and limiters act on the voltage error $V_c^*-V_c$, corroborated with the PSS output. A voltage load compensator is added. It is beyond our scope here to explain the whole, multilevel, control system of an SG into a large power system ([1], Chapter 2).

We will concentrate a little only on SG exciters and on the automatic voltage regulator (AVR).

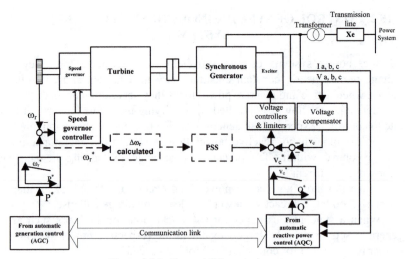

Figure 15.1. General SG control system

15.2.1. Exciters for SG

The exciter is an electric power source that is capable of supplying in a controllable manner the field winding of the SG such that the SG terminal voltage is satisfactorily controlled under designed active and reactive power load variation range.

There are two main types of exciters in fabrication today:

- a.c. brushless exciters
- static exciters

15.2.2. The a.c. brushless exciter

The standard a.c. brushless exciter (Figure 15.2) is made of an inside-out synchronous machine whose stator holds a d.c. fed field winding while its rotor contains the three phase (armature) winding and a diode rectifier which supplies directly the SG field winding. The stator-placed exciter d.c. field winding current is controlled through a small power static power converter, to provide controlled voltage at SG terminals.

It is evident that the time constant and transfer functions of both, SG and the a.c. brushless exciter (basically an inverted, small, 3%, rating SG), stand between the command voltage V_{con} at the output of the voltage regulator and the regulated terminal voltage of SG.

There is no easy way to measure any variable such as the SG field current or voltage and thus a very robust AVR is needed.

The time allowed for the field winding voltage V_f to rise from 0 to 95 % rated value is limited by standards to a few tens of milliseconds. On the other hand, forcing the voltage V_f to (1.5 - 3) times rated value V_{fn} is needed to produce acceptably fast terminal voltage response to perturbations.

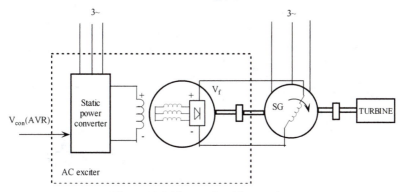

Figure 15.2. The a.c. brushless exciter

The a.c. brushless exciter has the inherent advantage that it is rather immune to SG faults, when the SG terminal voltage decreases notably, if the small power static converter (Figure 15.2) has a back-up source during SG faults or is fed from a rechargeable battery.

An approximate model of the a.c. exciter (alternator without a rotor cage) is shown in Figure 15.3.

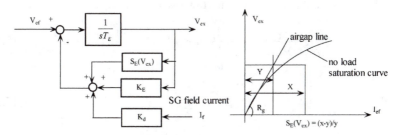

Figure 15.3. A.c. exciter alternator structural diagram for control

The load of the a.c. exciter alternator comes into play only through the SG field current I_f produced armature reaction in the a.c exciter. As the diode rectifier provides for almost unity power factor, I_f influence on V_{ex} regulation is rather straightforward (Figure 15.3).

S_E (V_{ex}) introduces the influence of the magnetic saturation in the a.c. exciter on the V_{ex} limitation. T_E is the a.c. exciter alternator excitation time constant.

The diode rectifier (Figure 15.4) introduces an additional voltage regulation and its model is needed also.

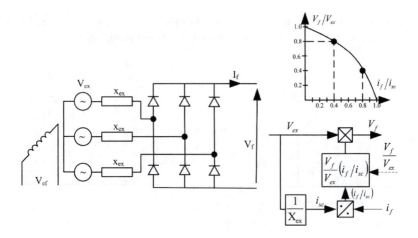

Figure 15.4. A.c. exciter alternator + diode rectifier

The V_f (I_f) output characteristic is rather nonlinear and depends heavily on diode commutation overlapping.

The a.c. exciter reactance (X_{ex}) plays a major role in the diode commutation process.

Three main operation modes may be identified:

Stage 1: two diodes conducting (low load):

$$\frac{V_f}{V_{ex}} \approx 1 - \frac{1}{\sqrt{3}}\frac{I_f}{I_{sc}} \quad \text{for } \frac{I_f}{I_{sc}} < \left(1 - \frac{1}{\sqrt{3}}\right) \tag{15.1}$$

with
$$I_{sc} = \frac{V_{ex}\sqrt{2}}{X_{ex}} \tag{15.2}$$

Stage 2: each diode conducts only when the other one on the same leg ended conduction (medium load):

$$\frac{V_f}{V_{ex}} = \sqrt{\frac{3}{4} - \left(\frac{I_f}{I_{sc}}\right)^2} \quad \text{for } \left(1 - \frac{1}{\sqrt{3}}\right) \le \frac{I_f}{I_{sc}} \le \frac{3}{4} \tag{15.3}$$

Stage 3: four diodes are conducting at any time:

$$\frac{V_f}{V_{ex}} = \sqrt{3}\left(1 - \frac{I_f}{I_{sc}}\right) \quad \text{for } \frac{3}{4} \le \frac{I_f}{I_{sc}} \le 1 \tag{15.4}$$

Figure 15.4 illustrates eqns (15.1) – (15.4) and introduces the nonlinear structural diagram that represents the diode rectifier.

15.2.3. The static exciter

The standard static exciter for SGs is the fully controlled phase rectifier (Chapter 5) — Figure 15.5 where V_{ex}, X_{ex} represent the voltage and internal (transient) reactance of the power source (a transformer in general) that supplies the static exciter. The static exciter is placed on ground and thus it transmits the power to the SG field winding through brushes and copper slip rings.

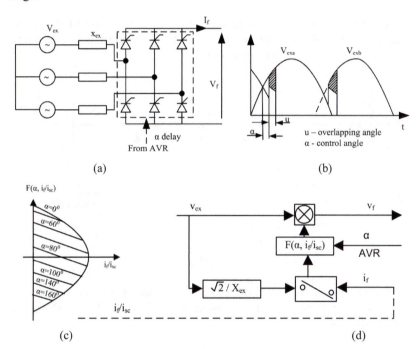

Figure 15.5. a.) The controlled rectifier as static exciter; b.)its output voltage waveform, c.) voltage/current curve, d.) nonlinear structural diagram

As shown in Chapter 5, the controlled rectifier output voltage V_f is:

$$V_f' = \frac{3\sqrt{2}V_{ex}\sqrt{3}\cos\alpha}{\pi} - \frac{3}{\pi}X_{ex}I_f; \quad I_{sc} = \frac{V_{ex}\sqrt{2}}{X_{ex}} \qquad (15.5)$$

Consequently:

$$\frac{V_f}{V_{ex}} = \cos\alpha - \frac{1}{\sqrt{3}}\frac{I_f}{I_{sc}}; \quad V_f = \frac{V_f'\pi}{3\sqrt{6}} \qquad (15.6)$$

For $\alpha = 0$ – zero phase delay angle – (15.6) degenerates into (15.1): the diode rectifier.

With the exciter main parts already modeled here, only the automated voltage regulator (AVR) is still needed to complete the system.

Apart from standardized excitation — AVR systems such as IEEE 1992 AC1A (with a.c. brushless exciter) and IEEE 1992 ST1A (with static exciter) which are basically analog systems — let us consider here a digital PID AVR system.

15.2.4. A digital PID AVR system

The a.c. exciter is simplified (armature reaction is neglected) and the diode rectifier model is a gain, as in Figure 15.6. A proportional — derivative — integral (PID) voltage regulator is used.

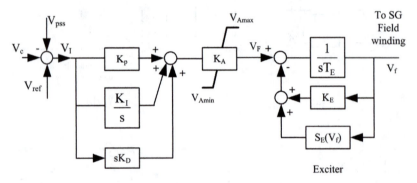

Figure 15.6. PID voltage regulator for SG

With a sampling frequency more than 20 times the damped frequency of the close loop system we may design the PID controller as if it were continuous, with:

$$G_c(s) = K_P + \frac{K_I}{s} + sK_D \qquad (15.7)$$

with K_P – proportional gain, K_I integral gain, K_D – derivative gain.

As the exciter was considered here as a first order model (only one time constant, T_E), the SG may be modeled also as a first order system with the excitation time constant T'_{d0}, at constant speed and for small deviation transients.

Consequently, the a.c. exciter plus the SG exhibits a simplified second order transfer function $G(s)$:

$$G(s) = \frac{l_{dm}\big/r_f(S_E + K_E)}{(1 + sT'_{d0}) \cdot (1 + sT_e)}; \quad T_e = \frac{T_E}{S_E + K_E}; \quad T'_{d0} = \frac{l_{dm} + l_{fl}}{\omega_b \cdot r_f} \qquad (15.8)$$

Where l_{dm} is the SG magnetization inductance in P.U. and r_f is the SG field winding resistance in P.U.; l_{fl} is the SG field winding leakage inductance in P.U.; ω_b – the base (rated) angular frequency of the SG.

The closed loop system, which includes the PID controller, has the known characteristic equation:

$$G(s) \cdot G_c(s) + 1 = 0 \qquad (15.9)$$

Let us consider $l_{dm}/(r_f(S_E + K_E)) = 1$ for simplicity.

Making use now of (15.7) and (15.8) in (15.9) we obtain:

$$K_D^2 s^2 + K_P s + K_I = -s(1 + sT_{d0}')(1 + sT_e) \qquad (15.10)$$

As it would be desirable to deal with a second order system, we select from the start a negative pole $s_3 = c$ in the left half-plane. The other two poles are chosen as complex and conjugate: $s_{1,2} = a \pm jb$.

The peak overshoot and settling time, or other method of pole placement, may be used to find the two remaining unknowns $s_{1,2}$ in (15.10) and thus to find the controller gains K_P, K_I, K_D.

The controller gains give rise to two zeros that might affect the transients unfavorably and thus, by trial and error, the final values of K_P, K_I, K_D are settled.

As an example, for $T'_{d0} = 1.5$ s, $T_e = 0.3$ s, $f_1 = 60$ Hz, settling time = 1.5 s, peak overshoot = 10 %, a satisfactory analog PID controller gain set would be: $K_P = 39.3$, $K_I = 76.5$, $K_D = 5.4$ [3].

The conversion into discrete form may use the trapezoidal integration method:

$$s \to \frac{1 - z^{-1}}{T}; \quad \frac{1}{s} \to \frac{T}{2} \frac{1 + z^{-1}}{1 - z^{-1}} \qquad (15.11)$$

z^{-1} is the unit delay.

The PID controller discrete transfer function $G_c(z)$ is thus:

$$G_c(z) = \left[K_{PD} + \frac{K_{ID}}{1 - z^{-1}} + K_{DD}(1 - z^{-1}) \right] \cdot K_{AA} = \frac{\Delta V_f(z)}{\Delta V_I(z)} \qquad (15.12)$$

with $K_{PD} = K_P - K_I(T/2)$, $K_{ID} = K_I T$, $K_{DD} = K_D/T$.

The K_{AA} gain was added in (15.12).

Making use of the property: $z^{-1} X(k) = X(k-1)$ leads to the discrete form of the PID voltage controller output $\Delta F(k)$:

$$\Delta F(k) = \Delta F(k-1) + (K_{PD} + K_{ID} + K_{DD}) \Delta V_I(k) - \\ - (K_{PD} + 2K_{DD}) \Delta V_I(k-1) + K_{DD} \Delta V_I(k-2) \qquad (15.13)$$

with ΔV_I as the SG voltage error variation (Figure 15.7).

For a 75KVA, 208 V, 0.8 PF (lagging) SG with T=12.5 ms, K_{PD}=777, K_{ID}=19, K_{DD}=8640, K_{AA}=7.00, a 50 KVAR reactive load application and rejection response is shown in (Figure 15.7a). A step reference voltage response (up and down) is shown in Figure 15.7b.

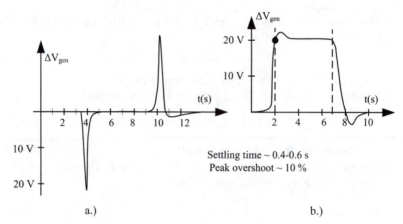

Figure 15.7. SG voltage response with PID AVR

Stable voltage response under severe reactive power surge is evident.

Note: The PID voltage regulator is a mere example of AVR and many other more robust configurations have been introduced. Among them the ones based on sliding mode and fuzzy logic stand out as apparently more practical.

15.3. CONTROL OF WOUND-ROTOR INDUCTION GENERATORS (WRIGs) WITH LIMITED SPEED RANGE

Wound-rotor induction motors have been treated in Chapter 14 for large power drive applications with limited speed control range.

The stator and rotor active power balance for steady state still holds (from Chapter 14):

$$P_s \approx -P_{Cos} + P_{elm}; \quad P_{elm} = \frac{T_e \omega_1}{p} \tag{15.14}$$

$$P_r \approx -P_{Cor} - SP_{elm}; \quad S = \frac{\omega_1 - \omega_r}{\omega_r} \tag{15.15}$$

P_{Cos} and P_{Cor} are the stator and rotor winding losses.

P_s, P_r positive means generated (delivered) electric powers; S is slip and is positive in sub-synchronous ($\omega_r < \omega_1$)and negative in super-synchronous ($\omega_r > \omega_1$) operation.

P_{elm} is the electromagnetic power, which again is positive for the generator operation mode. The torque T_e is considered positive for the generator mode ($\omega_1 = 2\pi f_1$, f_1 – stator frequency, p - pole pairs).

For generator mode in sub-synchronous operation, $P_s > 0, P_r < 0$, while for super - synchronous operation, $P_s > 0, P_r > 0$ (Figure 15.8).

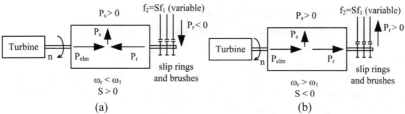

(a)   (b)

Figure 15.8. WRIG power balance: a.) sub-synchronous generator mode,
b.) super-synchronous generator mode

The core and and mechanical losses have been neglected here, for more clarity.

It should be noted that the stator power P_s comes at frequency f_1 and the rotor electric power P_r at frequency $f_2 = Sf_1$ with $|f_2| < 0.2f_1$, in general.

Also f_1 = constant and f_2 is variable with speed:

$$f_1 = f_2 + np; \quad n - speed \ (rps) \tag{15.16}$$

For super-synchronous operation $(n > f_1/p)$, $f_2 < 0$, which means that the sequence of rotor phases (supplied through a static power converter) is changed form abc to acb.

For super-synchronous operation, say at $S_{max} = -0.25$, with copper losses neglected, $P_r > 0$ and thus the total power of WRIG is:

$$P_{max} = P_{elm} + |S_{max}| \cdot P_{elm} \tag{15.17}$$

The machine is electromagnetically designed for P_{elm}, at synchronous speed $n_1 = f_1/p$ but, because it runs at larger speed $n_{max} = n(1 + |S_{max}|)$, it produces additional electric power P_r through the rotor $P_r = |S_{max}| \ P_{elm}$. This is considered a notable cost advantage of WRIG.

The reactive power flow in WRIGs is not dependent on slip S sign, as expected, and we may magnetize the machine from the rotor side (through the static power converter) or from the stator. In addition, it may be feasible to work with reactive power delivery through the stator or the rotor — if the rotor-side converter is capable to provide it. Fortunately the ratio of the rotor reactive power Q_r (at f_2), necessary to produce a reactive power Q_s (at f_1), is:

$$\frac{Q_r}{Q_s} = \frac{f_2}{f_1} = S \tag{15.18}$$

This is so because the magnetic energy, which is conserved, does not depend on frequency.

To limit the costs of the rotor side converter, WRIGs in general are operated at best at unity power factor in the stator. In this case the current oversizing of the rotor static power converter over $|S_{max}|$ P_{elm} is not larger than 10 – 15 %, to provide for the reactive power required to magnetize the machine.

Typical static power converters used for the scope are:
- Back to back voltage source two (or multi) level PWM converters;
- Cycloconverters;
- Matrix converters;

They all provide 4 quadrant operation (Figure 15.9) but only back to back PWM converters and matrix converters have enough secondary frequency range to provide the starting of the machine as a motor (in pump storage applications) from the rotor side with short-circuited stator.

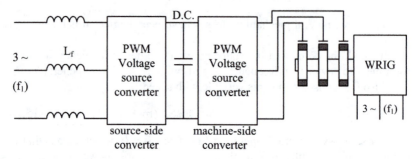

Figure 15.9. Bidirectional back to back PWM converter for the rotor of WRIGs

The two converters in Figure 15.9 are practically identical and constitute the typical two (or multi) level voltage source PWM converter used for a.c. drives.

There is a filter L_f at the source side but the voltage-matching transformer to the a.c. source is missing. It is feasible to leave it out if we design the WRIG such that at maximum slip ($|S_{max}|$) the required rotor voltage V_{rmax} is equal to the constant stator voltage V_s:

$$V_{r\,max} \approx V_s \qquad (15.19)$$

E_s and E_r are the stator and rotor e.m.f.s:

$$S_{max} E_s \frac{W_r K_{Wr}}{W_s K_{Ws}} = E_{r\,max} = V_{r\,max} \qquad (15.20)$$

W_s, W_r are the numbers of turns per current path in the stator and, respectively, in the rotor, while K_{Ws}, K_{Wr} are the corresponding winding factors. For example, with $|S_{max}|=0.25$ and identical winding factors $K_{Ws}=K_{Wr}$ $W_r/W_s=4/1$. This situation leads to a corresponding reduction of rotor rated current I_r: $I_r/I_s \cong 1/4$, as expected.

15.3.1. The space phasor model of WRIG

WRIG is a typical IM and thus the space phasor model developed in Chapter 8 stands here valid:

$$\bar{I}_s R_s + \bar{V}_s = -\frac{d\bar{\lambda}_s}{dt} - j\omega_b \bar{\lambda}_s; \quad \bar{\lambda}_s = L_{sl}\bar{I}_s + L_m \bar{I}_m; \quad \bar{I}_m = \bar{I}_s + \bar{I}_r$$

$$\bar{I}_r R_r + \bar{V}_r = -\frac{d\bar{\lambda}_r}{dt} - j(\omega_b - \omega_r)\bar{\lambda}_r; \quad \bar{\lambda}_r = L_{rl}\bar{I}_r + L_m \bar{I}_m \tag{15.21}$$

The electromagnetic torque T_e ($T_e > 0$ for generating) is:

$$T_e = \frac{3}{2} p I_{mag}[\bar{\lambda}_s \bar{I}_s^*] = \frac{3}{2} p\left(\lambda_d I_q - \lambda_q I_d\right) \tag{15.22}$$

With d/dt=s, eqns (15.21) become:

$$\left(R_s + (s + j\omega_b)L_{sl}\right)\cdot\bar{I}_s + \bar{V}_s = -L_{mt}s(\bar{I}_s + \bar{I}_r) - j\omega_b L_m(\bar{I}_s + \bar{I}_r)$$

$$\left(R_r + (s + j(\omega_b - \omega_r))L_{rl}\right)\cdot\bar{I}_r + \bar{V}_r = -L_{mt}s(\bar{I}_s + \bar{I}_r) - j(\omega_b - \omega_r)L_m(\bar{I}_s + \bar{I}_r) \tag{15.23}$$

These equations lead to the equivalent circuit of Figure 15.10.

L_{mt} is the transient magnetization inductance while L_m is the steady state one, both depend on magnetization current I_m, due to magnetic saturation.

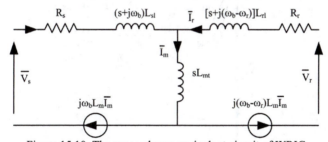

Figure 15.10. The space phasor equivalent circuit of WRIG

The speed of the reference system ω_b is free to choose but stator coordinates ($\omega_b = 0$), rotor coordinates ($\omega_b = \omega_r$) and synchronous coordinates ($\omega_s = \omega_1$) are favored, depending on the application.

For steady state $s \rightarrow j(\omega_1 - \omega_b)$ with ω_1 the frequency of actual stator variables (currents and voltages).

We may add two equations for the stator and rotor homopolar components, which are however independent of the space phasor components (Chapter 8):

$$I_{s0} R_s + V_{s0} \approx -L_{sl}\frac{dI_{s0}}{dt}$$

$$I_{r0} R_r + V_{r0} \approx -L_{rl}\frac{dI_{r0}}{dt} \tag{15.24}$$

Note: In all equations above the rotor variables are reduced to the stator ones.

For steady state, the stator and rotor voltages in their phase coordinates, when symmetric, are:

$$V_{abc}(t) = V_s \sqrt{2}\cos(\omega_1 t - (i-1)\frac{2\pi}{3})$$

$$V_{ar\,br\,cr}(t) = V_r \sqrt{2}\cos((\omega_1 - \omega_r)t + \gamma_v - (i-1)\frac{2\pi}{3})$$

(15.25)

Also for steady state a simplified vector diagram is obtained by using eqn (15.23) with $s \rightarrow j(\omega_1 - \omega_b)$.

In Figure 15.11 such a vector diagram is presented for unity power factor in the stator and sub-synchronous operation.

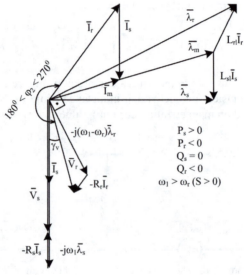

$P_s > 0$
$P_r < 0$
$Q_s = 0$
$Q_r < 0$
$\omega_1 > \omega_r \, (S > 0)$

Figure 15.11. WRIG steady state vector diagram for sub-synchronous operation

$((\omega_1 - \omega_b) > 0)$ at unity stator power factor

A few remarks are in order:

- As the stator power factor is unity, the machine magnetization comes from the rotor and thus the rotor flux amplitude λ_r is larger than the stator flux amplitude λ_s.
- Also, the rotor power factor angle φ_2 is between 180° and 270° to provide for negative (drained) active and reactive rotor electric power.
- At zero slip (conventional synchronous speed, $\omega_1 = \omega_r$) the vector diagram still holds but then $\overline{V}_r = \overline{I}_r R_r$ and the rotor flux λ_r loses its paramount role in explaining the machine behavior.

As the stator flux voltage and frequency (or stator flux) are kept rather constant, the reference system may be tied to stator voltage $\overline{V}_s$ or stator flux $\overline{\lambda}_s$, for vector control.

15.3.2. Vector control principles

Let us align the reference system to stator flux $\overline{\lambda}_s$:

$$\overline{\lambda}_s = \lambda_s = \lambda_d; \ \lambda_q = 0; \ \frac{d\lambda_q}{dt} = 0 \tag{15.26}$$

In dq coordinates, the stator equation in (15.21), with $\lambda_s \sim$ constant and $R_s \sim 0$, becomes:

$$V_d = 0; \ \lambda_q = L_s I_q + L_m I_{qr} = 0$$
$$V_q = -\omega_r \lambda_d; \ \lambda_d = L_s I_d + L_m I_{dr} \tag{15.27}$$

Consequently, the stator active and reactive powers P_s, Q_s are:

$$P_s = \frac{3}{2}(V_d I_d + V_q I_q) = \frac{3}{2} V_q I_q \approx \frac{3}{2} \omega_1 \lambda_d \frac{L_m I_{qr}}{L_s}$$
$$Q_s = \frac{3}{2}(V_d I_q - V_q I_d) = \frac{3}{2} \omega_1 \lambda_d I_d = \frac{3}{2} \omega_1 \frac{\lambda_d}{L_s}(\lambda_d - L_m I_{dr}) \tag{15.28}$$

If is now evident that, with constant stator flux amplitude (λ_d), the stator active power, P_s, may be controlled by I_{qr} (rotor q axis) current control, while stator reactive power, Q_s, control is performed through I_{dr} (rotor d axis) current control. This is in fact the principle of current vector control.

The rotor voltage equations, for steady state, in these conditions are simply:

$$V_{dr} = -R_r I_{dr} + L_{sc} S \omega_1 I_{qr} \tag{15.29}$$

$$V_{qr} = -R_r I_{qr} - S \omega_1 \left(\frac{L_m}{L_r} \lambda_d + L_{sc} I_{dr} \right) \tag{15.30}$$

Now eqns (15.29) - (15.30) pave the way for combined voltage / current vector control as they represent the voltage decoupling conditions.

15.3.3. Vector control of the machine side converter

Combined voltage/current vector control is straightforward using (15.28) - (15.30). Such a scheme is illustrated in Figure 15.12 where P_s and Q_s power controllers are added, and θ_s is the stator flux position angle with respects to phase a in the stator and θ_{er} is the rotor phase a_r axis position with respect to stator phase a, in electrical degrees.

Dual Park transformation is required first to bring the rotor currents into stator flux coordinates and then, once again, to "translate" the rotor voltages V_{dr}, V_{qr} into rotor coordinates.

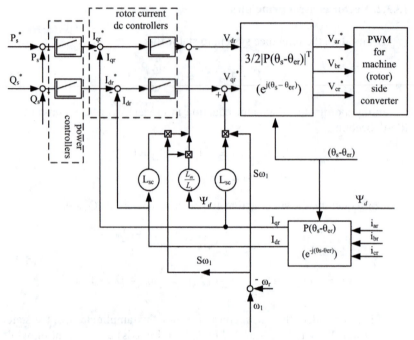

Figure 15.12. P_s, Q_s vector control of WRIG side converter

It is evident the stator voltages, stator currents and rotor currents have to be measured. The stator flux estimator is straightforward:

$$\bar{\lambda}_s = \bar{\lambda}_{s0} - \int (V_s + R_s I_s) \cdot dt \qquad (15.31)$$

The offset in the integrator has to be taken care of. A first order delay replacement of the integrator might do it. This way the stator flux amplitude $\lambda_s = \lambda_d$ and angle θ_s are obtained.

The rotor position θ_{er} has to be either measured by a robust and precise encoder or may be estimated to provide for motion-sensorless control.

15.3.4. Rotor position estimation

For rotor position (θ_{er}) estimation we first have to investigate the angle relationships between rotor axis angle θ_{er}, stator flux angle θ_s and the rotor current vector (Figure 15.13).

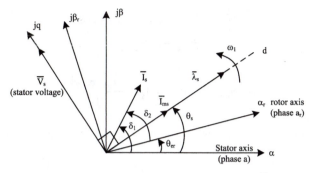

Figure 15.13. Location of rotor current vector $\bar{I}_r$

Let us consider that we know the magnetization curve of the machine, say $I_{ms}(\lambda_s)$ and start with a given value of I_{ms}.

The $\alpha\beta$ components of I_{ms} are evidently (Figure 15.13).

$$
\begin{aligned}
I_{ms\alpha} &= I_{ms} \cos\theta_s \\
I_{ms\beta} &= I_{ms} \sin\theta_s
\end{aligned}
\tag{15.32}
$$

Then, in stator coordinates, the rotor current components are (from $\bar{I}_r = \bar{I}_{ms} - \bar{I}_s$):

$$
\begin{aligned}
I_{r\alpha} &= \left(I_{ms\alpha} - I_{s\alpha}\right)\frac{L_s}{L_m}; \quad I_{s\alpha} = I_a \\
I_{r\beta} &= \left(I_{ms\beta} - I_{s\beta}\right)\frac{L_s}{L_m}; \quad I_{s\beta} = \frac{1}{\sqrt{3}}(2I_b + I_a) \\
I_r &= \sqrt{I_{r\alpha}^2 + I_{r\beta}^2}
\end{aligned}
\tag{15.33}
$$

The rotor currents are, however, measured in rotor coordinates, $I_{r\alpha r}$, $I_{r\beta r}$, and:

$$
\begin{aligned}
\cos\delta_2 &= \frac{I_{r\alpha r}}{I_r}, \quad \sin\delta_2 = \frac{I_{r\beta r}}{I_r} \\
I_{r\alpha r} &= I_{ar}; \quad I_{r\beta r} = \frac{1}{\sqrt{3}}(2I_{br} + I_{ar})
\end{aligned}
\tag{15.34}
$$

But $\hat{\theta}_{er} = \delta_1 - \delta_2$ and thus:

$$\sin\hat{\theta}_{er} = \sin(\delta_1 - \delta_2) = \sin\delta_1\cos\delta_2 - \cos\delta_1\sin\delta_2 =$$

$$= \frac{I_{r\beta}I_{r\alpha r} - I_{r\alpha}I_{r\beta r}}{I_r^2} \qquad (15.35)$$

$$\cos\hat{\theta}_{er} = \cos(\delta_1 - \delta_2) = \frac{I_{r\alpha}I_{r\alpha r} + I_{r\beta}I_{r\beta r}}{I_r^2}$$

Knowing $\cos\hat{\theta}_{er}$ and $\sin\hat{\theta}_{er}$ the rotor speed $\hat{\omega}_r$ is estimated by digital filtering because:

$$\frac{d\hat{\theta}_{er}}{dt} = \hat{\omega}_r = -\sin\theta_{er}\frac{d}{dt}(\cos\theta_{er}) + \cos\theta_{er}\frac{d}{dt}(\sin\theta_{er}) \qquad (15.36)$$

To account for magnetic saturation state change (especially during faults when the stator voltage and flux vary notably), after starting the online computation cycle by an initial value of I_{ms}, we will recalculate it as $I'_{ms\alpha}(k)$, $I'_{ms\beta}(k)$, after every computation cycle, in two stages [4]. The magnetization current computation is one step behind but this is acceptable.

$$I'_{ms\alpha}(k) = I_{s\alpha}(k) + \frac{L_m}{L_s}I'_{r\alpha}(k)$$

$$I'_{ms\beta}(k) = I_{s\beta}(k) + \frac{L_m}{L_s}I'_{r\beta}(k) \qquad (15.37)$$

With:

$$I'_{r\alpha}(k) = I_{r\alpha r}(k)\cos\hat{\theta}_{er}(k-1) - I_{r\beta r}(k)\sin\hat{\theta}_{er}(k-1)$$

$$I'_{r\beta}(k) = I_{r\beta r}(k)\cos\hat{\theta}_{er}(k-1) + I_{r\alpha r}(k)\sin\hat{\theta}_{er}(k-1) \qquad (15.38)$$

15.3.5. The control of the source-side converter

The source-side PWM converter (Figure 15.14) includes at least an LR filter, in order to reduce the source-side current harmonics.

We start with the LR filter equations:

$$\begin{vmatrix} V_a \\ V_b \\ V_c \end{vmatrix} = R\begin{vmatrix} I_{as} \\ I_{bs} \\ I_{cs} \end{vmatrix} + L\frac{d}{dt}\begin{vmatrix} I_{as} \\ I_{bs} \\ I_{cs} \end{vmatrix} + \begin{vmatrix} V_{as} \\ V_{bs} \\ V_{cs} \end{vmatrix} \qquad (15.39)$$

After translation into dq synchronous coordinates, aligned with d axis voltage ($V_d = V_s, V_q = 0$), equations (15.39) become:

$$V_d = RI_{ds} + L\frac{dI_{ds}}{dt} - \omega_1 LI_{qs} + V_{ds}$$

$$V_q = RI_{qs} + L\frac{dI_{qs}}{dt} + \omega_1 LI_{ds} + V_{qs}$$

(15.40)

ω_1 is the frequency of the a.c. supply voltages.

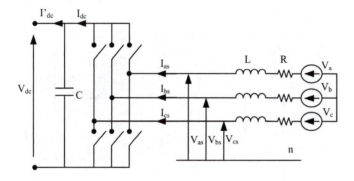

Figure 15.14. The source-side converter with LR filter

Neglecting all losses in the machine and in the converter:

$$V_{dc}I_{dc} = \frac{3}{2}V_d I_d = P_r; V_q = 0$$

(15.41)

With the PWM depth m_1:

$$V_d = \frac{m_1}{2\sqrt{2}}V_{dc}$$

(15.42)

Making use of (15.41) in (15.42):

$$I_{dc} = \frac{3m_1 I_d}{4\sqrt{2}}$$

(15.43)

The d.c. link equation writes:

$$C\frac{dV_{dc}}{dt} = I_{dc} - I'_{dc} = \frac{3m_1 I_d}{4\sqrt{2}} - I'_{dc}$$

(15.44)

If is thus evident that the d.c. link voltage may be controlled by controlling the current I_d.

The reactive power Q_r at rotor terminals is:

$$Q_r = \frac{3}{2}\left(V_d I_q - V_q I_d\right) = V_d I_q; V_q = 0$$

(15.45)

Consequently, the reactive rotor power flow may be controlled by controlling the current I_q. We may add the voltage decoupler:

$$V'_{ds} = V_d + \omega_1 L I_q$$
$$V'_{qs} = -\omega_1 L I_q$$

(15.46)

This way a combined voltage/current vector control system ([1], Chapter 9) is born (Figure 15.15)

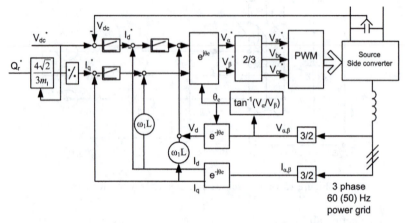

Figure 15.15. Vector control of the source-side converter

For operation at power grid the d.c. voltage V_{dc} is controlled to remain constant. The Park transformation is operated into two stages (abc to $\alpha\beta$, $\alpha\beta$ to dq) and I_d, I_q, V_d, and the voltage vector angle θ_e are calculated from measured I_a, I_b, V_a, V_b. A rather complete design of such a vector control system is given in Ref. 5.

15.3.6. WRIG control – a case study

The case of a wind turbine driving a 2.0 MW WRIG with the following parameters is considered:

V_{SN} (line, RMS) = 690 V, 2p = 4, f_N = 50 Hz, l_m = 3.658P.U., l_{sl} = 0.0634 P.U., l_{rl} = 0.08466 P.U., r_s = 4.694x10^{-3} P.U., r_r = 4.86x10^{-3} P.U., H = 3.611 seconds.

A ±25% maximum slip is considered and, to eliminate the voltage matching transformer from the rotor circuit, the rotor/stator turns ratio $W_r/W_s = 1/|S_{max}| = 4/1$.

The wind turbine is simulated by a 2D – look-up table that produces the mechanical torque T_M as a function of wind speed U (m/s) and generator speed (ω_r) such that to extract maximum power, up to the speed limit.

The control of the source-side converter:

The d-q axis current controllers' design is obtained from the system's transfer function in z-domain with a sampling time T = 0.45 ms. The closed loop natural frequency is 125Hz and the damping factor is 0.8. In these conditions the dc voltage loop PI controller transfer function is 18.69(z-1335.4)/(z-1).

The current controllers' loops are much faster than the d.c. voltage (V_{dc}) control loop and their transfer function is 17.95 (z-2640) / (z-1).

The generator side converter control:

Again, by imposing the natural close-loop frequency and the damping factor of the plant, the d,q PI current controllers' transfer functions are: 12(z-0.995)/(z-1). The power control loops are much slower with their transfer functions as 0.00009 (z-0.9) / (z-1).

Space vector PWM is used in both converters to generate their reference a.c. voltages.

Simulation results are obtained through a Matlab/Simulink dedicated program (WRIG.sim – see the attached CD).

At 1.5 seconds in time the wind speed is increased from 7 to 11 m/s, then, at 6 s, the reference value of active power is increased from 0 to 1.2 MW for constant 0.3 MVAR reactive power.

The rotor currents waveforms (Figure 15.16a) and the generator speed (Figure 15.16b) show clearly the smooth passing of the machine through the synchronous speed around 2.1 seconds in time.

The active and reactive power fast transients are shown in Figure 15.16c, d.

A three phase short-circuit at the power grid for same WRIG is illustrated in Figure 15.17. There are two main current peaks, one at the initiation of the short - circuit at half length of a symmetric power line (Figure 15.16e) and the other after the short-circuit.

Decreasing these peaks may be accomplished by simply limiting the output of the current controllers at the machine-side converter at, say 150 % rated value.

This way the WRIG may remain connected to the power grid during the short-circuit, to be ready to contribute to the voltage restoration once the short-circuit is cleared.

15.4. AUTONOMOUS D.C. EXCITED SYNCHRONOUS GENERATOR CONTROL AT VARIABLE SPEED

The d.c. excited synchronous generator may perform as an autonomous power source at constant speed (and voltage) such as in standard standby emergency or gensets, many driven by Diesel engines. In this case, however, the speed (frequency) and voltage are controlled separately such as for power grid operation.

The speed is controlled to remain constant by the speed governor of the prime-mover (Diesel engine) while the voltage is controlled, to stay constant, by the field current control.

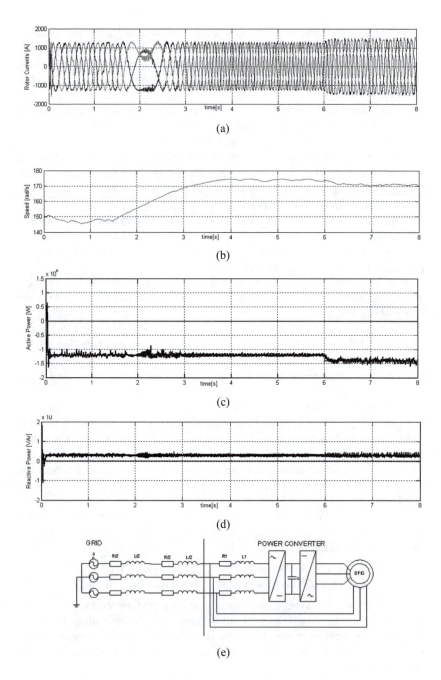

Figure 15.16. 2MW WRIG transients: a) rotor currents, b) generator speed,
c) active power, d) reactive power, e) the system structure

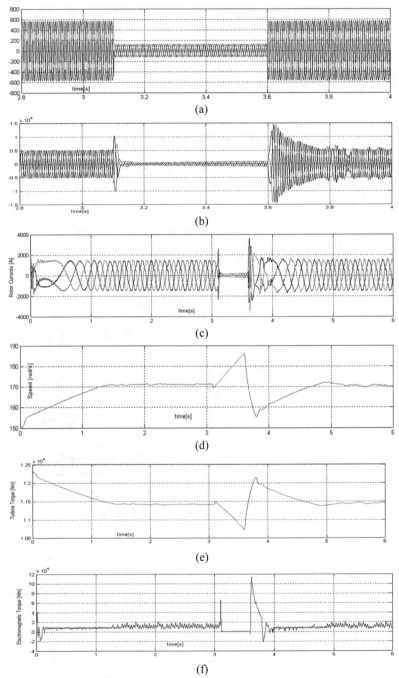

Figure 15.17. Three phase short-circuit at power grid: a) stator voltage,
b) stator currents, c) rotor currents, d) speed, e) turbine torque,
f) electromagnetic torque

As this case has been treated in paragraph 15.2, we will dwell here only on variable speed control of d.c. excited autonomous SG.

Even here there are two main situations:

- With battery back-up and d.c. controlled output (as for vehicle alternators).
- With a.c. constant voltage and frequency controlled output.

We will treat both cases in some detail.

15.4.1. Control of the automotive alternator

The automotive industry's alternator is of a single topology: the Lundell - Rice configuration. It has a single ring-shape rotor coil which is d.c. fed through slip rings and brushes to produce a multipolar ($2p = 10, 12, 14, 16, 18$) magnetic field in the airgap. The ring-shape rotor coil is surrounded by solid-iron claws to help configure the multipolar magnetic field in the airgap.

The stator is made of standard rotary machine silicon iron laminations with uniform slots that hold a q=1 slot/pole/phase single layer three phase a.c. winding (Figure 15.18).

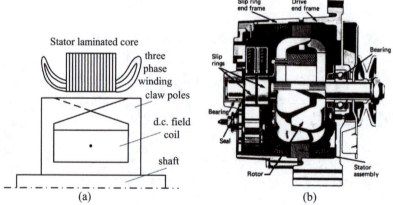

Figure 15.18. Lundell-Rice automotive alternator:
a) cross-section, b) general view

The rotor claw-poles are made of solid iron; so they act as a rather weak damper cage, to be neglected, to a first approximation.

The dq axis reactances X_d and X_q differ from each other ($X_d > X_q$).

The claw-pole alternator is to deliver d.c. power to the battery – back - up d.c. loads on automobiles.

To do so, it uses a full power diode rectifier connected between the stator terminals and the battery and a d.c. field current controller.

The field winding is fed from the alternator terminals through a 3 diode (half) rectifier and a d.c. - d.c. converter (Figure 15.19).

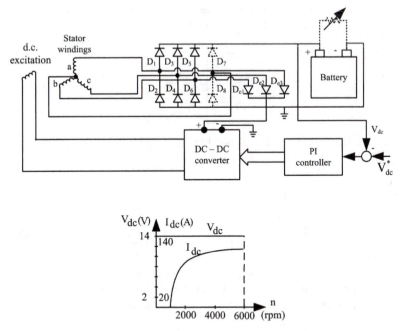

Figure 15.19. Typical automotive alternator control system

Though Figure 15.19 depicts a control system for automotive alternators any variable speed alternator with battery back-up d.c. output will fit into the category.

A voltage close loop controller is included. For maximum field current the d.c. output current I_{dc} increases with speed as in Figure 15.19. Slightly below the engine idle speed the d.c. current is zero because the alternator e.m.f. is insufficient so that the diode rectifier does not open up the energy flow.

The rather large machine transient inductances L'_d, L'_q, due to the weak solid iron rotor claw-poles damping effect, makes the response in d.c. current, to a step load power increase, not very fast.

As the diode rectifier provides for almost unity power factor, the phasor diagram of the alternator with $L_d \sim L_q$ gets simplified (Figure 15.20).

$$\underline{I}_1 R_s + \underline{V}_1 = \underline{E}_1 - jX_s \underline{I}_1 \tag{15.47}$$

$$\underline{E}_1 = -j\omega_1 X_{dm} \underline{I}_F \tag{15.48}$$

Equations (15.47) – (15.48) have been documented in Chapter 10 but they are straightforward as $\underline{E}_1$ is the e.m.f. and R_s, X_s resistance and cyclic synchronous reactances. $\underline{I}_F$ is the field current equivalent phasor seen from the stator as equations (15.47) are written in stator coordinates.

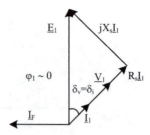

Figure 15.20. The simplified phasor diagram of claw-pole alternator with diode rectifier and $L_d \sim L_q$.

From the power balance:

$$V_1 \approx \frac{V_{dc}I_{dc} + P_{diode}}{3I_1 \cos \varphi_1}; K_i = \frac{I_{dc}}{I_1} \approx \frac{3\sqrt{3}}{\pi} \cos \varphi_1 \qquad (15.49)$$

As the speed increases, E_1 (for a given field current value) and $X_s = \omega_1 L_s$ increase and thus the current I_1 (and I_{dc}) tends to be limited.

From the phasor diagram (with $R_s \sim 0$) the current I_1 is approximately:

$$I_1 = \frac{\sqrt{E_1^2 - V_1^2}}{\omega_1 L_s} \qquad (15.50)$$

But, for constant field current I_F:

$$\begin{aligned} E_1 &= K_E (I_F) \omega_1 W_1 \\ X_s &= K_x \omega_1 W_1^2 \end{aligned} \qquad (15.51)$$

W_1 is the number of turns/phase and ω_1 is the stator frequency.

Making use of (15.51) in (15.50), the maximum current I_1 is obtained for W_{1opt}:

$$W_{1opt} = \frac{V_1 \sqrt{2}}{K_e \omega_1} \text{ or } E_{1opt} = V_1 \sqrt{2} \qquad (15.52)$$

Consequently, $X_s I_{1opt} = V_1$ for maximum current ($\delta_v = \delta_i = \pi/4$ in Figure 15.20).

It is evident that W_{1opt} should decrease with speed (ω_1) to maintain the maximum optimum current with increasing speed.

Winding tapping or operating at larger voltage are thus practical solutions to increase alternator output current at high speeds. Alternatively, an alternator designed for given power P_n at 14Vdc with a switched mode diode rectifier (as voltage booster) can produce almost twice as much power at high speeds [7] (Figure 15.21) at 42 Vdc.

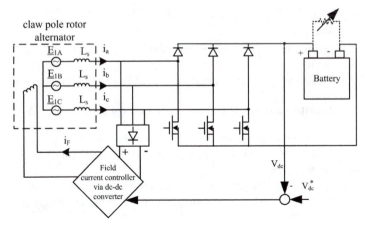

Figure 15.21. Alternator with switched-mode diode rectifier and voltage controller

By operating the automotive alternator at larger d.c. voltage, not only the output is doubled, but implicitly, the efficiency was improved from less than 50% to 70% [7].

When the load is dumped the maximum voltage has to be limited to 80 Vdc by adequate field current control (protection) provisions, to meet the current standards in automotive industry. Results in Figure 15.22 [7] prove such good performance with the switched-mode rectifier (Figure 15.21).

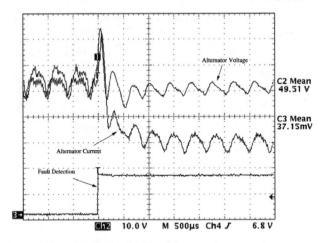

Figure 15.22. Limited load dump voltage transients

with the switched-mode diode rectifier

As the electric energy on board the vehicle is on the rise, the adaptation of 42 Vdc bus becomes necessary and beneficial in the sense that today's alternators, designed for 14 Vdc at maximum power P_n, may be used with a switched-mode rectifier and produce $2P_n$ maximum power at 42 Vdc, for increased efficiency.

While efforts to improve the output to efficiency of claw-pole alternators continue, other d.c. excited alternators such as those with smaller voltage regulation (smaller L_q, with PMs in rotor q axis) have been proposed [8].

15.4.2. A.c. output autonomous alternator control at variable speed

The autonomous alternator may also provide for constant voltage and frequency output at variable speed (Figure 15.23). A full power PWM inverter is required.

The field circuit is designed and controlled to provide for constant voltage amplitude V_1 (at variable frequency) at alternator terminals for the whole speed range.

Further on, a standard full power diode rectifier with a PWM voltage source inverter will provide for constant frequency and voltage a.c. output at power grid or for autonomous (separate) loads.

Only unidirectional power flow — from alternator to power grid or to loads — is possible but at moderate costs, as the standard PWM inverter at power grid is similar to an active front end rectifier [9].

V/f standard control of latter is adaptable for autonomous loads when V and f have to be controlled to remain constant with load, up to a point.

It is feasible to use a single d.c. voltage link for a cluster of alternators and thus a single PWM inverter is required for all of them.

The desired power sharing of each alternator is tailored through its field current controller.

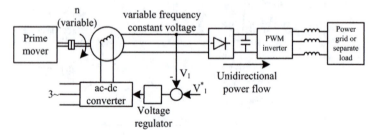

Figure 15.23. Alternator with full power PWM converter control at variable speed

15.5. CAGE-ROTOR INDUCTION GENERATOR CONTROL

The cage-rotor induction machine is known for its ruggedness, moderate costs and rather good performance.

In some applications, such as wind or small hydro energy conversion (up to 1 – 2 MW / unit), variable speed is required to track most of the available wind and hydro energy when the wind speed, and respectively, the water head vary. Also, variable speed with power increasing with generator speed, results in better system efficiency.

One drawback of cage-rotor induction machines is the need for an external source to magnetize them, corroborated with reactive power "production" for voltage control, at constant frequency.

The cage-rotor induction generator (called simply IG) has to be associated with a bidirectional a.c.-a.c. power electronic converter in the stator in order to produce active and reactive power control, both at power grid and in stand alone operation. For $\pm100\%$ P_1 and Q_1 control a back to back PWM voltage source converter (Figure 15.24) is required.

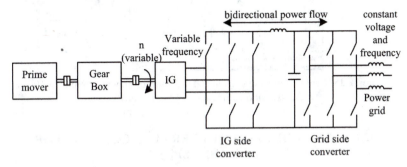

Figure 15.23. IG with back to back PWM converter to power grid:

$\pm100\%$ P_1 and Q_1 control

The bidirectional back to back PWM inverter contains the grid-side converter and the IG side converter. The source-side converter control is the same as for WRIG (see paragraph 13.3) where this converter was connected to the rotor. In essence, for power grid operation, the vector control in grid voltage coordinates is used such that the active power flow is controlled by controlling the d.c. link voltage along axis d and the reactive power Q_1 along axis q. But in this case the d.c. link capacitor has to be designed for $\pm100\%$ reactive power control. This way the IG behaves towards the loads as a synchronous generator, with faster $\pm100\%$ active and reactive power control.

The IG side converter may be either vector or direct torque and flux control type (DTFC), for motoring and generating, as widely discussed in Chapter 9 on induction motor drives.

When the IG works in autonomous load mode, the load side converter may be either vector or V/f controlled, for constant or drooped voltage amplitude control and constant frequency output control (for more details see Reference 1, Chapter 13). Typical a.c. grid side voltage and current waveforms for 100% active power and zero reactive power control of an 11KW IG (connected to an off the shelf bidirectional back to back PWM converter designed for fast braking electric drives) is shown in Figure 15.25 ([1], Chapter 13).

More filtering of the current is required to meet the today's strict harmonics content limitation standards.

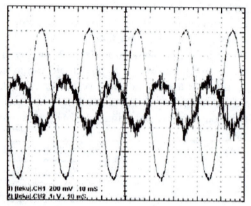

Figure 15.25. Grid voltage and current at 1500 rpm for generator mode at 100 % active and zero reactive power control

15.6. PM SYNCHRONOUS GENERATOR CONTROL FOR VARIABLE SPEED

PM synchronous generators (PMSG) are built either with surface-pole PM or interior PM rotor, with radial or axial airgap, with distributed or non - overlapping windings, just as PM synchronous motors (Chapter 10).

For convenience we will show here only the radial-airgap, distributed stator winding configuration with surface-pole-PM and, respectively, inset-pole-PM rotor (Figure 15.26).

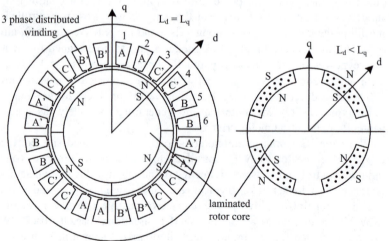

Figure 15.26. PMSG with distributed single layer stator winding ($2p_1=4$, $N_c=24$ slots): a) surface PM-pole rotor, b) inset-PM-pole rotor

In essence, the distributed three phase winding in Figure 15.26 is characterized by rather sinusoidal stator phase e.m.f.s. However, as a single layer winding is chosen, the stator current m.m.f. produces notable 5 and 7 space harmonics which, in turn, may produce notable eddy current losses in the rotor PMs.

Surface PM pole rotors are characterized by the zero magnetic saliency ($L_d = L_q$) while inset-PMs may lead to a special design when, for rated resistive load, the rated PMSG voltage V_{1N} is equal to no load voltage E_1, for given speed (Figure 15.27).

The phasor diagram for the two rotors (Figure 15.27) is based on the phase equations:

$$\underline{I}_1 R_s + \underline{V}_1 = \underline{E}_1 - jX_d \underline{I}_d - jX_q \underline{I}_q; \underline{I}_1 = \underline{I}_d + \underline{I}_q$$
$$\underline{E}_1 = -j\omega_1 \lambda_{PM} \tag{15.53}$$

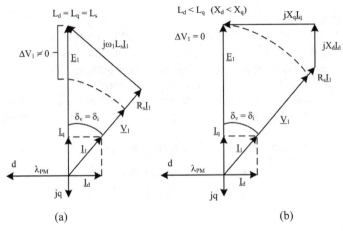

Figure 15.27. The phasor diagram of PMSG on resistive (or diode rectifier) load:

a) with surface–PM–pole rotor, b) with inset-PM-pole rotor

The dq model of the PMSG is practically the same as for PMSM (Chapter 10):

$$I_d R_s + V_d = -L_d \frac{dI_d}{dt} + \omega_r \lambda_q; \quad \lambda_d = L_d I_d + \lambda_{PM}$$
$$I_q R_s + V_q = -L_q \frac{dI_q}{dt} - \omega_r \lambda_d; \quad \lambda_q = L_q I_q \tag{15.54}$$

$$V_d = \frac{2}{3}\left(V_a \cos(-\theta_{er}) + V_b \cos(-\theta_{er} + \frac{2\pi}{3}) + V_c \cos(-\theta_{er} - \frac{2\pi}{3})\right)$$
$$V_q = \frac{2}{3}\left(V_a \sin(-\theta_{er}) + V_b \sin(-\theta_{er} + \frac{2\pi}{3}) + V_c \sin(-\theta_{er} - \frac{2\pi}{3})\right) \tag{15.55}$$

The electromagnetic torque writes (Chapter 10):

$$T_e = \frac{3}{2}p(\lambda_d I_q - \lambda_q I_d) = \frac{3}{2}p(\lambda_{PM} + (L_d - L_q)I_d)\cdot I_q \qquad (15.56)$$

As the equations are written for generator association of signs $T_e>0$ for generator. Negative torque (motoring) is obtained with negative I_q.

The motion equations have to be added:

$$\frac{J}{p}\frac{d\omega_r}{dt} = T_{mec} - T_e; \frac{d\theta_{er}}{dt} = \omega_r \qquad (15.57)$$

The prime mover torque $T_{mec}>0$ for PMSG generator operation mode and negative for PMSG motoring operation.

Let us take a numerical example to calculate PMSG steady state performance.

Example 15.1.

Let us consider a surface-PM-pole rotor PMSG with the data: $R_s=0.1\Omega$, $L_s=L_d=L_q=0.005H$, $\lambda_{PM}=0.5Wb$, $p_1=2$ pole pairs, that has to deliver power into a three phase resistance $R_L=3\Omega$/phase at $n_1=1800$rpm:

a) Calculate the phase current, voltage and power.

b) For same phase current, calculate the load resistance R'_L and power and $L_q>L_d$ (inset-PM-pole rotor), when the load voltage $V_{1r}=E_1$ (e.m.f.).

Solution:

a) The equations (15.54) apply directly for $V_d=R_L I_d$, $V_q=R_L I_q$ and $d/dt=0$:

$$(R_s + R_L)I_d = \omega_r L_q I_q$$
$$(R_s + R_L)I_q = -\omega_r(\lambda_{PM} + L_d I_d)$$

Or:

$$(0.1+3)I_d = 2\pi\cdot 30\cdot 2\cdot 0.005\cdot I_q \qquad I_d = -19.5766A$$
$$(0.1+3)I_q = -2\pi\cdot 30\cdot 2(0.5+0.005\cdot I_d) \xrightarrow{} I_q = -53.664A$$

The phase current I_1 is:

$$I_1 = \sqrt{\frac{1}{2}(I_d^2 + I_q^2)} = 40.3922A(rms)$$

Consequently, the delivered power P_1 writes:

$$P_1 = 3R_L I_1^2 = 14.683KW$$

The phase voltage V_1 is:

$$V_1 = R_L I_1 = 121.1766V(rms/phase)$$

b) For zero voltage regulation:

$$V_{1r} = E_1 = 2\pi \cdot 30 \cdot 2 \cdot 0.5 = 188.5 V(rms/phase)$$

If the current I_1 is kept the same, the power is delivered on a new load resistance R'_L:

$$R'_L = \frac{V_{1r}}{I_1} = 4.6667 \ \Omega/phase$$

We return to the above equations with known R_s, R'_L, L_d and I_1, and need to find out L_q (or I_d):

$$(0.1 + 4.6667)I_q = -377(0.5 + 0.005I_d)$$
$$4.7667I_d = 377L_qI_q$$

From the first equation with $I_q = \sqrt{3264 - I_d^2}$, we finally calculate:

$$I_d = -31.128A$$

Consequently:

$$L_q = \frac{4.7667 \cdot (-31.128)}{377 \cdot (-47.906)} = 0.00821H$$

$L_q/L_d = 0.00821/0.005 = 1.643$, a ratio which may be reached with inset-PMs on the rotor if only with a slightly smaller mechanical airgap.

The power delivered now is:

$$P'_1 = 3R'_L I_1^2 = 23.153W$$

Consequently, 57.69 % more power is gained from basically same machine and stator windings losses, just by using inset magnets on the rotor poles. Definitely the inverse saliency seems to pay off.

15.6.1. Control schemes for PMSG

The PMSG may work alone or at power grid. Let us consider the power grid operation, at variable speed, by two essential power electronics control configurations (Figure 15.28).

The voltage booster active switch T_1 (Figure 15.28a) makes use of machine inductances and the additional L_{dc} to boost the voltage in the d.c. link in a controllable manner.

A constant d.c. voltage link may be secured from ω_{rmin} to ω_{rmax}, provided there is still room at ω_{rmax} to boost a little the d.c. link voltage, to keep T_1 active.

The grid side PWM inverter is controlled as for the WRIG (paragraph 15.3).

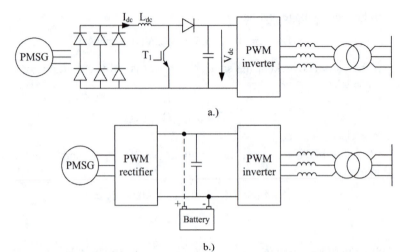

Figure 15.28. Variable speed PMSG with constant voltage and frequency a.c. output:
a) with hard - switched diode – rectifier and PWM inverter;
b) with active front rectifier and PWM inverter

On the other hand, if motoring (for starting the prime mover) is required, an active front end rectifier is required on the machine side (Figure 15.28b). This PWM converter is to be controlled as for motoring and regenerative braking of PMSM. Vector or direct torque (or power) and flux control (DTFC), as in Chapter 11 for PMSM drives, is applicable with or without motion sensors.

On the other hand, the PMSG may work as a stand alone generator driven by a medium speed Diesel engine or by a high speed gas turbine. Standby emergency or remote-area electric energy generation require such solutions. Again, the load side PWM converter may be V/f or vector controlled with an output harmonics power filter as for the IG.

An up to date Diesel engine PMSG system with multiple output frequency, single phase and three phase a.c. output, for variable speed, with fuel saving, is presented in Reference 10.

15.7. SWITCHED RELUCTANCE GENERATOR (SRG) CONTROL

The switched reluctance machine (SRM), which can be built in single phase or multiphase configurations, may work as a motor or as a generator.

Motoring is required if the prime-mover starting or assistance is required as in hybrid electric vehicles [1, 11-12].

If only generator operation is necessary, single phase (for low power) and three phase configurations seem favorite.

The circuit mathematical model of SRM developed in Chapter 12 holds valid here and thus will not be derived again.

We will rather concentrate on typical current waveforms for generator mode, typical PWM converters and a voltage controller for d.c. output at variable speed.

The SRG current waveforms

The turn on θ_{on} and θ_{off} angles for generator mode occur at the beginning, and respectively long before the end of negative inductance slope of the conducting phase (Figure 15.28a).

The phase i voltage equation is:

$$V_i = R_s I_i + L_{ti}(I_i, \theta_r)\frac{dI_i}{dt} + K_E(I_i, \theta_r) \cdot 2\pi n \qquad (15.58)$$

K_E is the pseudo e.m.f. coefficient.

$$E = K_E 2\pi n; \quad K_E(I_i, \theta_r) = I_i \frac{\partial L_i}{\partial \theta_r} < 0, \text{for generating} \qquad (15.59)$$

We may consider $L_{ti} \cong L_u$ (unaligned) = constant for the heavily saturated core machine.

Voltage $V_i = +V_{dc}$ for phase energization and $V_i = -V_{dc}$ during power delivery (phase de-energization).

There are three generator operation modes as illustrated in Figure 15.29a, b.

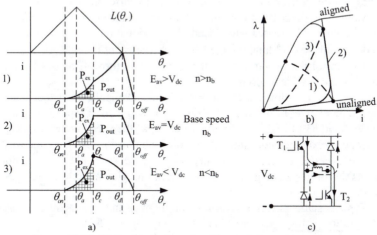

Figure 15.29. SRG: a) inductance/position and current waveforms, b) corresponding energy cycles, c) typical phase PWM converter.

As during generator mode $V_i = -V_{dc} < 0$ and $E < 0$, the current which remains positive increases as long as $|E| > |V_{dc}|$. This case is typical for high speeds when the torque (energy cycle area) is smaller (Figure 15.29b, cycle 1).

For case a2 in Figure 15.29a, at the turn off angle θ_c, $|E| = V_{dc}$ and thus, with $R_s \sim 0, di/dt=0$, the current stays constant until the phase inductance reaches its minimum at angle θ_d (Energy cycle 2 in Figure 15.28b).

In case a3 in Figure 15.29a the maximum current is reached at θ_c and, after that, the current decreases monotonously because $|E| < V_{dc}$, which corresponds to low speeds (energy cycle 3 in Figure 15.29b).

For constant d.c. voltage power delivery, the excitation (energization) energy per cycle W_{exc} is:

$$W_{exc} \approx \frac{V_{dc}}{2\pi n} \int_{\theta_{on}}^{\theta_c} i d\theta_r \qquad (15.60)$$

while the energy delivered per cycle W_{out} is:

$$W_{out} \approx \frac{V_{dc}}{2\pi n} \int_{\theta_c}^{\theta_{off}} i d\theta_r \qquad (15.61)$$

We may define the excitation penalty ε as:

$$\varepsilon = \frac{W_{exc}}{W_{out}} \qquad (15.62)$$

$|E| > V_{dc}$ leads to a smaller excitation penalty but not necessarily to the highest energy conversion ratio, which seems to correspond to $|E| \sim V_{dc}$.

Maintaining this latter condition, however, it means to increase V_{dc} with speed because E increases with speed. A step down d.c.-d.c. converter has to be added as the d.c. output voltage has to stay constant for ω_{rmin} to ω_{rmax}.

To control the power output, the turn on and turn off angles θ_{on} and θ_c should be changed in a certain manner.

If such a step-down d.c. converter is not available, then the generator mode will evolve from case a1 ($|E| < V_{dc}$), below base speed, to case a2 ($|E| = V_{dc}$) and a3 ($|E| > V_{dc}$) — Figure 15.29 — as the speed increases. A rather moderate speed range constant output voltage control is provided this way.

An average excitation penalty $\varepsilon = 0.3$-0.4 should be considered acceptable for good energy conversion ratio, though a notable energy reserve is lost.

Typical static power converters for $|E| = V_{dc}$ and, respectively, variable $|E|/V_{dc}$ ratio (with speed) control are shown in Figure 15.30.

The additional boost capability, to the buck converter is required when motoring is needed, as for starter-alternators on hybrid vehicles (Figure 15.30a).

To stabilize the output voltage of stand-alone loads it may be suitable to supply the excitation power from a separate battery (Figure 15.30b, left). The diode D_{oe} allows for battery assistance during generating mode and thus the battery cost is rather small.

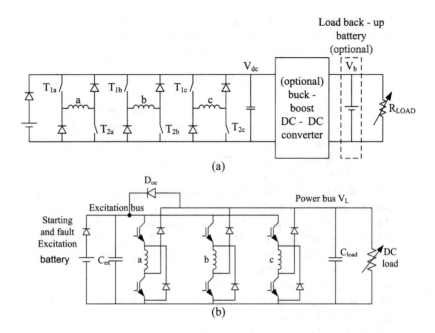

(a)

(b)

Figure 15.30. Three SRG: a) with asymmetrical PWM converter, self excitation
(from battery) or load back-up battery and d.c.-d.c. converter;
b) with separate excitation power bus and fault clearing capability

The excitation battery also allows SRG to operate during load faults and
clear them quickly (Figure 15.31) [15].

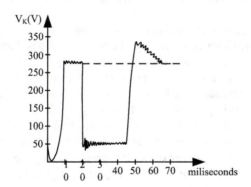

Figure 15.31. SRG voltage rise-up, experiencing a fault
(by sharp load resistance reduction) and voltage swift recovery

Intricate SRG control at variable speed for electric vehicles is
demonstrated in Reference [14] with feedforward torque control and in
Reference [15-17] with direct torque control.

The ruggedness of SRG and its wide speed range constant power capability at constant d.c. voltage are strong assets in its favor, especially in battery back-up d.c. loads, typical for automobiles, ships and aircraft.

15.8. SUMMARY

- Practically all electric energy is produced (from primary fuel energy) through electric generators driven by prime-movers (turbines).
- In today's standard electric power systems many synchronous electric generators are connected in parallel.
- Their voltage (reactive power) and speed (active power) is controlled with a small droop to allow for desired power sharing between paralleled synchronous generators.
- The voltage (reactive power) control of SG is performed through the SG excitation current; an exciter source is required; it may be a brushless a.c. exciter or a static exciter. Fast voltage recovery response is required; various automatic voltage regulators have been proposed and a PID digital version is illustrated in this chapter.
- While SGs in power systems are controlled with a small speed droop, when they act alone on a few loads their speed is controlled either to stay constant or is left to vary (to reduce speed with power decrease) to save fuel in the prime-mover, when a full power converter is needed to yield constant voltage and frequency output over the entire speed range.
- SGs are present on automobiles as alternators with full power diode rectifier output to a d.c. battery. The alternator provides d.c. current in the battery, increasing with speed through proper d.c. excitation current control via voltage close loop regulator and a low power electronics converter.
- For wind and small hydro energy conversion in the MW/unit range variable speed for best primary energy tapping and better stability, the doubly-fed (wound rotor) induction generator (WRIG) is used.
- WRIG contains a dual (a.c.-d.c.-a.c.) PWM converter connected to the wound rotor through brushes and slip-rings. This converter is feeding the rotor with a voltage $V_r \leq V_s$ at a frequency $f_2 = f_1 - np >< 0$; (V_r, V_s, rotor and stator voltages, f_2, f_1 – rotor and stator frequency, n – speed rps, and p – pole pairs). The speed range is defined by the maximum slip $S_{max} = \pm(5\text{-}30)\% = f_2/f_1$; smaller values are typical for larger powers.
- The PWM bidirectional converter is practically rated at $|S_{max}|P_n < 30\%P_n$ and thus lower system costs are obtained.
- WRIG has been implemented up to 400 MW/unit in pump storage hydro - generators. The fast active and reactive power control within the speed range from 70 % to 130 % of rated speed is a flexibility asset that fixed speed SGs are lacking. WRIG is practically an SG at all speeds within the speed range and its synchronization to the grid is very quick and safe.

- Cage rotor induction generators with full power bidirectional PWM converters connected to the stators seem the favorite solution in wind, small hydro and even gensets up to 1-2 MW/unit as they provide fast smooth connection to the power grid. From the grid point of view they behave like very fast controlled SGs, but at variable speed.

- Permanent magnet synchronous generators are characterized by low losses and, for inset (or interior) PM – rotor, they may exhibit even zero voltage regulation on resistive rated load.

- PMSG may be controlled for variable speed with constant d.c. voltage output or with constant frequency and voltage output, when a full power electronic converter is required. PMSG may work in stand alone applications (starter/alternators on cars) or in power grid applications, driven by high speed gas turbines to produce cogeneration or emergency, or peak power in various applications.

- Switched reluctance generators have been proposed especially for vehicle applications (automobile and aircraft starter alternators) at variable speed, when they deliver power on d.c. battery back-up loads.

- Electric generator control is needed both in power systems, now still standard but more distributed in the future, and also in stand alone vehicular or on ground applications. Decentralized, controlled, electric energy production requires variable speed in general and thus power electronics control of active and reactive power comes into play.

- As for electric motor operation mode, there is a myriad of electric generators with their digital control via power electronics for better energy conversion and higher power quality.

- Control of electric generators at variable speed can collect from the immense heritage of electric drives control techniques, and should constitute a major technology of the future.

- The present chapter should be considered only a tiny introduction to EG control; see Reference 1 to start your own journey into this field.

15.9. SELECTED REFERENCES

1. **I. Boldea,** Electric generators Handbook, Part 1/2 and 2/2, Taylor and Francis, 2005.
2. **P. Kundur,** Power systems stability and control, MacGraw Hill, 1994.
3. **A. Godhwani, M.J. Bassler,** "A digital excitation control system for use on brushless excited synchronous generators", IEEE – Trans, Vol. EC – 11, no. 3, 1996, pp. 616-620.
4. **R. Datta, V.T. Ranganathan,** "A simple position – sensorless algorithm for rotor-side field oriented control of wound rotor induction machine", IEEE-Trans. Vol. IE – 48, no. 4, 2001, pp. 786-793.
5. **R. Rena, J.C. Clare, G.M. Asher,** "Doubly fed induction generator using back to back PWM converters and its application to variable speed wind-energy generation", Proc. IEE, Vol. EPA – 143, no. 3, 1996, pp. 231-241.
6. **I. Serban, F. Blaabjerg, I. Boldea, Z. Chen,** "A study of the double fed wind power generator under power system faults", Record of EPE-2003, Toulouse, France.
7. **D.J. Perreault, V. Caliskan,** "Automotive power generation and control", IEEE Trans.

Vol. PE – 19, no. 3, 2004, pp. 618-630.

8. **S. Scridon, I. Boldea, L. Tutelea, F. Blaabjerg, E. Ritchie,** "BEGA – a biaxial excitation generator for automobiles: comprehensive characterization and test results", Record of IEEE-IAS-2004, Seattle, USA.

9. **M. Kazmierkowski, R. Krishnan, F. Blaabjerg (editors),** Control in power electronics, Academic Press, Amsterdam, 2002.

10. **L.M. Tolbert, W.A. Peterson, T.J. Theiss, M.B. Sardiere,** "Gen-sets", IEEE-IA Magazine, vol. 9, no. 2, 2003, pp. 48-54.

11. **J. Hussain,** Electric and hybrid vehicles, CRC Press, Florida, 2003.

12. **A. Emadi, M. Ehsani, J.M. Miller,** Vehicular electric power systems, Marcel Dekker, New York, 2004.

13. **A.V. Radun, C.A. Ferreira, E. Richter,** "Two channel switched reluctance starter-generator results", IEEE Trans. Vol. IA-34, no. 5, 1998, pp. 1106-1109.

14. **H. Bausch, A. Grief, K. Kanelis, A. Mickel,** "Torque control battery supplied switched reluctance drives for electrical vehicles", Record of ICEM – 1998, vol. 1, pp. 229-234.

15. **R.B. Inderka, R.W. De Doncker,** "Simple average torque estimation for control of switched reluctance machines", Record of EPE-PEMC-2000, Kosice, Vol. 5, pp. 176-181.

16. **S. Dixon, B. Fahimi,** "Enhancement of output electric power in SRGs", Record of IEEE-IEMDC-2003, Madison, Wi, Vol. 2, pp. 849-856.

INDEX